Topics in Applied Physics
Volume 77

Now also Available Online

Starting with Volume 77, Topics in Applied Physics is part of the Springer LINK service. For all customers with standing orders for Topics in Applied Physics we offer the full text in electronic form via LINK free of charge. Please contact your librarian who can receive a password for free access to the full articles by registration at:

http://link.springer.de/series/tap/reg_form.htm

If you do not have a standing order you can nevertheless browse through the table of contents of the volumes and the abstracts of each article at:

http://link.springer.de/series/tap/

There you will also find more information about the series.

Springer
Berlin
Heidelberg
New York
Barcelona
Hong Kong
London
Milan
Paris
Singapore
Tokyo

Physics and Astronomy

ONLINE LIBRARY

http://www.springer.de/phys/

Topics in Applied Physics

Topics in Applied Physics is a well-established series of review books, each of which presents a comprehensive survey of a selected topic within the broad area of applied physics. Edited and written by leading research scientists in the field concerned, each volume contains review contributions covering the various aspects of the topic. Together these provide an overview of the state of the art in the respective field, extending from an introduction to the subject right up to the frontiers of contemporary research.

Topics in Applied Physics is addressed to all scientists at universities and in industry who wish to obtain an overview and to keep abreast of advances in applied physics. The series also provides easy but comprehensive access to the fields for newcomers starting research.

Contributions are specially commissioned. The Managing Editors are open to any suggestions for topics coming from the community of applied physicists no matter what the field and encourage prospective editors to approach them with ideas.

See also: http://www.springer.de/phys/series/TAP

Managing Editors

Dr. Claus E. Ascheron

Springer-Verlag Heidelberg
Topics in Applied Physics
Tiergartenstr. 17
69121 Heidelberg
Germany
Email: ascheron@springer.de

Dr. Hans J. Kölsch

Springer-Verlag Heidelberg
Topics in Applied Physics
Tiergartenstr. 17
69121 Heidelberg
Germany
Email: koelsch@springer.de

Assistant Editor

Dr. Werner Skolaut

Springer-Verlag Heidelberg
Topics in Applied Physics
Tiergartenstr. 17
69121 Heidelberg
Germany
Email: skolaut@springer.de

Pramod K. Rastogi (Ed.)

Photomechanics

With 314 Figures

Springer

Dr. Pramod K. Rastogi (Ed.)
Stress Analysis Laboratory IMAC-DGC
Swiss Federal Institute of Technology
1015 Lausanne
Switzerland
Email: rastogi@imac.epfl.ch

Library of Congress Cataloging-in-Publication Data

Photomechanics / Pramod K. Rastogi (ed.).
    p. cm. -- (Topics in applied physics, ISSN 03034216 ; v. 77)
   Includes bibliographical references and index.
   ISBN 3540659900 (hardcover : alk. paper)
   1. Non-destructive testing. 2. Photonics. 3. Holographic interferometry. 4. Moirâ
method. 5. Optoelectronic devices. I. Rastogi, P. K. (Pramod K.) II. Series.

ta417.2 .P483 1999
620.1'127--dc21

99-052204

Physics and Astronomy Classification Scheme (PACS):
06.30.-k; 06.20.-f; 42.62.Eh

ISSN print edition: 0303-4216
ISSN electronic edition: 1437-0859
ISBN 3-540-65990-0 Springer-Verlag Berlin Heidelberg New York

This work is subject to copyright. All rights are reserved, whether the whole or part of the material is concerned, specifically the rights of translation, reprinting, reuse of illustrations, recitation, broadcasting, reproduction on microfilm or in any other way, and storage in data banks. Duplication of this publication or parts thereof is permitted only under the provisions of the German Copyright Law of September 9, 1965, in its current version, and permission for use must always be obtained from Springer-Verlag. Violations are liable for prosecution under the German Copyright Law.

Springer-Verlag is a company in the specialist publishing group BertelsmannSpringer
© Springer-Verlag Berlin Heidelberg 2000
Printed in Germany

The use of general descriptive names, registered names, trademarks, etc. in this publication does not imply, even in the absence of a specific statement, that such names are exempt from the relevant protective laws and regulations and therefore free for general use.

Typesetting: Data conversion by DA-TeX, Blumenstein, Seidel GbR, Leipzig
Cover design: *design & production* GmbH, Heidelberg
Computer-to-plate and printing: Saladruck, Berlin
Binding: Buchbinderei Lüderitz & Bauer, Berlin

SPIN: 10664377   57/3144/mf - 5 4 3 2 1 0 – Printed on acid-free paper

The important thing in science is not so much to obtain new facts as to discover new ways to thinking about them.

Sir William Lawrence Bragg.
in *Beyond Reductionism*, A. Koestler and J. R. Smithies (London: Hutchinson) p. 115.

Preface

The word "photomechanics" describes the synergy between the fields of photonics and mechanics. Photonics is the technology of generating and detecting radiant energy whose basic unit is the photon. Mechanics, on the other hand, is the science that deals with objects (solids or fluids) at rest or in motion and under the action of external loads. Photomechanics thus refers to the technology that uses photonics for measurements in mechanics. In other words, photomechanics forms the basis of a whole range of photonics-related techniques used in the measurement of variations in certain important physical quantities such as displacements, strains, densities, etc. in solid mechanics and flow problems. While photonics by definition is supposed to encompass the entire spectral width, the scope of the present book has been intentionally limited to cover a broad selection of topics in the visible range of the spectrum in order to keep the length of the book within reasonable limits.

The application of photomechanics has experienced constant expansion over the last several years to just about every field of engineering sciences. This has been driven, first, by the extreme flexibility and the unique features associated to optical measurement methods; second, by an improved understanding of the physics behind these methods, permitting their performance to be optimized; third, by advances in the rapid acquisition and analysis of data through the availability of refined information-technology tools; and last but not the least, by a growing awareness among the engineers and scientists of the real utility of these methods in diverse areas of science and engineering. There is every reason to believe that this strong growth will continue in the future.

The book is broadly targeted to meet the requirements of beginners and of all those scientists and engineers who want to use, or are involved in, optical nondestructive measurements. The first two Chapters are of an introductory character. The first Chapter, by D. Malacara-Hernández, provides a brief insight into relevant optical principles for engineers, while the introduction to engineering mechanics by A. Asundi in the next chapter is addressed specifically to physicists involved in the development of photomechanics methods. The contribution, by Y. Surrel, treats fringe analysis using digital image processing. The next chapter , by R. K. Rastogi, introduces the principles of holographic interferometry and speckle metrology and describes the experimental methods needed to perform in situ measurements. In the fifth contribution D. Post, B. Han and P. G. Ifju present several Moiré approaches with emphasis on applications. The contribution, by T. Y. Chen, offers an introduction to digital photoelasticity and the following contribution, by J. S. Sirkis and D. Inaudi, respectively, consider concepts involved in fiber-optic sensing, with special emphasis on their use in an engineering environment. The contribution, by S. Krishnaswamy, describes the use of shearing interferometer and caustics for experimental stress analysis. The contribution, by M. A. Sutton, S. R. McNeill, J. D. Helm and Y. J. Chao, details the use of two- and three-dimensional computer-vision methods for the measurement of surface shapes and surface deformations in simple and complex objects. The contribution, by J. M. Coupland which follows, describes the underlying principles of laser-Doppler and pulsed-laser velocimetries for the characterization of fluid flow. The last contribution, by D. J. Whitehouse, gives a representative view of the techniques useful for surface characterization and roughness measurement in engineering.

Thus, the book brings within the scope of graduate students as well as practicing scientists and engineers engaged in diverse fields, a compilation of the underlying principles, methods, and a variety of application possibilities of a broad range of topics in photomechanics, otherwise not readily available in a single textbook. A list of selected references at the end of each chapter will help readers interested in particular topics to pursue their interests in more detail.

My very special thanks are due to Professor Leopold Pflug. Without his encouragement the final goal would not have been achieved. His generosity is most appreciated. I wish to thank Jean-Louis Guignard for modifying and drawing many of the figures that are reproduced in this volume. I am thankful to Dr. Werner Skolaut for his cooperation and prompt handling of the book. I wish to thank Dr. Claus Ascheron for his motivation and support throughout the development of the book.

Lausanne, *Pramod K. Rastogi*
October 1999

Contents

Optics for Engineers
Daniel Malacara-Hernández 1

1. Introduction 1
2. Geometrical Optics Principles 1
   2.1. Fermat's Principle and the Law of Refraction 1
   2.2. First-Order Optics 3
   2.3. Aberrations 7
3. Wave-Optics Fundamentals 12
   3.1. Young's Double Slit 12
   3.2. Michelson Interferometer 13
   3.3. Coherence and Light Sources 14
4. Main Interferometers Used in Optical Engineering 15
   4.1. Fizeau Interferometer and Newtons Rings 16
   4.2. Twyman-Green Interferometer 19
   4.3. Ronchi and Lateral-Shear Interferometers 22
   4.4. Talbot Interferometer and Moiré Deflectometry 26
   4.5. Foucault Test and Schlieren Techniques 27
   4.6. Two-Wavelength Interferometry 28
   4.7. Holographic Principles 29
5. Summary 31
References 31

Introduction to Engineering Mechanics
Anand Asundi 33

1. Introduction 33
2. Basic Concepts 34
   2.1. Stress 34
   2.2. Stress Transformation 35
   2.3. Stress Transformation Using Mohr's Circle 36
   2.4. Displacement and Strain 37
   2.5. Strain Transformation 39
   2.6. Stress-Strain Relations 40
   2.7. Boundary Conditions 41
3. Theory of Elasticity Approach 42
   3.1. Plane Stress Formulation 44
   3.2. Plane Strain Formulation 45

4. Strength of Materials Approach 46
5. Examples 47
   5.1. Axial Loading 47
   5.2. Bending of Beams 49
   5.3. Combined Bending and Axial Load 52
6. Conclusion 53
Further Reading 54

Fringe Analysis
Yves Surrel 55

1. Introduction 55
2. Phase Evaluation 57
   2.1. Local and Global Strategies 57
   2.2. Local Phase detection 58
   2.3. Global Phase Detection 77
   2.4. Residual Phase Errors 81
   2.5. Effect of Additive Noise 84
3. Phase Unwrapping 86
   3.1. Spatial Approach 87
   3.2. Temporal Approach 89
4. Conclusion 94
References 99

Principles of Holographic Interferometry and Speckle Metrology
Pramod K. Rastogi 103

1. Holographic Interferometry 103
   1.1. Types of Holographic Interferometry 104
   1.2. Thermoplastic Camera 107
   1.3. Mapping of the Resolved Part of Displacement 109
   1.4. Determination of the Wavefront Phase
        Using Phase-shifting Interferometry 110
   1.5. Out-of-plane Displacement Measurement 111
   1.6. In-plane Displacement Measurement 112
   1.7. Holographic Shearing Interferometry 115
   1.8. Comparative Holographic Interferometry 117
   1.9. Vibration Analysis 119
2. Speckle Metrology 124
   2.1. Focused Speckle Photography 125
   2.2. Defocused Speckle Photography 128
   2.3. Speckle Shearing Photography 129
   2.4. Speckle Interferometry 131
   2.5. Speckle Shearing Interferometry 140

3. Conclusions 145
References 145

Moiré Methods for Engineering and Science – Moiré Interferometry and Shadow Moiré
Daniel Post, Bongtae Han, and Peter G. Ifju 151

1. Introduction 151
2. Moiré Interferometry 152
    2.1. Basic Concepts 152
    2.2. Equipment 156
    2.3. Specimen Gratings 157
    2.4. Bithermal Loading (Isothermal Loading) 158
    2.5. Fringe Counting 158
    2.6. Strain Analysis 160
    2.7. Carrier Fringes 160
    2.8. Out-of-Plane Deformation 162
3. Advanced Techniques in Moiré Interferometry 164
    3.1. Microscopic Moiré Interferometry 164
    3.2. Curved Surfaces 166
    3.3. Replication of Deformed Gratings 170
4. Diverse Applications of Moiré Interferometry 171
    4.1. Thermal Deformations of Microelectronics Devices 171
    4.2. Textile Composites 173
    4.3. Fracture Mechanics 176
    4.4. Micromechanics: Grain Deformations 177
5. Shadow Moiré 180
    5.1. Basic Concepts 180
    5.2. Additional Considerations 183
6. Increased Sensitivity, Shadow Moiré 184
    6.1. Phase Stepping (or Quasi-heterodyne) Method 184
    6.2. Optical/Digital Fringe Multiplication (O/DFM) 185
7. Applications of Shadow Moiré 185
    7.1. Post-buckling Behavior of a Composite Column 185
    7.2. Pre-buckling Behavior of an Aluminum Channel 189
    7.3. Warpage of Electronic Devices 191
References 194

Digital Photoelasticity
Terry Y. Chen 197

1. Basic Principles of Photoelasticity 197
    1.1. Light and Complex Notation 197
    1.2. Polarization of Light 198
    1.3. Retardation 200
    1.4. Optical Media 200
    1.5. The Stress-Optic Law 202

    1.6. Plane Polariscope 204
    1.7. Circular Polariscope 207
2. Computer Evaluation of Photoelastic Fringe Patterns 210
    2.1. Digital Image Processing 211
    2.2. Extraction of Fringe Point 213
    2.3. Fringe Multiplication 213
    2.4. Determination of the Isochromatic Fringe Order 215
    2.5. Determination of Principal Stress Direction 222
3. Applications of Evaluated Data 226
    3.1. Stress Analysis 226
    3.2. Examples 227
4. System and Error Assessment 227
5. Conclusions 230
References 230

Optical Fiber Strain Sensing in Engineering Mechanics
James S. Sirkis 233

1. Intrinsic Fabry-Perot Sensor 234
    1.1. Optical Arrangement 234
    1.2. Response to Strain and Temperature 236
    1.3. Sensor Multiplexing 239
2. Extrinsic Fabry-Perot Strain Sensor 239
    2.1. Optical Arrangement 241
    2.2. Response to Strain and Temperature 242
    2.3. Self-temperature Compensation 243
    2.4. EFP Sensor Variants 244
    2.5. Sensor Multiplexing 246
3. Bragg Grating Strain Sensor 246
    3.1. Bragg Grating Fabrication 247
    3.2. Strain and Temperature Response 250
    3.3. Effects of Gradients 250
    3.4. Bragg Grating Reliability 251
    3.5. Sensor Multiplexing 251
4. Fiber-Optic Coatings and Cables 253
5. Commercial Packaging 256
6. Sensor Bonding Techniques 257
7. Demodulation 258
    7.1. Serrodyne Fringe Counting (IFP Sensors) 259
    7.2. White Light Cross-Correlator (EFP Sensors) 260
    7.3. Scanning Fabry-Perot Filter (Bragg Grating Sensors) 262
    7.4. Spectral Interrogation (EFP Sensors) 263
    7.5. Other Demodulators 266
8. Sensor Comparison 266
References 269

Long-Gage Fiber-Optic Sensors for Structural Monitoring
Daniele Inaudi 273

1. Introduction: Monitoring versus Measuring Structures 273
2. Long-Gage versus Short-Gage Sensors 274
3. Interferometric Sensors 275
   3.1. Optical Arrangements 276
   3.2. Strain and Temperature Sensitivity 276
   3.3. Demodulation Techniques 277
   3.4. Sensor Packaging 280
   3.5. Application Example: Monitoring the Versoix Bridge 283
4. Intensity-Based Sensors 284
5. Brillouin-Scattering Sensors 286
   5.1. Principles 286
   5.2. Measurement Techniques 287
   5.3. Sensor Packaging 289
   5.4. Application Example: Temperature Monitoring
        of the Luzzone Dam 290
6. Outlook 291
7. Conclusions 291
References 292

Techniques for Non-Birefringent Objects:
Coherent Shearing Interferometry and Caustics
Sridhar Krishnaswamy 295

1. Introduction 295
2. Elasto-Optic Relations 297
   2.1. Optical Phase Shift due to Transmission through an
        Optically Isotropic Linear Elastic Medium 297
   2.2. Optical Phase-Shift due to Reflection from a Polished Surface 298
3. Shearing Interferometry 299
   3.1. A Dual-Grating Shearing Interferometer –
        The Coherent Gradient Sensor (CGS) 300
   3.2. A Polarization-Based Shearing Interferometer:
        Compact Polariscope / Shearing Interferometer (PSI) 304
4. Optical Caustics 309
5. Applications 311
   5.1. Compressive Line Load on the Edge of a Plate 312
   5.2. Fracture Mechanics 315
6. Conclusion 319
References 319

Advances in Two-Dimensional and Three-Dimensional Computer Vision
Michael A. Sutton, Stephen R. McNeill, Jeffrey D. Helm, and Yuh J. Chao 323

1. Introduction 323
2. Theory and Numerical Implementation 326
   2.1. Two-Dimensional Video Image Correlation 326
   2.2. Three-Dimensional Video Image Correlation 334
3. Applications 343
   3.1. Two-Dimensional Video Image Correlation 343
   3.2. Three-Dimensional Video Image Correlation 354
4. Discussion 366
5. Summary 367
References 368

Laser Doppler and Pulsed Laser Velocimetry in Fluid Mechanics
Jeremy M. Coupland 373

1. Introduction 373
   1.1. The Scattering and Dynamic Properties of Seeding Particles . 375
2. Laser Doppler Velocimetry (LDV) 376
   2.1. Fundamentals of LDV 376
   2.2. Fourier Optics Model of LDV 381
   2.3. The Doppler Signal and Signal Processing 386
   2.4. LDV Measurements in Practice 389
3. Planar Doppler Velocimetry (PDV) 391
4. Pulsed Laser Velocimetry 395
   4.1. Particle Image Velocimetry (PIV) 395
   4.2. Removal of Directional Ambiguity 399
   4.3. PIV Measurements in Practice 400
   4.4. Three-Dimensional PIV 402
   4.5. Holographic Particle Image Velocimetry (HPIV) 402
5. Conclusions, Discussion and Future Development 408
References 410

Surface Characterization and Roughness Measurement in Engineering
David J. Whitehouse 413

1. General 413
   1.1. Historical 413
   1.2. Nature and Importance of Surfaces 414
   1.3. Trends 416
2. Instrumentation 419
   2.1. General Points 419

2.2. The Stylus	422
2.3. Basic Instrument	425
2.4. The Stylus Method (Mechanical)	427
2.5. The Optical Methods	428
2.6. Other Conventional Methods	433
2.7. Non-conventional Methods	433
3. Pre-Processing and Filtering	433
3.1. Levelling	433
3.2. Curve Fitting for Form	433
3.3. Filtering for Waviness	434
4. Parameters	439
4.1. General	439
4.2. Height Parameters (see [36] for example)	440
4.3. Peak Parameters	441
4.4. Spacing Parameters	442
4.5. Peak Parameters (Statistical)	443
4.6. Hybrid Parameters	445
4.7. Effects of Filtering on Parameter Values	445
5. Random Process Analysis in Surface Metrology	446
5.1. General	446
5.2. Height Information	446
5.3. Spacing Information	447
6. Areal (or Three-Dimensional) Parameters	453
6.1. General	453
6.2. Comments on Areal Measurement	454
7. Conclusions	458
References	459
Index	463

Optics for Engineers

Daniel Malacara-Hernández

Centro de Investigaciónes en Optica,
Lomas del Bosque No. 115, Lomas del Campestre,
León, Gto, Mexico 37150
dmalacara@foton.cio.mx

Abstract. Optical metrology techniques are fundamental tools in optical engineering. They use the wavelength of light as the basic scale for most measurements and thus their accuracy is much higher than that of mechanical metrology methods. A good understanding of the nature of light and of optical phenomena is necessary when working with optical metrology methods. Here a review of the main metrological instruments and their associated optical phenomena will be made.

1 Introduction

To study how light behaves in optical instruments, quantum theory is not generally needed. Most phenomena can be explained using the geometric or wave models of light. Geometrical optics considers a beam of light formed by rays of light, which are the geometric trajectories followed by the light energy. This model assumes that the media in which light propagates are isotropic, but not necessarily homogeneous.

If the aperture through which the beam of light passes is smaller than a small fraction of a millimeter, diffraction effects appear and the geometric model of light is not valid. If two or more beams are superimposed in some places in space, interference effects may arise. If the material is not isotropic, polarization effects appear. In all of these cases a wave model of light becomes necessary. Thus, light has to be considered as a transverse wave, namely an electromagnetic wave.

2 Geometrical Optics Principles

2.1 Fermat's Principle and the Law of Refraction

The Fermat principle is the foundation of all geometrical optics theory. It can be expressed as follows:

The path followed by the light when traveling from one point in space to another is such that the traveling time is an extremum. In other words, this time is a minimum, a maximum or stationary with respect to neighboring paths.

In order to understand Fermat's principle we first need to be aware that the speed of light is different in different media. The speed of light is

$c = 299,792$ km/s, in vacuum, but is smaller in any other medium. The refractive index n is defined as the ratio of the speed of light in vacuum to the speed of light in the medium. Thus, the refractive index n is

$$n = \frac{c}{v}, \qquad (1)$$

where v is the velocity of light in the medium, and is always greater than one. Table 1 shows the refractive indices for some transparent media.

Table 1. Refractive indices for some transparent materials

Material	Refractive index
Vacuum	1.0000
Air	1.0003
Water	1.33
Fused silica	1.46
Plexiglass	1.49
Borosilicate crown	1.51
Ordinary crown	1.52
Canada Balsam	1.53
Light flint	1.57
Extra dense barium crown	1.62
Extra dense flint	1.72
Diamond	2.42

Another quantity frequently used is the optical path OP, from one point P1 to another point P2 in space which is defined by

$$\text{OP} = \int_{P_1}^{P_2} n \, dx \, . \qquad (2)$$

Fermat's can therefore be expressed by saying that the optical path when light travels from the point P1 to the principle for a reflecting circle and a reflecting ellipse point P2 should an extremum. Figure 1 illustrates Fermat's principle.

The laws of refraction and reflection can then be derived, as shown in almost any introductory optics book, as a consequence of Fermat's principle. If light travels from a point P1 in a medium with refractive index n to a point P2 in a medium with refractive index n' separated by a plane interface, as in Fig. 2, we have

$$n \sin I = n' \sin I' \, , \qquad (3)$$

where I and I' are the angles forming the incident and refracted rays, respectively with respect to the normal to the flat interface. The incident and the refracted rays are in a common plane with the interface normal. This is the well-known refraction law or Snell's law, illustrated in Fig. 2.

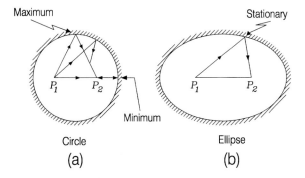

Fig. 1a,b. Fermat's principle in a reflecting sphere and a reflecting ellipsoid

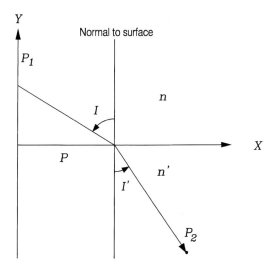

Fig. 2. Snell's law of reflection

The reflection law is

$$I = -I', \tag{4}$$

which may be considered as a special case of Snell's law if $n = -n'$.

Using the laws of refraction and reflection all of geometrical optics theory can be derived, as will be shown in the following sections.

2.2 First-Order Optics

Nearly all reflective or refractive surfaces in optical systems are either spherical or flat. On the other hand, a flat surface may be considered as a special kind of spherical surface, with an infinite radius of curvature. Non-spheric, also called aspheric, surfaces are sometimes used, but their deviation from the spherical shape is not generally large. So, it is natural to begin with the study of the refraction of light rays at a spherical surface.

The spherical surface in Fig. 3 is a circular section of a sphere. The radius of the sphere is the radius of curvature r and c its center of curvature. The center V of the circular section is the vertex. A straight line passing through the vertex V and the center of curvature C is the optical axis.

A light ray can follow any straight trajectory in the first medium, to arrive at the optical axis. If it is not in a common plane with the optical axis it is said to be an oblique ray. Unfortunately the geometry involved in the refraction of oblique rays is a little complicated [1].

If the ray of light is in a common plane with the optical axis, it is a meridional ray and the plane containing both rays is the meridional plane, which normally is the drawing plane. Meridional rays are much simpler than general oblique rays, and their study provides most of the important properties of optical systems. Let us consider a meridional ray with the parameters indicated in Fig. 3. The incident ray has two degrees of freedom (U and L) and the refracted ray also two (U' and L'). Thus, only four equations are necessary, as follows

$$\frac{\sin I}{L-r} = -\frac{\sin U}{r}, \tag{5}$$

$$\frac{\sin I'}{L'-r} = -\frac{\sin U'}{r}, \tag{6}$$

$$-U + I = -U' + I', \tag{7}$$

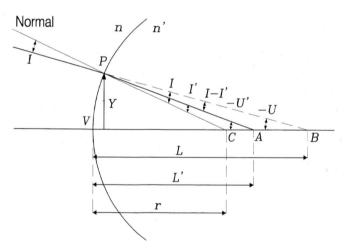

Fig. 3. Refraction of a ray of light at a spherical surface

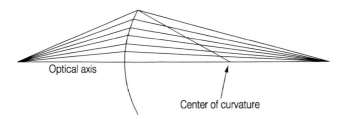

Fig. 4. Convergence of a beam of rays by a spherical refracting surface

and

$$n \sin I = n' \sin I' . \tag{8}$$

With these four expressions we can show that if a beam of rays starts from a common point on the optical axis (object), they pass very close to each other at another point on the optical axis (object), as illustrated in Fig. 4. We see that if the diameter of the surface is sufficiently reduced, we can consider all rays to be refracted to a common point. These meridional rays, which are not far from the optical axis, are the paraxial rays.

The study of paraxial rays is called paraxial or first-order optics. Thus, we can replace the sine of the angles by the angles measured in radians. Then, it is possible to obtain the following useful relation known as the Gaussian equation.

$$\frac{n' - n}{r} = \frac{n'}{l'} - \frac{n}{l} . \tag{9}$$

With two spherical surfaces aligned with a common optical axis we have a lens. If the separation between the surfaces is very small compared to the radii of curvature of the surfaces we have a thin lens. Applying the gaussian equation to the two surfaces of a thin lens we may obtain

$$\frac{l}{f} = (n - 1) \left(\frac{1}{r_1} - \frac{1}{r_2} \right) , \tag{10}$$

where f is the focal length, defined as the distance from the lens to the point where the paraxial rays converge when the lens is illuminated with a point source (object) at infinity. This point of convergence of the rays (image) is the focus. If the focus and the point light source are on opposite sides of the lens, it is convergent (positive focal lengths). Figure 5 shows convergent and divergent lenses.

If the object is at the right-hand side of the lens, the distance l from the lens to the object is positive and the object is virtual. If the object is to the left of the lens, the distance l is negative and the object is real (see Fig. 6). If the image is to the right of the lens, the distance l' from the lens to the

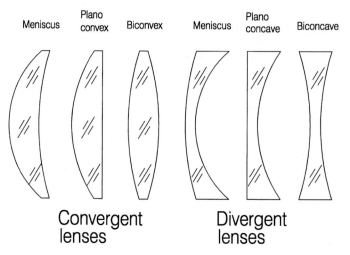

Fig. 5. Converging and diverging lenses

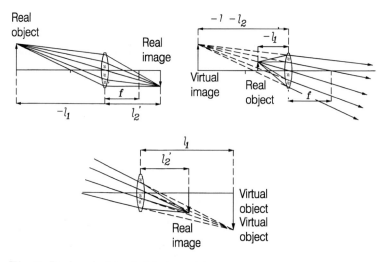

Fig. 6. Real and virtual objects and images

object is positive and the image is real. If this image is to the left of the lens, the distance l' is negative and the image is virtual (see Fig. 6).

The object and image distances for a thin lens are related by

$$\frac{1}{f} = \frac{1}{l'_2} - \frac{1}{l_1} \,. \tag{11}$$

Optical systems with spherical surfaces with a common axis of symmetry can also be described with first-order optics, as in many geometrical optics textbooks [1].

In thick optical systems formed by several thin and/or thick lenses, the principal points $P1$ and $P2$ and the nodal points $N1$ and $N2$ are defined as graphically shown in Fig. 7a. The distance from the principal point $P2$ to the focus $F2$ is the effective focal length. This focal length can be experimentally measured with the property of the principal points illustrated in Fig. 7b.

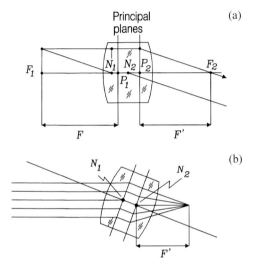

Fig. 7. (a) Illustration of the principal and nodal points in an optical system (b) Rotation of a lens about the second nodal point

2.3 Aberrations

We have seen that, strictly speaking, if we have a point light source, only paraxial rays converge at a single point called the image. If the diameter of the lens is not very small compared with the radii of curvature of the surfaces, the rays will not cross the axis at the same point, as shown in Fig. 8. Marginal rays will have a different point of convergence to that of the paraxial rays. This is the spherical aberration of a lens. It can be canceled with a proper lens design, so that some surfaces compensate the spherical aberration of other surfaces. However, given a thin lens, some rules of thumb can be found for reducing as much as possible the spherical aberration. The approximate condition is that the angle of incidence of the ray entering the lens and the angle of refraction of the ray exiting the lens should have magnitudes as close as possible to each other, as shown in the examples in Fig. 9.

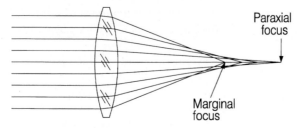

Fig. 8. Spherical aberration

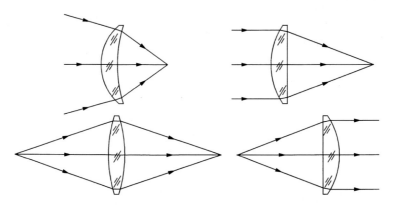

Fig. 9. Proper lens orientations to reduce the spherical aberration in a lens

Even if the spherical aberration is perfectly eliminated in an optical system, if the point object is laterally shifted away from the optical axis, the image also shifts, but a single converging point for the rays no longer exists. Instead of a point, the image has the shape illustrated in Fig. 10 due to the coma aberration. Fortunately, if in a single thin lens without any stop in front of or behind the lens the spherical aberration is minimized, the coma is also minimized. In more complicated lens systems, with several lenses and a stop, the spherical aberration and coma are completely independent and their analysis is quite complicated. The magnitude of the comatic image grows linearly with the image separation from the optical axis (image height).

It is interesting and useful to know that symmetric optical systems, with the object and image also placed symmetrically with respect to the lens, automatically eliminate the coma aberration, as in Fig. 11.

There is another off-axis aberration called astigmatism. In this aberration rays in the meridional plane (or tangential plane) converge to a different point to that for rays in a perpendicular plane (sagittal), as illustrated in Fig. 12. Astigmatism makes the image of a point appear elongated as an ellipse or even a small line, instead of a point. The magnitude of this aberration increases with the square of the image height.

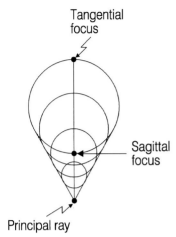

Fig. 10. Image of coma aberration

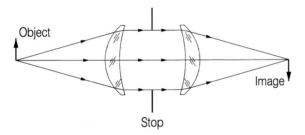

Fig. 11. Symmetric system without coma

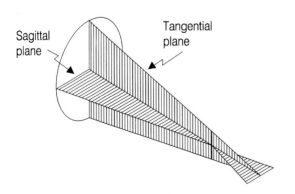

Fig. 12. An optical system with astigmatism

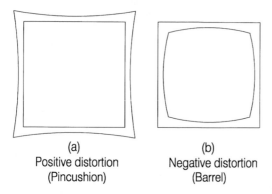

(a) Positive distortion (Pincushion)   (b) Negative distortion (Barrel)

Fig. 13a,b. Image of a square with distortion

Coma and astigmatism degrade the quality of the off-axis-images. Since coma increases linearly with the off-axis image displacement, while astigmatism increases as with the square of this displacement, a practical consequence is that coma is more important than astigmatism for small off-axis displacements. On the other hand, for large off-axis displacements astigmatism is more important than coma. Symmetrical optical systems are free of coma, but not free of astigmatism.

If the image is a perfect point, but its lateral off-axis displacement is not linear with the object displacement, an aberration known as distortion appears. The image of a square then looks as illustrated in Fig. 13. In general, the analysis of distortion is quite complicated. The distortion aberration, like coma, is automatically canceled in a completely symmetrical optical system.

Another aberration appears if for different point objects in a plane the corresponding point images are not in a plane but in a sphere, as in Fig. 14. This is probably the most complicated of all aberrations, because it is associated with astigmatism. In the presence of astigmatism, there are four different image surfaces, with different curvatures. These are: (a) the tangential surface, where the tangential rays are focused, (b) the sagittal surface,

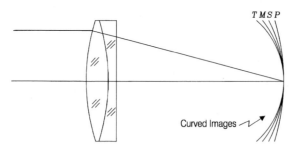

Fig. 14. Image surface with field curvature

where the sagittal rays are focused, (c) the best focus surface, where neither the sagittal nor the tangential rays are focused, but we have an average focus between these two and, (d) the Petzval surface, where the image is clearly focused if the astigmatism is zero. These surfaces are illustrated in Fig. 14.

The refractive index is a function of the wavelength of light. So, images with different colors may be located at different planes and with different magnifications. If the magnification is the same for different colors, but the images are at different axial locations, we have axial chromatic aberration. If the images for different colors are at the same axial locations, but have different magnifications, we have magnification chromatic aberration. These chromatic aberrations are illustrated in Fig. 15 and Fig. 16. A thin lens is corrected for both aberrations simultaneously if a doublet with two different glasses is designed.

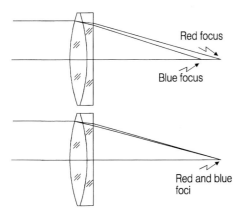

Fig. 15. Nature of the axial chromatic aberration

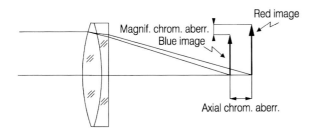

Fig. 16. Images with axial and magnification chromatic aberrations

3 Wave-Optics Fundamentals

Many optics experiments cannot be explained using the concept of a ray of light. Instead, the wave nature of light has to be considered, as will be seen in this section.

3.1 Young's Double Slit

A typical example of an experiment where light is clearly a wave is the Young's double slit in Fig. 17. A point source of light emits a spherical wave that illuminates both slits. At each slit the light is diffracted, producing waves diverging from these slits. Thus, any point on the screen is illuminated by the waves emerging from the two slits. Since the total paths from the point source to a point on the observing screen are different, the phases of the two waves are not the same. Then, constructive or destructive interference takes place at different points on the screen, forming interference fringes. The basic relation to explain the formation of the fringes is

$$(AB + BC) - (AD + DC) = m\lambda, \tag{12}$$

where m is the order of interference. When m is an integer we have at the center of a bright fringe.

In this experiment the fringes are formed even if the light source is not monochromatic. Since the fringe separation is different for different colors, the fringe contrast is high only near a central line on the screen, parallel to the slits.

If the light source is extended, each point of the light source forms its own set of fringes, mutually displaced with respect to each other. If the light source is large, the fringe contrast is reduced or even disappears.

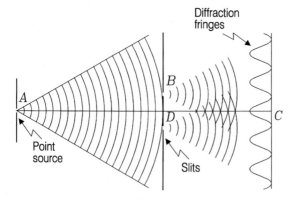

Fig. 17. Diffraction by a Young's double slit

3.2 Michelson Interferometer

A Michelson Interferometer, shown in Fig. 18, is illuminated with an extended light source. The emitted beam of light is separated into two beams with smaller amplitudes, by the plane parallel glass plate, called the beam splitter. The two beams are then reflected back to the beam splitter by two mirrors. At the beam splitter the two beams are recombined along a common path to the observer's eye.

Interference between these two waves takes place. The observer's eye sees two virtual images of the extended light source, one on the top of the other, but separated by a certain distance. The reason is that the two arms of the interferometer may have different lengths.

To observe interference fringes with a non-monochromatic or white light source, the two optical paths must be identical for all wavelengths. This is possible only if the optical path is the same for the two interfering beams, at all wavelengths. This is possible only if two conditions are met:

a) the optical path is the same for both beams, for any wavelength,
b) the two observed virtual images of the light source must coincide at the same plane in space.

If the length of one of the arms of the interferometer is changed by moving one of the mirrors along its perpendicular, either one of these two conditions is satisfied but not necessarily both. Both conditions can be simultaneously fulfilled only if the two interferometer beams travel in the same materials, glass and air. It can be easily seen that this is possible only if a compensation plate is introduced in one of the interferometer arms, as shown in Fig. 18. If the interferometer is compensated and the two optical paths are exactly equal, interference fringe with white light can be observed.

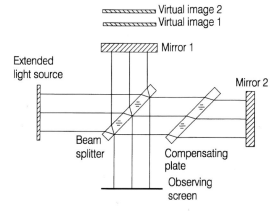

Fig. 18. Michelson interferometer

3.3 Coherence and Light Sources

An ideal light source for many optics experiments is a point source with only one pure wavelength (color). However, in practice most light sources are not a point, but have a certain finite size and emit several wavelengths simultaneously.

A point source is said be spatially coherent if, when used to illuminate a system of two slits, interference fringes are produced. If an extended light source is used to illuminate the two slits no interference fringes are observed. Thus, an extended light source is said to be spatially incoherent.

A monochromatic light source has an infinite sinusoidal wavetrain (long coherence length). On the other hand, a non-monochromatic light source has a short wavetrain (short coherence length). A light source with a short wave train is said to be temporally incoherent. A monochromatic light source is temporally coherent.

The most common light source in interferometry is the helium-neon laser, which has a long wavetrain coherence length and monochromaticity. However, these characteristics are sometimes a problem, because many spurious fringes are formed, unless great precautions must be taken to avoid them.

When a laser light source is used, extremely large optical path differences can be introduced in the interferometer without losing fringe contrast. The light emitted by a gas laser consists of several equally spaced spectral lines, called longitudinal modes, with a frequency separation

$$\Delta \nu = \frac{c}{2L}, \tag{13}$$

where L is the laser cavity length.

If the cavity length L of a laser is modified for some reason, for example by thermal expansion or contraction or mechanical vibrations, the spectral lines (longitudinal modes) move, preserving their relative separations, but with their intensities inside a gaussian dotted envelope (power gain curve) as shown in Fig. 19.

Single mode or single frequency lasers are designed and made to produce a perfectly monochromatic wavetrain, but because of instabilities in the cavity length, the frequency may be unstable. Single frequency lasers with extremely

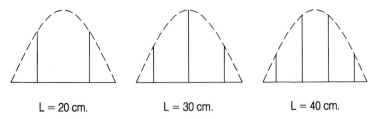

Fig. 19. Longitudinal modes in a helium-neon laser

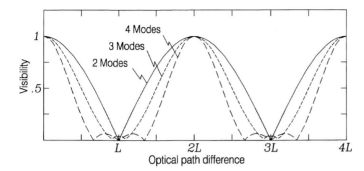

Fig. 20. Contrast variation as a function of the optical path difference in a Twyman-Green or Fizeau interferometer when illuminated with a helium-neon laser

stable frequencies are commercially produced. They are the ideal source for interferometry with large optical path differences without reducing the fringe contrast.

The fringe visibility in an interferometer using a laser source with several longitudinal modes is a function of the optical path difference (OPD), as shown in Fig. 20. To have a good contrast, the OPD has to be an integral multiple of $2L$. A laser with two longitudinal modes is sometimes stabilized to avoid contrast changes.

A laser diode is also frequently used in interferometers. *Creath* [2], and *Ning* et al. [3] have described the most important coherence characteristics of these lasers for use in interferometers. The low coherence length, of the order of one millimeter, is a great advantage in many applications, with the additional advantage of their low price and small size.

4 Main Interferometers Used in Optical Engineering

Two-wave interferometers produce an interferogram by superimposing two wavefronts: a flat reference wavefront and a distorted wavefront whose distortions are to be measured. An interferometer measures small wavefront deformations of the order of a small fraction of the wavelength of light. The main interferometers used in optical engineering have been described in several places in the literature [4].

To understand how interferometers work, consider a two-wave interferogram with a flat wavefront, whose deformations are given by $W(x, y)$ interfering with a flat wavefront as in Fig. 21. The amplitude $E_1(x, y)$ in the observing plane is the sum of the of the two waves $A_1(x, y)$ and $A_2(x, y)$, given by

$$E_1(x, y) = A_1(x, y) \exp i[(kW(x, y)] + A_2(x, y) \exp i(kx \sin \theta) , \qquad (14)$$

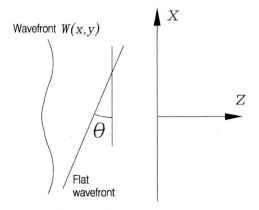

Fig. 21. Two interfering wavefronts

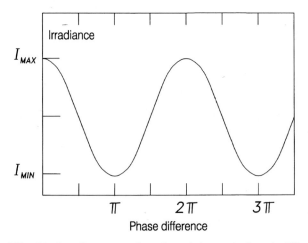

Fig. 22. Irradiance as a function of the optical path difference in a Twyman-Green or Fizeau interferometer

where $k = 2\pi/\lambda$. The irradiance function $I(x,y)$ may then be written as

$$E_1(x,y)E_1^*(x,y) = A_1^2(x,y) + A_2^2(x,y)$$
$$+ 2A_1(x,y)A_2(x,y)\cos k[x\sin\theta - W(x,y)] \,, \quad (15)$$

where the symbol $*$ denotes the complex conjugate. This function is plotted in Fig. 22.

4.1 Fizeau Interferometer and Newtons Rings

The Fizeau interferometer, shown in Fig. 23, like the Michelson interferometer described in the preceding section, produces the two interfering beams with an amplitude beam splitter. However, unlike the Michelson interferometer the illuminating light source is a point source and monochromatic. The spherical wavefront produced by the light source is transformed into a flat wavefront (collimated) by means of a lens. This wavefront is reflected back onto one of the faces of the beam splitter glass plate, which is partially reflecting. The transmitted beam goes to the optical element under test and then gets reflected back to the beam splitter.

Different kinds of optical elements can be tested with this interferometer, for example, a glass plate, which is also used as the reference beam splitter, as shown in Fig. 24. Then, the optical path difference OPD can be written as

$$\text{OPD} = nt \;, \tag{16}$$

where t is the glass plate thickness. So, a field with no fringes means that nt is a constant, but n and t are not measured separately.

When testing a convex surface, the reference surface can be flat or concave, as in Fig. 25. The quality of an optical surface can be measured as in Fig. 23. In this case the optical path difference is equal to $2t$. Let us imagine now that the point light source is laterally displaced by a small amoun s. Then, the collimated wavefront will be tilted by an angle θ given by

$$\theta = \frac{s}{f} \;, \tag{17}$$

where f is the focal length of the collimator. With this tilted wavefront the optical path difference OPD is equal to the distance $AB + BC$ in Fig. 26 or

$$\text{OPD} = 2t \cos \theta \;. \tag{18}$$

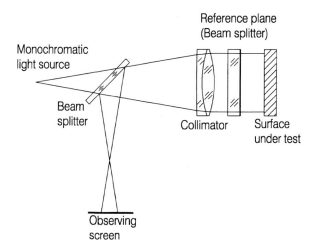

Fig. 23. Fizeau interferometer

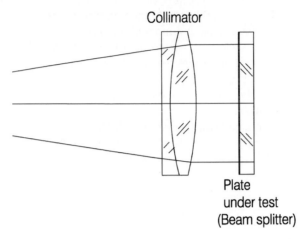

Fig. 24. Fizeau interferometer testing a glass plate used as a reference beam splitter

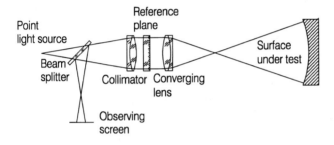

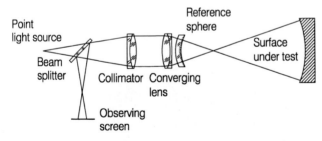

Fig. 25. Fizeau interferometer to test a concave surface

Thus, OPD at a small angle θ minus the optical path difference on-axis is approximately given by

$$\Delta \text{OPD} = t\theta^2 = \frac{ts^2}{f^2} . \tag{19}$$

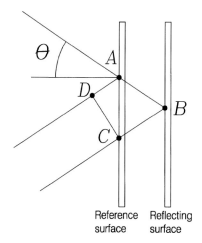

Fig. 26. Optical path difference for an angle of incidence θ, in a Fizeau interferometer

If we have a small extended light source with semidiameter s, the fringes will have a reasonably high contrast as long as

$$\Delta \text{OPD} = \frac{ts^2}{f^2} \leq \frac{\lambda}{4} . \tag{20}$$

We can see that the light source can increase its size s if the air gap thickness t is reduced.

If the collimator lens has spherical aberration, the refracted wavefront will not be flat. Thus, the maximum transverse aberration TA in this lens can be interpreted as the semidiameter s of the light source. The conclusion is that the quality requirements for the collimator lens increase as the optical path difference is also increased. If the optical path difference is zero, the collimator lens can have any magnitude of spherical aberration.

The so-called Newton interferometer is a Fizeau interferometer in which the air gap is reduced so much that a large extended source can be used. This high tolerance in the magnitude of the angle θ therefore allows the collimator to be eliminated, if a reasonably large observing distance is used. Figure 27 shows a Newton interferometer, with a collimator, so that the effective observing distance is always infinite.

A Newton interferometer is quite frequently used to test plane, concave spherical or convex spherical surfaces by means of measuring test plates with the opposite curvature that are placed over the surface under test.

4.2 Twyman-Green Interferometer

A Twyman-Green interferometer is really a modification of the Michelson interferometer described previously, as shown in Fig. 28. The basic modification

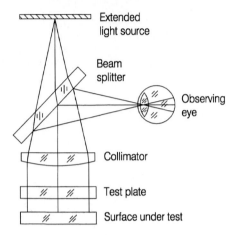

Fig. 27. Newton interferometer

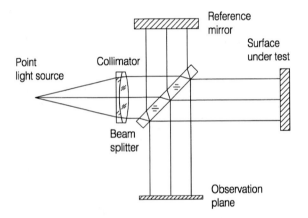

Fig. 28. Twyman-Green interferometer

is to replace the extended light source by a point source and a collimator, as in the Fizeau interferometer. Thus, the wavefront is illuminated with a flat wavefront. Hence, the fringes in a Twyman-Green interferometer are of the equal thickness type.

As in the Michelson interferometer, white light fringes are observed only if the instrument is compensated with the compensating plate. However, normally a monochromatic light source is used, eliminating the need for the compensating plate.

The beam splitter must have extremely flat surfaces and its material must be highly homogeneous. The best surface must be the reflecting one. The non-reflecting surface must not reflect any light, to avoid spurious interference fringes. Thus the non-reflecting face must be coated with an antireflection multilayer coating. Another possibility is to have an angle of incidence at the beam splitter with a magnitude equal to the Brewster angle and thus, completely polarizing the incident light beam.

The size of the light source can be slightly increased to a small finite size if the optical path difference between the two interferometer arms is small, following the same principles used for the Fizeau interferometer.

The arrangements in Fig. 29 can be used to test a glass plate or a convergent lens. When testing a glass plate, the optical path difference is given by

$$\text{OPD} = (n-1)\,t\ . \tag{21}$$

When no fringes are present, we can conclude that $(n-1)t$ is a constant, but not n or t independently. If we compare this expression with the equivalent for the Fizeau interferometer (16), we see that n and t can independently be measured if Fizeau and Twyman-Green interferometers are both used.

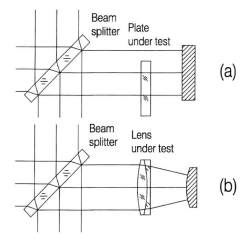

Fig. 29. Testing a glass plate (**a**) and a convergent lens (**b**) in a Twyman-Green interferometer

A convex spherical mirror with its center of curvature at the focus of the lens is used to test lenses with long focal lengths, and a concave spherical mirror to test lenses with short focal lengths. A small, flat mirror at the focus of the lens can also be employed. The small region used on the flat mirror is so small that its surface does not need to be very accurate. However, the wavefront is rotated 180° making the spatial coherence requirements (a gas laser is necessary) higher and canceling odd aberrations like coma.

Concave or convex optical surfaces are tested using a Twyman-Green interferometer, as shown in Fig. 30. Large astronomical mirrors can also be tested with an unequal path interferometer as described by *Houston* et al. [5].

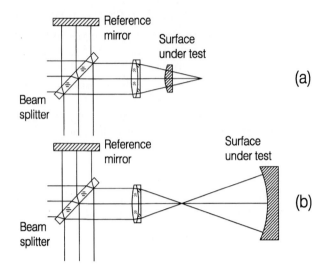

Fig. 30a,b. Testing convex and concave surfaces in a Twyman-Green interferometer

However, it must be remembered that if the collimator does not produce a perfectly flat wavefront, the optical path difference must not be very large, as in the Fizeau interferometer.

The Twyman-Green and Fizeau interferograms produce the same interferogram if the same aberration is present. The interferograms produced by the Seidel primary aberrations have been described by *Kingslake* [6], and their associated wavefront deformations can be expressed by

$$W(x,y) = A(x^2+y^2)^2 + By(x^2+y^2)$$
$$+ C(x^2+y^2) + D(x^2+y^2) + Ex + Fy + G \qquad (22)$$

where

A = spherical aberration coefficient
B = coma coefficient
C = astigmatism coefficient
D = defocusing coefficient
E = tilt about the y-axis coefficient (image displacement along the x-axis)
F = tilt about the x-axis coefficient (image displacement along the y-axis)
G = piston or constant term.

4.3 Ronchi and Lateral-Shear Interferometers

Lateral shear interferometers produce identical wavefronts, one laterally sheared with respect to the other, as shown in Fig. 31. The advantage is that a perfect reference wavefront is not needed. The optical path difference in these interferometers can be written as

$$\text{OPD} = W(x, y) - W(x - S, y) , \tag{23}$$

where S is the lateral shear. If this lateral shear is very small compared with the aperture diameter, we obtain

$$\text{OPD} = S \frac{\delta W(x, y)}{\delta x} . \tag{24}$$

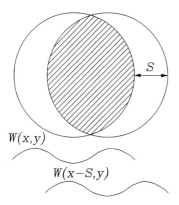

Fig. 31. Wavefronts in a lateral shear interferometer

This interferometer can become quite simple, but the practical problem they have is that the interferogram represents the wavefront slopes in the shear direction, not the wavefront shape. Thus, to obtain the wavefront deformations a numerical integration of the slopes have to be performed. Besides, two laterally sheared interferograms in mutually perpendicular directions are needed.

If the shear is not small enough, the interferogram does not represent the wavefront slope. Then, another procedure to obtain the wavefront deformation has to be employed. One of these possible methods has been proposed by *Saunders* [7] and it is described in Fig. 32. To begin, let us assume that $W_1 = 0$. Then, we may write

$$\begin{aligned}
W_1 &= 0 \\
W_2 &= \Delta W_1 + W_1 \\
W_3 &= \Delta W_2 + W_2 \\
&\dots\dots\dots\dots\dots\dots \\
W_n &= \Delta W_{n-1} + W_{n-1} .
\end{aligned} \tag{25}$$

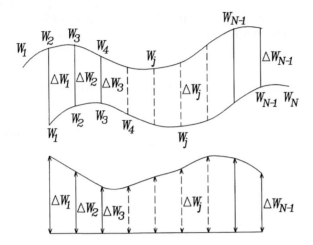

Fig. 32. Saunders method for finding the wavefront in a lateral shear interferometer

A disadvantage of this method is that the wavefront is evaluated only at points separated by a distance S. Intermediate values are not measured and have to be interpolated.

The simplest lateral shear interferometer is the one due to *Murty* [8] and shown in Fig. 33. The great practical advantages of this instrument are simplicity, low price and fringe stability. The only small disadvantage is that it is not compensated and thus has to be illuminated by laser light.

The lateral shear interferograms for the primary aberrations may be obtained from the expression for the primary aberrations, as will now be described.

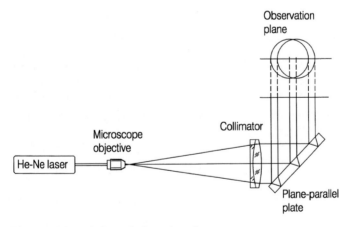

Fig. 33. Murty's lateral shear interferometer

The interferogram with a defocused wavefront is given by

$$2DxS = m\lambda . \tag{26}$$

This is a system of straight, parallel and equidistant fringes. These fringes are perpendicular to the lateral shear direction. When the defocusing is large, the spacing between the fringes is small. On the other hand, when there is no defocus, there are no fringes in the field.

In the case of spherical aberration the interferogram is given by

$$4A(x^2 + y^2)xS = m\lambda , \tag{27}$$

if this aberration is combined with defocus we may write instead

$$\left[4A(x^2 + y^2)x + 2Dx\right] S = m\lambda , \tag{28}$$

then, the interference fringes are cubic curves.

When coma aberration is present, the interferogram is given by

$$2BxyS = m\lambda , \tag{29}$$

where the lateral shear is S in the *sagittal* direction. If the lateral shear is T in the tangential y direction, the fringes are given by

$$B(x^2 + 3y^2)T = m\lambda . \tag{30}$$

In the case of astigmatism, when the lateral shear is S in the sagittal x direction, the fringes are given by

$$(2Dx + 2Cx)S = m\lambda , \tag{31}$$

and for the lateral shear T in the tangential y direction we have

$$(2Dy - 2Cy)T = m\lambda . \tag{32}$$

The fringes are straight and parallel as in the case of defocus, but with a different separation for both interferograms.

The well-known and venerable Ronchi test [9], illustrated in Fig. 34, can be considered a geometrical test but also a lateral shear interferometer. In

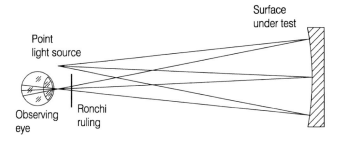

Fig. 34. Ronchi test

the geometrical model the fringes are the projected shadows of the Ronchi-ruling dark lines. However, the wave model assumes that several laterally sheared wavefronts are produced by diffraction. Thus the Ronchi test is a multiple-wavefront lateral-shear interferometer.

4.4 Talbot Interferometer and Moiré Deflectometry

If we project the shadow of a Ronchi ruling with a collimated beam of light, as in Fig. 35, due to diffraction the shadows of the dark and clear lines are not clearly defined. Intuitively, we might expect to get sharp and well-defined shadows for extremely short distances from the ruling to the observing screen, which is easily confirmed to be true. When we gradually increase the observing distance, we see that the fringe sharpness also decreases, until at a certain distance the fringes completely disappear. It is surprising, however, as discovered by *Talbot* [10], that by further increasing the observing distance the fringes become again sharp and then disappear again in an alternating manner. Figure 36 shows how these fringes change their contrast with the observation distance. Negative contrast means that a clear fringe appears where there should be a dark fringe and vice versa. Talbot was not able to explain this phenomenon. Later, *Rayleigh* [11] explained it. The period of this contrast variation is called the Rayleigh distance L_R, which can be written as

$$L_R = \frac{2d^2}{\lambda}, \tag{33}$$

where d is the spatial period (line separation) of the ruling and λ is the wavelength of the light.

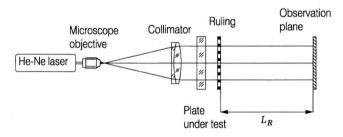

Fig. 35. Observation of the Talbot effect

If a distorted glass plate is placed in the path of the collimated light beam, the illuminating light beam is no longer flat but distorted, and the observed projected fringes will also be distorted, instead of straight and parallel.

A simple interpretation is analogous to the Ronchi test, for both the geometrical and the wave models. The geometrical model interprets the fringe

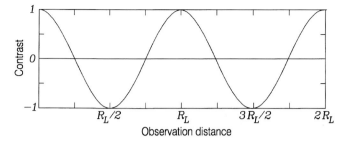

Fig. 36. Contrast variation distance in the Talbot effect

deformation as due to the different local wavefront slopes producing different illumination directions. Hence this method is frequently given the name of deflectometry [12,13]. The wave model interprets the fringes as due to the interference between multiple diffracted and laterally sheared wavefronts. Talbot interferometry and their multiple applications have been described by many authors, for example by *Patorski* [14] and *Takeda* et al. [15].

The fringes produced are of such a high spatial frequency that the linear carrier should be removed by a moiré effect with another identical Ronchi ruling at the observation plane.

4.5 Foucault Test and Schlieren Techniques

Leon Foucault [16] proposed an extremely simple method to evaluate the shape of concave optical surfaces. A point light source is located slightly off-axis, near the center of curvature. If the optical surface is perfectly spherical, a point image will be formed by the reflected light also near the center of curvature, as shown in Fig. 37.

A knife edge then cuts the converging reflected beam of light. Let us consider three possible planes for the knife edge. If the knife is inside the focus the projected shadow of the knife will be projected on the optical surface on

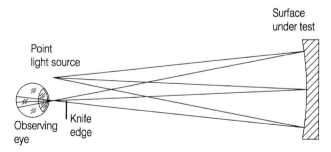

Fig. 37. Foucault test

the same side as the knife. On the other hand, if the knife is outside of focus, the shadow will be on the opposite side to the knife. If the knife is at the image plane, nearly all light is intercepted with even a small movement of the knife.

If the wavefront is not spherical the shadow of the knife will create a light pattern on the mirror where the darkness or lightness will be directly proportional to the wavefront slope in the direction perpendicular to the knife edge. The intuitive impression is a picture of the wavefront topography. With this test even small amounts of wavefront deformation a fraction of the wavefront can be detected.

If a transparent fluid or gas is placed in the optical light path between the lens or mirror used to produce the spherical wavefront and the knife edge, a good sensitivity to the refractive index gradients in the direction perpendicular to the knife edge is obtained. For example, any air turbulence can thus be detected and measured. This is the working principle of the Schlieren techniques used in atmospheric turbulence studies [17].

4.6 Two-Wavelength Interferometry

When the asphericity of the wavefront being measured is quite strong, the slope of the wavefront may be so large that the Nyquist limit is exceeded. In other words, the fringe spacing becomes smaller than twice the pixel separation, as required by the sampling theorem. Under these circumstances a phase unwrapping of the wavefront is almost imposible. To solve this problem, *Wyant* [18] proposed an interferometric method in which two different wavelengths are used simultaneously. If two wavelengths λ_1 and λ_2 are used, the wavetrain is modulated as in Fig. 38, with a group length given by

$$\lambda_{eq} = \frac{\lambda_1 \lambda_2}{|\lambda_2 - \lambda_1|}. \tag{34}$$

In one approach two different interferograms are taken for each of the two wavelengths. Then a moiré pattern is formed by the superposition of the two interferograms. In another approach, the interferometer is simultaneously illuminated with the two wavelengths to directly form the moiré pattern. With this two-wavelength method, the precision of the smaller wavelength

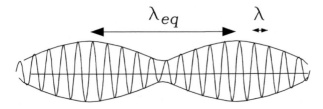

Fig. 38. Wavetrain produced by the combination of two wavelengths

components is retained, while increasing the wavefront slope range to that given by the equivalent wavelength.

4.7 Holographic Principles

The basic principle of holography was discovered by *Gabor* [19] as a procedure to form images without lenses, using diffraction. However, his attempts were not completely successful because he could not separate in the final image the first order of diffraction producing the image from the undesired zero and minus one orders of diffraction. A few years later *Leith* and *Upatnieks* [20] discovered a method for separating these diffracted beams by introducing a large angle between them.

We give next, a brief description of the holographic principles and their applications.

4.7.1 Thin Holograms

By means of holography a wavefront can be stored with amplitude and phase (shape) in a photographic plate. Then, the original wavefront can be reconstructed. Consider, as in Fig. 39a, a wavefront with deformations $W(x,y)$ interfering at a photographic plate with a flat reference wavefront with an angle between them. Then, the resultant amplitude can be written as follows

$$E(x,y) = A(x,y)\exp \mathrm{i}[kW(x,y)] + A_0 \exp \mathrm{i}(kx\sin\theta) \ . \tag{35}$$

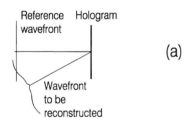

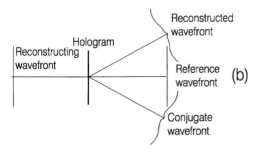

Fig. 39. (a) Forming and (b) reconstructing a hologram

After being developed, this photographic plate (hologram) is illuminated with the reference wavefront used during the hologram formation, as in Fig. 39b. If we assume, as a first approximation, that the optical amplitude transmission of the developed hologram is linear with the irradiance that formed it, we may write the transmitted amplitude in the reconstruction process as

$$[E(x,y)E^*(x,y)]A_0 \exp i(kx \sin \theta) = [A^2(x,y) + A_0^2]A_0 \exp i(kx \sin \theta)$$
$$+ A(x,y)A_0^2 \exp[ikW(x,y)]$$
$$+ A(x,y)A_0^2 \exp\{ik[2x \sin \theta - W(x,y)]\} \ . \qquad (36)$$

The first term in this expression is the reference or reconstructing wavefront. The second term is the original wavefront to be reconstructed. The third term is a wavefront with identical deformations to these on the original wavefront, but with opposite sign (conjugate wavefront).

The hologram may be interpreted as a diffraction grating with distorted (non straight) fringes. The, first term in (36) is the direct non-diffracted wave (zero'th order). The second term in (36) is the reconstructed wavefront (first order) and the last term is the minus-one order of diffraction.

4.7.2 Thick Holograms

The preceding theory assumes that the recording medium (photographic emulsion) is very thin compared with the fringe spacing in the hologram. However, if the fringe spacing becomes small enough, of the same order as the recording medium thickness, Bragg's law will come out of the hologram after diffraction, as in Fig. 40.

If properly designed, a thick hologram will produce only the reconstructed wave, and the other orders of diffraction will be absent. This effect has many important practical applications. One example is the fabrication of a color hologram.

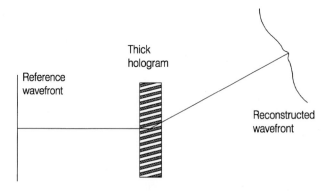

Fig. 40. Thick hologram and Bragg's law

4.7.3 Interferometric Holography

Let us assume that we make a hologram to reconstruct a wavefront $W(x,y)$. Then, if this wavefront changes its shape due to a modification in the system producing it, the change can be detected by making the new wavefront $W'(x,y)$ with the original one $W(x,y)$ stored in the hologram. This process is called holographic interferometry or interferometric holography.

With this basic principle, we can superimpose in space the image of an object produced by a hologram on the object used to form it. If the object changed its position or shape at any time between the hologram formation and its reconstruction, the difference will be observed as interference fringes.

Holographic interference has many important applications, for example, in the detection of small fractures or distortions in machinery.

5 Summary

The main optics tools used in engineering have been briefly reviewed. The reader interested in any of the topics presented here can go to the references for additional information.

References

1. D. Malacara, Z. Malacara: *Handbook of Lens Design*, Dekker, New York (1994)
2. K. Creath: Interferometric Investigation of a Laser Diode, Appl. Opt. **24**, 1291–1293 (1985)
3. Y. Ning, K. T. V. Grattan, B. T. Meggitt, A. W. Palmer: Characteristics of Laser Diodes for Interferometric Use, Appl. Opt. **28**, 3657–3661 (1989)
4. D. Malacara (Ed.): *Optical Shop Testing*, 2nd edn., Wiley, New York (1992)
5. J. B. Houston Jr., C. J. Buccini, P. K. O'Neill: A Laser Unequal Path Interferometer for the Optical Shop, Appl. Opt. **6**, 1237–1242 (1967)
6. R. Kingslake: The Interferometer Patterns due to the Primary Aberrations, Trans. Opt. Soc. **27**, 94, 1925–1926
7. J. B. Saunders: Measurement of Wavefronts without a Reference Standard: The Wavefront Shearing Interferometer, J. Res. Natl. Bur. Stand. **65**, 239 (1961)
8. M. V. R. K. Murty: The Use of a Single Plane Parallel Plate as a Lateral Shearing Interferometer with a Visible Gas Laser Source, Appl. Opt. **3**, 531–351 (1964)
9. A. Cornejo: Ronchi Test, in D. Malacara (Ed.): *Optical Shop Testing*, 2nd edn., Wiley, New York (1992)
10. W. H. F. Talbot: Facts Relating to Optical Science, Philos. Mag. **9**, 401 (1836)
11. Lord Rayleigh: Philos. Mag. **11**, 196 (1881)
12. J. Stricker: Electronic Heterodyne Readout of Fringes in Moiré Deflectometry, Opt. Lett. **10**, 247–249 (1985)
13. I. Glatt, O. Kafri: Moiré Deflectometry - Ray Tracing Interferometry, Opt. Lasers Eng. **8**, 277–320 (1988)

14. K. Patorski: Moiré Methods in Interferometry, Opt. Lasers Eng. **8**, 147–170 (1988)
15. M. Takeda, S. Kobayashi: Lateral Aberration Measurements with a Digital Talbot Interferometer, Appl. Opt. **23**, 1760–1764 (1984)
16. L. M. Foucault: Description des Procédés Employés pou Reconnaitre la Configuration des Surfaces Optiques, C. R. Acad. Sci. Paris **47**, 958 (1858)
17. Y. I. Ostrovsky, M. M. Butusov, G. V. Ostrovskaya: *Interferometry by Holography*, Springer, Berlin, Heidelberg 1980
18. J. C. Wyant: Testing Aspherics Using Two Wavelength Holography, Appl. Opt. **10**, 2113–2118 (1971)
19. D. Gabor: A New Microscopic Principle, Nature **161**, 777 (1948)
20. E. N. Leith, J. Upatnieks: Reconstructed Wavefronts and Communication Theory, J. Opt. Soc. Am. **52**, 1123–1130 (1962)

Introduction to Engineering Mechanics

Anand Asundi

School of Mechanical & Production Engineering
Nanyang Technological University, Nanyang Avenue, Singapore
masundi@ntu.edu.sg

Abstract. This reviews introduces the basics of engineering mechanics of solids as relevant to practising engineers and optical metrologists. The treatment is elementary in breadth but at the same time comprehensive in depth so as to touch on most topics which would be encountered in practice. The chapter gives a basic introduction to the concepts of engineering solid mechanics, the more common relationships and the determination of the parameters of interest and concludes with some elementary examples.

1 Introduction

Engineering mechanics is a wide field encompassing both solids and fluids. However, in this chapter, the emphasis is primarily on solid mechanics, which is the area of application for most optical techniques. Engineering solid mechanics deals with the response of structures to external stimuli such as force, temperature, etc. The treatment can proceed in two ways - the strength of materials approach and the theory of elasticity approach. The former is more practice-based while the latter has a sound mathematical basis.

In the strength of materials approach, the entire structure is considered under the action of external forces. The various components of the structure are broken down in accordance with the equations of equilibrium. Each of these components, representing self-equilibrating free-body diagrams, can then be analyzed for the distribution of the internal forces and displacements. In order to achieve this, a few standard geometric and loading conditions are developed. Complex objects and loading are then split into these simpler geometries and analyzed.

The theory of elasticity, on the other hand, provides a generic analysis methodology. The theory progresses from a small internal element with initially no consideration of the geometry or loading to be analyzed. From the equilibrium of this generic element, the equilibrium equations are derived. These along with the compatibility equations, to ensure deformation consistency, and the stress-strain equations form the governing equations of elasticity for any deformable body subject to external loading. Solution of these equations and inclusion of the specific boundary conditions for a particular problem enables development of acceptable stress, strain and displacement distributions for the body being studied.

2 Basic Concepts

2.1 Stress

Stress is defined as the force per unit area. If the force is acting normal to the cross-sectional area, the stress is termed normal stress and will be denoted by σ. Subscripts are attached to this symbol to denote the direction of stress and the direction of the normal to the surface on which the force acts. Thus σ_{xx} or simply σ_x denotes normal stress acting in the x-direction on a surface whose normal is parallel to the x-direction. Shear stress, on the other hand is the tangential component of the force divided by the cross-sectional area. Shear stress is denoted by the symbol τ, with subscripts as before to signify the direction of stress and the normal to the surface on which the stress acts. Thus τ_{xy} signifies shear stress acting on a plane whose normal is the x-direction and the stress is directed in the y-direction. Thus, for an element in three dimensions, as shown in Fig. 1a, there are a total of nine stress components - three normal stresses and six shear stresses. From the figure it can be readily deduced that to satisfy moment equilibrium, i.e. to prevent rotation of the element under the action of the stresses, there needs to be only three independent shear stresses ($\tau_{xy} = \tau_{yx}, \tau_{xz} = \tau_{zx}$ and $\tau_{xz} = \tau_{zx}$). Incidentally, the directions shown in the figure are the positive directions for the stresses by convention. Thus tensile normal stresses are positive and shear stresses as represented are positive.

In general the stress components vary from point to point in a stressed body. However, since the whole object is in equilibrium, each element of it should also be in equilibrium. Thus the stresses acting on a generic element shown in Fig. 1b provide the basis for the equations of equilibrium. Summing the forces along the three Cartesian axes gives the following set of equations,

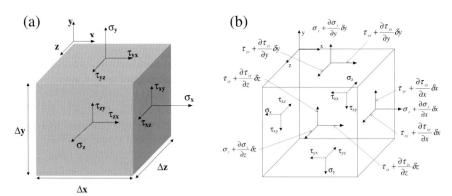

Fig. 1. Stress components for (a) a generic three-dimensional element, and (b) stress equilibrium of a generic three-dimensional element

when no body forces are present

$$\frac{\partial \sigma_x}{\partial x} + \frac{\partial \tau_{xy}}{\partial y} + \frac{\partial \tau_{xz}}{\partial z} = 0$$
$$\frac{\partial \tau_{xy}}{\partial x} + \frac{\partial \sigma_y}{\partial y} + \frac{\partial \tau_{yz}}{\partial z} = 0$$
$$\frac{\partial \tau_{xz}}{\partial z} + \frac{\partial \tau_{yz}}{\partial y} + \frac{\partial \sigma_z}{\partial z} = 0 \,. \tag{1}$$

2.2 Stress Transformation

Without loss of generality, consider the 2-D element shown in Fig. 2a. Since the choice of x-direction is completely arbitrary and based more on the geometry of the section and the loading directions, these axes might not coincide with the directions of maximum (or minimum) principal and/or shear stresses. Determination of the maximum stress components at a generic point is vital in structural design based on the strength of materials approach. Stress transformation equations permit the determination of stress at any point in an arbitrary coordinate system (Fig. 2b) from the knowledge of the stresses in one particular Cartesian system. The stress transformation equations for plane stress conditions are:

$$\sigma_{x'} = \sigma_x \cos^2 \theta + \sigma_y \sin^2 \theta + 2\tau_{xy} \sin \theta \cos \theta$$
$$\sigma_{y'} = \sigma_y \cos^2 \theta + \sigma_x \sin^2 \theta - 2\tau_{xy} \sin \theta \cos \theta$$
$$\tau_{x'y'} = (\sigma_y - \sigma_x) \sin \theta \cos \theta + \tau_{xy}(\cos^2 \theta - \sin^2 \theta) \,. \tag{2}$$

Graphical representation of the above equations in the form of the Mohr's circle provides more informative visualization of the stress transformation.

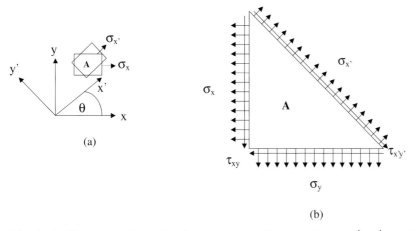

Fig. 2a,b. Stress transformation from $x-y$ coordinate system to $x'-y'$ coordinate system

The Mohr's circle is plotted with the normal stress on the horizontal axis and the shear stress on the vertical axis. From the stress transformation equations, given the normal and shear stress equations in one coordinate system, the following equation can be deduced

$$\left(\sigma - \frac{\sigma_x + \sigma_y}{2}\right)^2 + \tau^2 = \left(\frac{\sigma_x - \sigma_y}{2}\right)^2 + \tau_{xy}^2. \tag{3}$$

The equation represents a circle (Fig. 3) in the $\sigma - \tau$ plane with its center at $\{(\sigma_x + \sigma_y)/2, 0\}$ and with a radius equal to $\sqrt{\frac{(\sigma_x - \sigma_y)^2}{2} + \tau_{xy}^2}$.

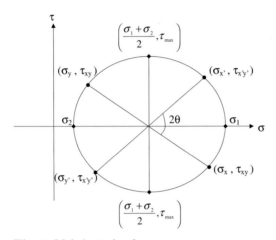

Fig. 3. Mohr's circle of stress

2.3 Stress Transformation Using Mohr's Circle

- The maximum and minimum normal stresses act on the same plane called the *principal plane* and the normal stresses are the *principal stresses* denoted as σ_1 and σ_2. Also the shear stress in this plane is zero. From the figure the principal stresses can be deduced as

$$\sigma_{1,2} = \left(\frac{\sigma_x + \sigma_y}{2}\right) \pm \sqrt{\frac{(\sigma_x - \sigma_y)^2}{2} + \tau_{xy}^2}. \tag{4}$$

- The plane of maximum shear stress is at an angle of 90° (or 45° in Cartesian coordinates) to the principal plane. The normal stresses on these planes are identical and equal to the mean stress $[(\sigma_1 + \sigma_2)/2 = (\sigma_x + \sigma_y)/2]$.
  The maximum shear stress is given as (assuming $\sigma_1 > \sigma_3 = 0 > \sigma_2$)

$$\tau_{max} = \left(\frac{\sigma_1 - \sigma_2}{2}\right) = \sqrt{\frac{(\sigma_x - \sigma_y)^2}{2} + \tau_{xy}^2}, \tag{5}$$

where σ_3 is the intermediate principal stress acting perpendicular to the $x-y$ plane.

Maximum normal stresses are generally used for analysis of brittle materials. Thus, when the maximum principal stress reaches a critical value, the object would fail. For ductile materials it is observed that the maximum shear stress is the more important parameter for the strength analysis approach. Note that photoelasticity provides the maximum shear stress contours and is a useful tool in design based on strength as both the locations and magnitude of the stress can be readily obtained.

2.4 Displacement and Strain

Displacements comprise rigid body displacement and deformation. Consider a cubic element within a body, as shown in Fig. 4a. The element can move as a rigid body in the three Cartesian directions as well as rotate about the three axes. In this case any two points of the element maintain their spatial relationship throughout the displacement. The displacement components along the three Cartesian (x, y, z) directions are denoted as u, v and w respectively. For deformation- or strain-induced displacement, the relative positions of points in the body changes. Deformation can cause a change in the length of the line or change in the angles between the two perpendicular lines or both (Fig. 4b). The change of length of a line segment divided by the original length is the normal strain, denoted by ε, in the direction of the initial line segment, while the change in angle from an initial right angle is a measure of the shear strain, written as γ. As for stresses, subscripts identify the direction in the case of normal strains and the planes in the case of shear strains. There are a total of three independent normal and shear strains.

Based on first principles the normal strain component in the x-direction can be derived as

$$\varepsilon_x = \frac{A'B' - AB}{AB} \tag{6}$$

and the shear strain is

$$\gamma_{xy} = \frac{\pi}{2} - \angle B'A'C'' \ . \tag{7}$$

For small strains, the following strain-displacement equations can be derived

$$\begin{aligned}
\varepsilon_x &= \frac{\partial u}{\partial x} & \gamma_{xy} &= \frac{\partial u}{\partial y} + \frac{\partial v}{\partial x} \\
\varepsilon_y &= \frac{\partial v}{\partial y} & \gamma_{yz} &= \frac{\partial v}{\partial z} + \frac{\partial w}{\partial y} \\
\varepsilon_z &= \frac{\partial w}{\partial z} & \gamma_{xz} &= \frac{\partial u}{\partial z} + \frac{\partial w}{\partial x}
\end{aligned} \tag{8}$$

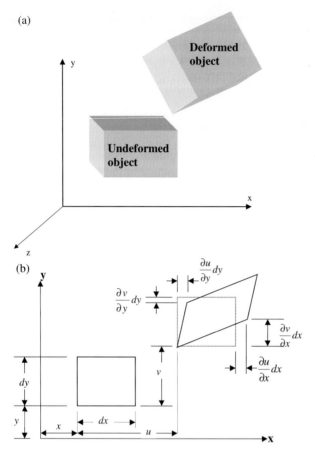

Fig. 4. (a) Deformation of an element subject to load, and (b) strain - displacement relations

These are also referred to as the engineering (Lagrangian) strain components. Other strain definitions are

- Eulerian strain, which is defined as the change in length divided by the final length. For the element above, this is $\varepsilon_x^E = \frac{A'B' - AB}{A'B'}$. For small strains this is identical to the Lagrangian strain, but for large strains it is significantly different. Displacement measuring techniques such as moiré measure the Eulerian strain since the measurements are made on the deformed object.
- True (incremental) strain, which is defined as the incremental change in length divided by the length prior to the increment; i.e. dl/l where l varies over the range from initial to final length. Thus integration gives the true strain as $\ln(A'B'/AB)$.

In (8) the six components of strain can be derived from the three displacement components. Alternately, there are six equations of strain for only three displacement components. This in general will produce a solution only if the strain components are related. From the above equations, one can readily obtain one of these relationships as

$$\frac{\partial^2 \varepsilon_x}{\partial y^2} + \frac{\partial^2 \varepsilon_y}{\partial x^2} = \frac{\partial^2 \gamma_{xy}}{\partial x \partial x} \ . \qquad (9)$$

This is one of the six compatibility equations. The strain components need to satisfy these equations for the proposed displacement solutions to exist. Thus, while there may be numerous stress distributions which will satisfy equilibrium conditions, the correct one must also maintain continuity of deformation in the body. This is ensured if the compatibility equations are satisfied.

2.5 Strain Transformation

As with stresses, strains are also associated with the choice of coordinate system and thus strain transformation equations need to be developed to enable conversion of strains from one system to the another. Consider the transformation of vector $\boldsymbol{A}$ from the $x - y$ plane to the $x' - y'$ plane as shown in Fig. 5. Since displacements are also vectors, the same transformation equation applies. Denoting strains in the transformed coordinate system with a prime, the strain transformation equation can be readily deduced as

$$\begin{aligned}
\varepsilon_{x'} &= \varepsilon_x \cos^2 \theta + \varepsilon_y \sin^2 \theta + \gamma_{xy} \sin \theta \cos \theta \ , \\
\gamma_{x'y'} &= 2(\varepsilon_y - \varepsilon_x) \sin \theta \cos \theta + \gamma_{xy}(\cos^2 \theta - \sin^2 \theta) \ , \\
\varepsilon_{y'} &= \varepsilon_y \cos^2 \theta - \varepsilon_x \sin^2 \theta - \gamma_{xy} \sin \theta \cos \theta \ ,
\end{aligned} \qquad (10)$$

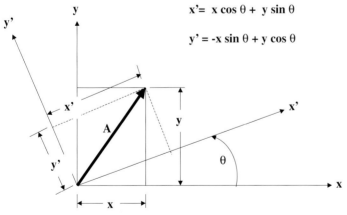

Fig. 5. Transformation of a line vector $\boldsymbol{A}$ from $x - y$ coordinate system to $x' - y'$ coordinate system

where $\varepsilon_{x'}, \varepsilon_{y'}, \gamma_{x'y'}$ are the strain components in the transformed coordinate system.

2.6 Stress-Strain Relations

Stress and strain are causal effects of the applied external loads and deformation. Thus, although their development above was independent of each other, they have a relationship with each other. These stress-strain relationships depend on the mechanical properties of the material. Solids can be broadly classified as *elastic* or *plastic*. For elastic bodies deformations of the body subject to external load are completely recovered when the load is removed. Within elasticity the materials can exhibit linear or non-linear stress-strain behaviour. Furthermore, the materials can be isotropic and homogeneous, wherein the mechanical properties are independent of orientation and location, or anisotropic and inhomogeneous, as with composite materials. Isotropic materials may exhibit anisotropic behaviour on the micro- and nano-scales. For isotropic materials, normal stresses give rise only to normal strains, and shear stresses are related to shear strains alone. Furthermore, shear stress in one plane generates shear strains in that plane only. This implies that there are only two *elastic constants* relating stresses to strains. These are the elastic or Young's modulus and the Poisson ratio.

From experimental observations, the following stress-strain equations (also called Hooke's Law) can be deduced as

$$\varepsilon_x = \frac{1}{E}[\sigma_x - \nu(\sigma_y + \sigma_z)] \qquad \tau_{xy} = \frac{\gamma_{xy}}{G}$$
$$\varepsilon_y = \frac{1}{E}[\sigma_y - \nu(\sigma_x + \sigma_z)] \qquad \tau_{yz} = \frac{\gamma_{yz}}{G}$$
$$\varepsilon_z = \frac{1}{E}[\sigma_z - \nu(\sigma_y + \sigma_x)] \qquad \tau_{xz} = \frac{\gamma_{xz}}{G} \qquad (11)$$

where E and ν are the Young's modulus and Poisson ratio, respectively, and $G = E/2(1+\nu)$ is the modulus of elasticity in shear or the modulus of rigidity.

These equations can be written in different forms depending on the stress and strain components of interest. Plane stress and plane strain (see Sect. 3.1) simplifications reduce the stress-strain equations accordingly. One form of equation, which is particularly useful in the extension of the theory of elasticity to plastic deformations, introduces some different strain and stress components. In plasticity, it has been observed that changes in volume do not contribute to plastic deformation. Thus, there is a need to separate the strains relating to volume changes from the total strain. Consider the case of hydrostatic pressure acting on a cubic element, as shown in Fig. 6. For this case the stress components are

$$\sigma_x = \sigma_y = \sigma_z = -p,$$
$$\tau_{xy} = \tau_{xz} = \tau_{zx} = 0, \qquad (12)$$

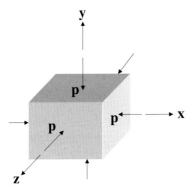

Fig. 6. Hydrostatic pressure acting on an element

and, from the stress-strain equations, the strain components are

$$\varepsilon_x = \varepsilon_y = \varepsilon_z = -(1-2\nu)p/E ,$$
$$\gamma_{xy} = \gamma_{yz} = \gamma_{xz} = 0 . \qquad (13)$$

Now the volume strain is defined as the change in the volume divided by the original volume, which can be derived as

$$\varepsilon = \varepsilon_x + \varepsilon_y + \varepsilon_z = -3(1-2\nu)p/E = -p/K . \qquad (14)$$

where $K = E/[3(1-2\nu)]$ is the bulk modulus of elasticity.

Defining the mean stress as the sum of the normal stresses divided by three, the following equation applies for any state of stress

$$\varepsilon = \frac{\sigma_m}{K} . \qquad (15)$$

The mean stress is also referred to as the hydrostatic component of stress or the spherical stress.

The total stress minus the hydrostatic stress is the deviatoric stress and contributes to plastic deformation. The deviatoric stress-strain relationship can thus be written as

$$\sigma'_{ij} = 2G\varepsilon_{ij} , \qquad (16)$$

where i,j take values x, y and z and ε_{ij} is the tensor notation of strain, i.e. ε_{ij} for i not equal to j, is half the engineering shear strain.

2.7 Boundary Conditions

A total of 15 unknown quantities, comprising six stress components, six strain components and three displacement components, are used to describe the problem. These are governed by the three equilibrium equations, six strain-displacement equations and six stress-strain equations. For an elastic body in equilibrium these equations must be satisfied at all points of the body. Note that no mention has been made of the compatibility equations. This is because it is inherent in the 15 governing equations, and if they are satisfied, so will the compatibility equations. These equations are generic and apply to all problems of elasticity. The differences between these problems arise from differences in the boundary conditions. Thus, in addition to satisfying the governing equations, the boundary conditions should also be satisfied. Boundary conditions can be prescribed in terms of forces or displacements. Equilibrium of a boundary element, shown in Fig. 7 provides the necessary stress boundary conditions. Besides forces, displacements on the boundaries can also be specified. In this case the displacement components obtained from the solution must satisfy the imposed boundary constraints.

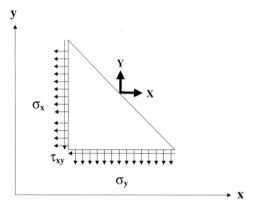

Fig. 7. Stress distribution on a boundary element

3 Theory of Elasticity Approach

To solve a particular problem in elasticity, one requires the following to be specified

- the geometry of the body,
- the boundary conditions,
- the mechanical properties of the material.

Various approaches to solving 3-D problems are detailed in Table 1. The straightforward process is to find solutions for the stress, strain and displacement unknowns which satisfy the equilibrium, stress-strain and strain-displacement equations and also the prescribed boundary conditions. However, due to the inherent relationship between the various unknowns, more elegant approaches can be found. One can, for example, write the three equilibrium equations in terms of displacements, called Navier's equations,

Introduction to Engineering Mechanics   43

Table 1. Three-dimensional equations of elasticity

Unknowns:

6 stress components	$\sigma_x, \sigma_y, \sigma_z, \tau_{xy}, \tau_{yz}, \tau_{zx}$
6 strain components	$\varepsilon_x, \varepsilon - \varepsilon_y, \varepsilon_z, \gamma_{xy}, \gamma_{yz}, \gamma_{zx}$
3 displacement components	u, v, w

Equations:

3 equilibrium equations	(1)
6 strain-displacement equations	(8)
6 stress-strain equations	(11)

Stress formulation
Beltrami-Mitchell equations

The six strain-displacement equations and the six stress-strain equations are combined to give three additional stress compatibility equations, which together with the equilibrium equations can be used to solve for the six unknown stress components. Strain and displacements are then derived from the stress components:

$$\nabla^2 \sigma_x + \frac{3}{1+\nu} \frac{\partial^2 \sigma_m}{\partial x^2} = 0$$

$$\nabla^2 \sigma_y + \frac{3}{1+\nu} \frac{\partial^2 \sigma_m}{\partial y^2} = 0$$

$$\nabla^2 \sigma_z + \frac{3}{1+\nu} \frac{\partial^2 \sigma_m}{\partial z^2} = 0$$

$$\nabla^2 \tau_{xy} + \frac{3}{1+\nu} \frac{\partial^2 \sigma_m}{\partial x \partial y} = 0$$

$$\nabla^2 \tau_{yz} + \frac{3}{1+\nu} \frac{\partial^2 \sigma_m}{\partial y \partial z} = 0$$

$$\nabla^2 \tau_{xz} + \frac{3}{1+\nu} \frac{\partial^2 \sigma_m}{\partial x \partial z} = 0$$

Displacement formulation
Navier's equations

The stress-strain and strain-displacement equations give stress-displacement equations. When substituted in the equilibrium equations those give three displacement equilibrium equations for the three unknown displacement components:

$$(\lambda + G)\frac{\partial \varepsilon}{\partial x} + G\nabla^2 u = 0$$

$$(\lambda + G)\frac{\partial \varepsilon}{\partial y} + G\nabla^2 v = 0$$

$$(\lambda + G)\frac{\partial \varepsilon}{\partial z} + G\nabla^2 w = 0$$

where

$$\varepsilon = \frac{\partial u}{\partial x} + \frac{\partial v}{\partial y} + \frac{\partial w}{\partial z}$$

$$\nabla^2 = \frac{\partial^2}{\partial x^2} + \frac{\partial^2}{\partial y^2} + \frac{\partial^2}{\partial z^2}$$

thus reducing the problem to one with three equations and three unknowns. The other alternative is to rewrite the compatibility equations in terms of stresses. These, along with the three stress equilibrium equations give the six required equations in terms of the unknown stress components. These are the Beltrami-Mitchell compatibility equations of stress. One of the features of the Beltrami-Mitchell equations is that the stresses are independent of the modulus of elasticity (E) and depend only on the Poisson ratio. This is strictly valid for singly-connected[1] bodies. This is the justification used in photoelasticity for modeling metal prototypes with birefringent plastics.

Plane stress and *plane strain* formulations provide further simplifications for practical problems.

3.1 Plane Stress Formulation

For plane stress, with $x-y$ as the Cartesian plane of investigation, the stresses satisfy the relations $\sigma_z = \tau_{xz} = \tau_{yz} = 0$, and thus only three non-zero stress components are present. Table 2 shows the simplified equilibrium, stress-strain and strain-displacement equations. Following the stress formulation approach, the two equilibrium equations and the compatibility equation in terms of stress can be written as

$$\frac{\partial \sigma_x}{\partial x} + \frac{\partial \tau_{xy}}{\partial y} = 0,$$
$$\frac{\partial \sigma_y}{\partial y} + \frac{\partial \tau_{xy}}{\partial x} = 0,$$
$$\nabla^2(\sigma_x + \sigma_y) = 0. \qquad (17)$$

Navier's equations (equilibrium equations in terms of displacement) for plane stress are

$$G\nabla^2 u + \frac{E}{2(1+\nu)} \frac{\partial}{\partial x}\left(\frac{\partial u}{\partial x} + \frac{\partial v}{\partial y}\right) = 0,$$
$$G\nabla^2 v + \frac{E}{2(1+\nu)} \frac{\partial}{\partial y}\left(\frac{\partial u}{\partial x} + \frac{\partial v}{\partial y}\right) = 0. \qquad (18)$$

Justification of the plane stress formulation requires that

- the body must be a thin plate;
- the two z surfaces of the plate are free from load;
- external forces do not have a component in the z-direction; and
- the external forces must either be uniformly distributed or symmetrically distributed through the thickness of the plate.

---

[1] Singly connected bodies are those where any cut from one boundary to another will split the body into two parts.

Table 2. Plane equations of elasticity

Plane Stress	**Plane Strain**
$\sigma_x, \sigma_y, \tau_{xy}$ non-zero stress components $\tau_{xy}, = \tau_{xz} = \sigma_z = 0$	$u = u(x,y), v = v(x,y), w = 0$
Equilibrium equations	Equilibrium equations
$\dfrac{\partial \sigma_x}{\partial x} + \dfrac{\partial \tau_{xy}}{\partial y} = 0$ $\dfrac{\partial \tau_{xy}}{\partial x} + \dfrac{\partial \sigma_y}{\partial y} = 0$	$\dfrac{\partial \sigma_x}{\partial x} + \dfrac{\partial \tau_{xy}}{\partial y} = 0$ $\dfrac{\partial \tau_{xy}}{\partial x} + \dfrac{\partial \sigma_y}{\partial y} = 0$
Compatibility equation	Compatibility equation
$\dfrac{\partial^2 \varepsilon_x}{\partial y^2} + \dfrac{\partial^2 \varepsilon_y}{\partial x^2} = \dfrac{\partial^2 \gamma_{xy}}{\partial x \partial x}$	$\dfrac{\partial^2 \varepsilon_x}{\partial y^2} + \dfrac{\partial^2 \varepsilon_y}{\partial x^2} = \dfrac{\partial^2 \gamma_{xy}}{\partial x \partial x}$
Stress-strain equations	Stress-strain equations
$\varepsilon_x = \dfrac{1}{E}(\sigma_x - \nu \sigma_y)$ $\varepsilon_y = \dfrac{1}{E}(\sigma_y - \nu \sigma_x)$ $\gamma_{xy} = \dfrac{\tau_{xy}}{G}$ $\varepsilon_z = -\dfrac{\nu}{E}(\sigma_y + \sigma_z)$	$\sigma_x = \lambda\left(\dfrac{\partial u}{\partial x} + \dfrac{\partial v}{\partial y}\right) + 2G\dfrac{\partial u}{\partial x}$ $\sigma_y = \lambda\left(\dfrac{\partial u}{\partial x} + \dfrac{\partial v}{\partial y}\right) + 2G\dfrac{\partial v}{\partial y}$ $\tau_{xy} = G\gamma_{xy}$ $\lambda = \dfrac{\nu E}{(1+\nu)(1-2\nu)}$

Biharmonic Equation

$$\nabla^4 \phi = 0$$

$$\sigma_x = \frac{\partial^2 \phi}{\partial y^2} \quad \sigma_y = \frac{\partial^2 \phi}{\partial x^2}, \quad \tau_{xy} = \frac{\partial^2 \phi}{\partial x \partial y}$$

3.2 Plane Strain Formulation

A state of plane strain can be assumed in the $x - y$ plane if the in-plane displacement components (u and v) are functions of x and y alone and the out-of-plane displacement component is zero (Table 2). For generalised plane strain, the out-of-plane displacement is a constant. The resulting equilibrium equations and stress compatibility equations are identical to those for plane stress in the absence of body forces. Navier's equations for plane strain can also be deduced in similar fashion or can be written from Navier's equations for plane stress by replacing the Young's modulus E by $E/(1-\nu^2)$ and the Poisson ratio, ν by $\nu/(1-\nu)$.

General requirements for validity of plane strain equations are:

- dimension of the object in the z direction is large; and
- prescribed surface forces are independent of z and have no component in the z–direction.

Despite these simplifications, direct solution of the equations is not always straightforward. One mathematical approach was to introduce the concept of stress functions ϕ (called the Airy's stress function) wherein the stress components are defined as:

$$\sigma_x = \frac{\partial^2 \phi}{\partial y^2},$$

$$\sigma_y = \frac{\partial^2 \phi}{\partial x^2},$$

$$\tau_{xy} = -\frac{\partial^2 \phi}{\partial x \partial y}. \tag{19}$$

Stress functions defined in this way clearly satisfy the equilibrium equations (17) and, in order to satisfy the stress compatibility equation, the stress function should also be a solution of the biharmonic equation

$$\nabla^4 \phi = 0. \tag{20}$$

Where $\nabla^4 = \frac{\partial^4}{\partial x^4} + 2\frac{\partial^4}{\partial x^2 \partial y^2} + \frac{\partial^4}{\partial y^4}$ is the biharmonic operator.

Use of polynomials is the simplest representation for the stress function. Thus polynomials of the second degree represent a uniform stress field; third order polynomials suggest a linearly varying stress field and so on. Boundary conditions make up the final form of the stress function. Choice of a particular polynomial or combination of polynomials can be guessed from the boundary conditions and an understanding of the problem at hand. Experimental methods discussed in this book can provide a better visual guide to selection of an appropriate stress function - a possible scheme for combined experimental and analytical methods.

4 Strength of Materials Approach

While the theory of elasticity approach discussed in the above section provides an elegant mathematical basis for the solution of engineering problems, there is a good chance that one may get too involved with the mathematics and may tend to forget the practical problem at hand. The strength of materials approach is the preferred choice with design engineers, while research and development engineers choose the theory of elasticity approach. While both are approximations to the exact solution, the strength of materials approach relies to a great extent on the experience and ingenuity of the designer to tackle new problems.

In the strength of materials approach the emphasis is more on the ability of the structure or system to withstand the operating loads that will be imposed on it. To aid this, certain standard loading configurations are analyzed, as shown in Table 3. Most practical problems can be simplified into components subject to one of these different loading conditions. If an element is subject to a combination of loading, then the stresses due to each loading system is separately evaluated and the principle of superposition invoked to evaluate the combined effect. As before, principal stresses and maximum shear stresses are determined at the critical areas to ensure that they are within allowable limits.

Table 3. Stresses and strains for some typical loading situations

Loading	Stress components	Strain components	Deformation
Axial loading	$\sigma_x = P/A$ $\sigma_y = 0$ $\tau_{xy} = 0$	$\varepsilon_x = P/AE$ $\varepsilon_y = -\nu P/AE$ $\gamma_{xy} = 0$	Displacement $u = \int \frac{P}{AE} dx$ $\nu = -\int \frac{\nu P}{AE} dx$
Bending Moment	$\sigma_x = My/I$ $\quad = Pxy/2I$ $\sigma_y = 0$ $\tau_{xy} = \frac{VQ}{It}$	$\varepsilon_x = Pxy/2EI$ $\varepsilon_y = -\nu Pxy/2EI$ $\gamma_{xy} = \frac{\tau_{xy}}{G_r}$	Deflection (y) $y = \iint \frac{M}{EI} dxdx$
Torsion	$\sigma_x = 0$ $\sigma_y = 0$ $\tau_{xy} = TR/J$ $\quad = 2Pcr/J$	$\varepsilon_x = 0$ $\varepsilon_y = 0$ $\gamma_{xy} = Tc/GJ$	Twist $\phi = \int \frac{\tau}{Gr} dz$

5 Examples

5.1 Axial Loading

Consider the large thin plate subject to an axial force P, as shown in Fig. 8a. The external loading might as well have been a uniformly distributed load as in Fig. 8b, which is statically equivalent to the point load P. Saint Venant's principle states that in such cases, while the loading can produce substantially different local effects, at distances greater than the width of the object the effects of the loading are identical. Thus, in the central part both systems

(a)
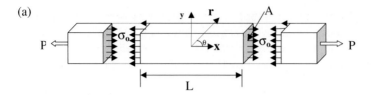

(b)
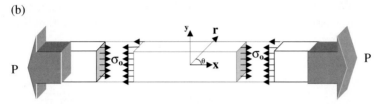

Fig. 8a,b. Stress analysis of a bar subject to a uniaxial load

produce the same uniform stress distributions as shown. Why should the stress distribution be uniform, when equilibrium suggests that any stress distribution symmetric about the centroidal axis would still satisfy equilibrium provided that $\int \sigma_x dA = P$. It is because only the uniform stress distribution satisfies compatibility of displacement and strain, the second criterion in elasticity theory and thus is the acceptable stress distribution.

For a stress distribution of the form $\sigma_x = \sigma_0$ and $\sigma_y = \tau_{xy} = 0$, the stress function $\phi = \sigma_0 y^2/2$ will satisfy the biharmonic equation (20). Using polar coordinates with $y = r \sin \theta$, the stress function can be transformed to

$$\phi = \frac{1}{4}\sigma_0 r^2 (1 - \cos 2\theta) . \tag{21}$$

The resulting stress components are thus

$$\sigma_r = \frac{1}{2}\sigma_0(1 + \cos 2\theta)$$

$$\sigma_\theta = \frac{1}{2}\sigma_0(1 - \cos 2\theta)$$

$$\tau_{r\theta} = -\frac{1}{2}\sigma_0 \sin 2\theta . \tag{22}$$

This is also the stress transformation equation for uniaxial loading, for which $\theta = 0°$ corresponds to the principal stress directions and $\theta = 45°$ corresponds to the maximum shear stress direction.

Displacements can be obtained using the strain components, which in turn can be determined from Hooke's law. Determination of displacements requires integration of the strain-displacement equations, which will provide a constant of integration whose value depends on any rigid body motion experienced by the object. If the object is fixed at one end and loaded from the other, as is usually the case, then the displacement at the fixed end is zero, which provides the necessary boundary condition. The strains and displacements along the principal directions are:

$$\varepsilon_x = \frac{\sigma_0}{E} = \frac{P}{AE} = \frac{\partial u}{\partial x},$$

$$\varepsilon_y = \varepsilon_z = -\nu\frac{\sigma_0}{E} = -\frac{\nu P}{AE},$$

$$u = \int \frac{P}{AE}\mathrm{d}x,$$

$$\nu = -\int \frac{\nu P}{AE}\mathrm{d}y,$$

$$w = -\int \frac{\nu P}{AE}\mathrm{d}z. \tag{23}$$

For the case when the load (P), the cross-section area (A) and the Young's modulus (E) are constant over a length (L), the u-displacement component can be written as $u = PL/AE = kP$; where k is the stiffness of the material. If the stiffness were to change either locally or globally, the effect would become noticeable in the reduced strain bearing capacity. Consider a bar with uniform circular cross-section of diameter D. Under the action of an axial load, the global or average strain over a length l is $4P/\pi D^2 E$. If the bar were tapered with the diameter varying from D at one end to d at the other, then the global strain becomes $4P/\pi D d E$, indicating a reduction in the load carrying capacity of the structure. Thus a global health-monitoring scheme for structures using long gauge length polarimetric fiber optic sensors is possible.

5.2 Bending of Beams

The bending of beams is another frequently encountered loading situation in practice. In the analysis of beams under bending the curvatures, being the significant physical feature, are used as the initial starting point when the strength of materials approach is to be formulated. Consider a beam, initially straight, which is subject to a pure bending moment (M). The deformed beam (greatly exaggerated) is shown in Fig. 9a. From the geometry, it can be deduced that the normal strain is $\varepsilon_x = -\kappa y$, where κ is the curvature and y is the distance from the neutral axis of the beam. From Hooke's law, the normal stress is $\sigma_x = E\varepsilon_x = -E\kappa y$.

Invoking equilibrium equations gives the following two conditions:
- the neutral axis passes through the centroid of the cross-section; and
- $\kappa = M/EI$ where M is the applied bending moment and I is the second moment of inertia.

The second condition, when incorporated into the stress equation, gives the familiar *flexural identities*:

$$\frac{\sigma_x}{y} = \frac{M}{I} = \frac{E}{R}, \tag{24}$$

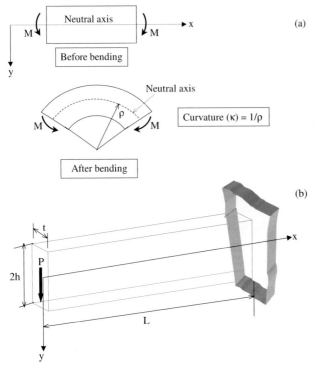

Fig. 9. Schematic of (**a**) a beam subject to pure bending, and (**b**) a cantilever beam subject to a point load at the free end

where $R = 1/\kappa$ is the radius of curvature of the bent beam. In addition to normal stress, shear stresses can also exist in beams subject to lateral loads. This shear stress leads to shear strains. These shear strains cause warping of the plane sections. However, it has been shown that warping of sections does not significantly affect the normal strains, and the flexure formula can be used for all loading situations.

Consider the ubiquitous cantilever beam (Fig. 9b) as an example to demonstrate the Airy stress function approach to beam-bending. The choice of the appropriate stress function relies to a great extent on the physical understanding of the problem. For the beam bending case, as discussed above, the normal stress varies linearly with the distance from the neutral axis and also linearly with the bending moment. Since the bending moment in this case is a linear function of distance from the free end, one can thus write

$$\sigma_x = \frac{\partial^2 \phi}{\partial y^2} = Axy \,, \tag{25}$$

which can be integrated twice to yield

$$\phi = \frac{Axy^3}{6} + yf_1(x) + f_2(x) \ . \tag{26}$$

On substitution into the biharmonic equation this gives

$$y\frac{d^4 f_1(x)}{dx^4} + \frac{d^4 f_2(x)}{dx^4} = 0 \ , \tag{27}$$

which can be then solved for $f_1(x)$ and $f_2(x)$ to give

$$f_1(x) = a_1 x^3 + a_2 x^2 + a_3 x + a_4 \ ,$$
$$f_2(x) = b_1 x^3 + b_2 x^2 + b_3 x + b_4 \ , \tag{28}$$

where $a1, a2, a3, a4, b1, b2, b3$ and $b4$ are constants of integration.

Thus the stress function becomes

$$\phi = \frac{Axy^3}{6} + y(a_1 x^3 + a_2 x^2 + a_3 x + a_4) + b_1 x^3 + b_2 x^2 + b_3 x + b_4 \ . \tag{29}$$

The boundary conditions are

$$(\tau_{xy})_{y=\pm h} = 0 = (\sigma_y)_{y=\pm h} \ ,$$

giving $a_1 = a_2 = b_1 = b_2 = 0$ and $a_3 = -1/2Ah^2$ and

$$P = \int_{-h}^{+h} \tau_{xy} t dy \ ,$$

giving $A = -3P/2th^3 = -P/I$ with I being the moment of inertia of the cross-section about the neutral axis.

The resulting stresses are

$$\sigma_x = \frac{Pxy}{I} \ ; \qquad \sigma_y = 0 \ ; \qquad \tau_{xy} = -\frac{P}{2I}(h^2 - y^2) \ . \tag{30}$$

Alternatively, one can assume that the normal stress can be deduced from the flexure formula for the case of pure bending, and with the shear stress τ_{xy} as the only other non-zero stress component the equilibrium equations give

$$\frac{\partial \tau_{xy}}{\partial x} = 0 \ ,$$

and thus τ_{xy} is independent of x, and

$$\frac{\partial \tau_{xy}}{\partial y} = -\frac{\partial \sigma_x}{\partial x} = \frac{Py}{I} \Rightarrow \tau_{xy} = \frac{Py^2}{2I} + B \ ,$$

where $B = -\frac{Ph^2}{2I}$. The integration constant B is determined from the shear boundary condition. The stress components thus obtained are identical to those in (30). These stresses are valid away from the loaded and free ends as per St. Venant's principle, since the boundary conditions specified at these ends are never achieved in practice.

Displacement can be determined as usual through the strain components, which in turn are obtained via the stress-strain equations. Thus

$$\varepsilon_x = \frac{\partial u}{\partial x} = \frac{\sigma_x}{E} = -\frac{Pxy}{EI},$$

$$\varepsilon_y = \frac{\partial v}{\partial y} = -\frac{\nu \sigma_x}{E} = -\frac{\nu Pxy}{EI},$$

$$\gamma_{xy} = \frac{\partial x}{\partial y} + \frac{\partial v}{\partial x} = \frac{\tau_{xy}}{G} = \frac{P}{8GI}(b^2 - 4y^2). \tag{31}$$

Integrating the first two equations and substituting for the u and v displacements in the third equation gives after mathematical manipulation

$$u = -\frac{Px^2y}{2EI} - \frac{\nu Py^3}{6EI} + \frac{Py^3}{6GI} + \left(\frac{Pl^2}{2EI} - \frac{Pb^2}{8GI}\right)y,$$

$$v = \frac{\nu Pxy^2}{2EI} + \frac{Px^3}{6EI} - \frac{Pl^2x}{2EI} + \frac{Pl^3}{3EI}, \tag{32}$$

where the constants of integration obtained from the displacement boundary conditions, namely the displacement components and the slope at the fixed end neutral axis are zero.

At the fixed-end, however, the beam can distort, and this distortion can be determined from the u displacement equations at $x = L$. The distortion of the cross-section is a result of the variation of the shear stress, which violates the plane section remaining plane assumption in the strength of materials approach. In most practical situations, this only occurs for short thick beams, while for long slender beams the effect of shear can be neglected.

5.3 Combined Bending and Axial Load

Consider the column shown in Fig. 10, subject to a load eccentric to the axis of the beam. Such situations are frequently encountered in bio-mechanics and civil engineering structures. At any cross-section, there is a resultant bending moment and an axial force. Superposition of stresses due to axial loading and those due to bending from Table 3 are used to determine the resultant stress distribution shown in Fig. 10. In such studies, the main concern is to determine the allowable range of eccentricities, which will give compressive stresses throughout the cross-section. Materials such as concrete and bone are stronger in compression and thus this is a necessity. For no tension in the column, it is clear that

$$-\frac{P}{A} + \frac{Pe}{z} \leq 0, \tag{33}$$

where $z = I/y_{\max}$ is the section modulus.

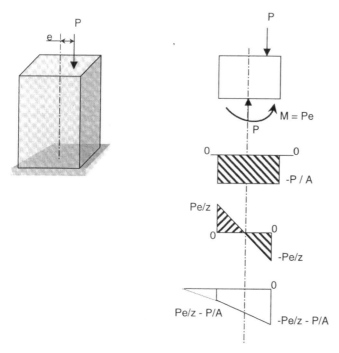

Fig. 10. Stresses due to eccentric loading on columns

6 Conclusion

This review gives a brief introduction to engineering mechanics with an emphasis on the stress and strain analysis of solid deformable bodies. Only the basic principles are covered, and a couple of simple examples demonstrate the application of these principles. As described in the chapter, there are two approaches to solving problems in stress/strain analysis - the strength of materials approach and the theory of elasticity approach. The former is more practically oriented, with intuition and experience providing the basis for problem solving, while the latter is more rigorous and used primarily to get a better understanding of the mechanics principles. However, both are approximations to the actual situation and any solution obtained would benefit from confirmation using the experimental techniques discussed in this book.

More advanced techniques such as plasticity, visco-elasticity, etc. are beyond the scope of this chapter. Engineering fluid mechanics is also an area that has not been discussed. The basic principles are similar to solid mechanics, except that in fluid mechanics one deals with velocities instead of displacement and thus their derivatives are strain rates rather than strains. More detailed treatments are given in the books listed in the bibliography, which should be available in most engineering libraries.

Further Reading

A. P. Boresi: *Elasticity in Engineering Mechanics*, Elsevier, New York (1987)

S. F. Borg: *Fundamentals of Engineering Elasticity*, World Scientific, Teaneck, NJ (1990)

C. R. Calladine : *Plasticity for Engineers*, Horwood, Chichester (1985)

J. Case, A. H. Chilver: *Strength of Materials and Structures*, Edward Arnold, London (1971)

J. Chakrabarty: *Theory of Plasticity*, McGraw-Hill, New York (1987)

J. F. Douglas, J. M. Gasiorek, J. A. Swaffield: *Fluid Mechanics*, Pitman, London (1979)

R. T. Fenner: *Engineering Elasticity: Application of Numerical and Analytical Techniques*, Halsted, New York (1986)

Y. C. Fung: *Foundations of Solid Mechanics*, Prentice Hall, Englewood Cliffs, NJ (1965)

J. M. Gere, S. P. Timoshenko: *Mechanics of Materials*, PWS-KENT, Boston, MA (1990)

V. Gopinathan: *Plasticity Theory and its Application in Metal Forming*, Wiley, New York (1982)

E. J. Hearn: *Mechanics of Materials*, Pergammon, Oxford (1985)

R. Hill: *The Mathematical Theory of Plasticity*, Clarendon, Oxford (1950)

W. Johnson, P. B. Mellor: *Engineering Plasticity*, Van Nostrand, London (1973)

S. G. Lekhnitskii: *Theory of Elasticity of an Anisotropic Body*, Mir, Moscow (1981)

A. E. H. Love: *A Treatise on the Mathematical Theory of Elasticity*, Dover, New York (1944)

A. Mendelson: *Plasticity Theory and Application*, Kreiger, Malabar, FL (1968)

E. P. Popov: *Introduction to Mechanics of Solids*, Prentice Hall, Englewood Cliffs, NJ (1968)

W. Prager, P. G. Hodge: *Theory of Perfectly Plastic Solids*, Wiley, New York (1951)

A. Pytel, F. L. Singer: *Strength of Materials*, Harper, New York (1987)

G. H. Ryder: *Strength of Materials*, ELBS Macmillan, London (1969)

I. H. Shames: *Elastic and Inelastic Stress Analysis*, Prentice Hall, Englewood Cliffs, NJ (1992)

I. H. Shames: *Applied Strength of Materials*, Prentice Hall, Englewood Cliffs, NJ (1989)

R. A. C. Slater: *Engineering Plasticity*, Wiley, New York (1977)

I. S. Sokolnikoff: *Mathematical Theory of Elasticity*, Kreiger, Malabar, FL (1956)

V. L. Streeter: *Fluid Mechanics*, McGraw-Hill, Ryerson, Toronto (1981)

S. P. Timoshenko, J. N. Goodier: *Theory of Elasticity*, McGraw-Hill, New York (1970)

S. P. Timoshenko, Young: *Elements of Strength of Materials*, Van Nostrand, New York (1968)

A. C. Ugural: *Advanced Strength and Applied Elasticity*, Arnold, London (1981)

C. T. Wang: *Applied Elasticity*, McGraw Hill, New York (1953)

Fringe Analysis

Yves Surrel

National Institute of Metrology, CNAM/INM,
292, rue St.Martin, F-75141 Paris Cedex 03, France
surrel@cnam.fr

Abstract. Fringe analysis is the process of extracting quantitative measurement data from fringe – or line – patterns. It usually consists of phase detection and phase unwrapping. Phase detection is the calculation of the fringes phase from the recorded intensity patterns, and this issue represents the major part of the material in this review. Different techniques for this phase calculation are presented, with special emphasis on the characteristic polynomial method, allows us permits to easily design customized algorithms coping with many error sources. For reference, a table presenting the properties of almost all algorithms which have been in recent years is provided in the Appendix. A generic method allowing us to quantitatively evaluate the phase errors and the effect of noise is presented here for the first time. Finally, some elements regarding the complex problem of phase unwrapping are given.

1 Introduction

Most optical methods for metrology require the processing of a fringe (or line) pattern. The intensity pattern which is provided by optical methods encodes the physical quantity which is measured, and this intensity varies as the cosine of a phase which is most often directly proportional to that physical quantity. Fringe processing is the extraction of the phase field from one or many intensity fields which have been acquired. The process of phase detection has received considerable attention during the last decade, as video acquisition and computer processing became more powerful and less expensive.

A typical example of a fringe pattern is presented in Fig. 1a. This example is a Michelson-type interferogram, and the fringe phase is directly proportional to the deviation of the measured surface from a reference plane. The problem is to extract the corresponding phase field represented in Fig. 1b, so that the profile of the surface can be obtained. The real fringe pattern exhibits usually variable bias and contrast, as can be seen when comparing Figs. 1a,c, and the phase extraction strategy has to take this into account. Finally, the phase which is obtained within the interval $[-\pi, +\pi]$ has to be unwrapped as in Fig. 1d to faithfully map the real profile. Note that very often the phase is only meaningful up to an additive constant.

This process should be compared to the historical approach using fringe orders, when fringes had to be counted manually throughout the field to

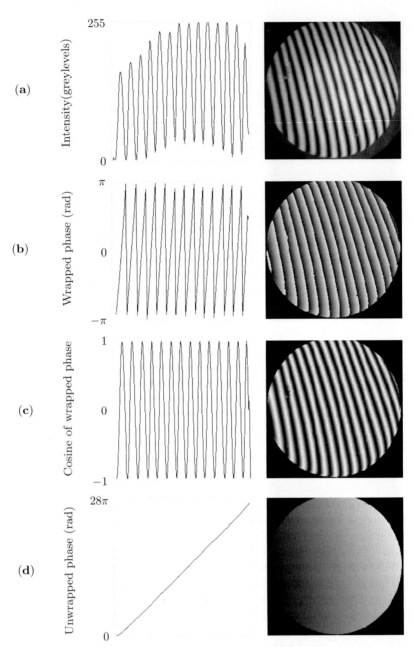

Fig. 1. Fringe processing (in all figures, the *left-hand side* displays a horizontal cross section across the center of the field): (**a**) fringe pattern, (**b**) corresponding wrapped phase, (**c**) ideal fringes, obtained from the cosine of the wrapped phase, and (**d**) unwrapped phase

make a quantitative measurement. The fringe order is the integer part of the division of the unwrapped phase by 2π. Actually, modern phase detection methods provide at any pixel not only the integer fringe order (once the phase is unwrapped) but also its fractional part, which can be written $\phi/2\pi$, where ϕ is the wrapped phase.

An important issue is the assessment of the relation between the phase and the physical quantity to be measured, when this relation is not linear. This is of particular importance in profilometry when using the fringe projection technique. However, this last topic is beyond the scope of this review.

2 Phase Evaluation

In this section, the problem of calculating a phase map from one or many fringe patterns is addressed. The phase detection process is first presented, and the discussion is centered on the role and properties of the detection "algorithm". The characteristic polynomial (CP) theory is used, as it provides on this topic a comprehensive set of results and requires only very simple calculations. Then the two major issues of phase error evaluation and noise effect are discussed.

2.1 Local and Global Strategies

Today, two main tracks can be followed for phase detection, which can be called "global" and "local" approaches.

- In the global approach, often called the Fourier transform technique [1–8], the whole intensity field is processed at the same time. The output phase data depend on the whole set of data in the intensity field. This technique cannot always be used: a carrier is necessary so that the phase modulation can be identified in the Fourier domain as the broadening of a well-defined frequency peak. This technique has two main advantages: first, only one intensity field is necessary, i.e. one image has to be grabbed, and dynamic events can be studied, and second, large phase modulations can be detected. This approach is presented at the end of this chapter. Using the vocabulary of signal processing, the procedure corresponds to extracting the 'analytic signal' from the intensity field. For metrological applications, the main disadvantage of this technique is that the error on the calculated phase is difficult to evaluate, especially near edges. Another disadvantage is that it is not easy to fully automate, as some filtering in the Fourier plane has to be done, that may depend on the pattern to be analyzed.
- The local approach is called phase-stepping or phase-shifting, and the phase is calculated from local intensity information. Two procedures exist, namely temporal and spatial phase-stepping. In temporal phase-stepping [9–14], a number of images are taken with a constant and known

phase shift δ introduced between successive images. There is no requirement on the fringe pattern, and the calculated phase at a given pixel depends only on the intensities recorded at the same pixel, unless spatial smoothing is used. Obviously, dynamic events cannot be studied as the phase-stepping takes some time. In spatial phase-stepping [15–17], a carrier is necessary so that a given number N of contiguous pixels sample a period of the intensity signal ("tuning" requirement). In this case, the values recorded by adjacent pixels play the role of the phase-stepped intensities and the phase can be calculated from one single frame. However, this tuning requirement is quite stringent and the method is often not applicable. It is especially suited to the grid method used for measuring displacements.

We can also mention an approach using wavelet analysis, which is also a kind of local detection. It will not be presented here, but a rather detailed analysis can be found in [18]. For completeness, a recent work concerning the use of adaptive quadrature filters must be cited [19].

In this review, we focus on the second (local) approach. The formula which is used to calculate the phase from the recorded intensities is usually called the 'algorithm'. Many algorithms have been proposed in the literature, and it is necessary to understand what are their respective advantages and disadvantages. In this chapter, we shall use the term 'phase-stepping' rather than 'phase-shifting'. The difference is that the shifter stops during image grabbing in the first case and not in the second one, thus decreasing the contrast. However, it is worthwhile mentioning that phase-shifting performs some amount of time averaging. This corresponds to a low-pass filtering which may reject part of the unwanted high frequency phase perturbations due to parasitic mechanical vibrations or air turbulence. The global approach will be shortly addressed in Sect. 2.3.

2.2 Local Phase detection

There are a number of systematic errors which can be introduced in the phase-stepping/phase detection process [20–23]. The main ones are listed below.

- There are many cases where the periodic intensity signal which encodes the physical quantity to measure is not sinusoidal. Many examples can be given. Moiré fringes have a profile which is rather triangular. In grid methods, the signal may be close to a binary one when black and white lines are used. Also, the Ronchi test provides fringes which have no reason to be sinusoidal, as they are the shadow of the Ronchi ruling placed in the beam path. In interferometry, some amount of multiple reflection may also lead to fringes which do not have an exact sine profile. Another very important example is when the detector exhibits some amount of

nonlinearity, in which case a perfect input sine is distorted and harmonics are generated.
- In case of a phase-step miscalibration, the phase variation between two images is not exactly equal to its nominal value δ. Two typical examples are: in interferometric techniques, a miscalibration of the piezoelectric transducer (PZT) which is used to move a mirror or rotate a parallel plate to vary the reference path length, and in the grid technique, the analysis of the strained state of a sample using spatial phase-stepping. There are also cases where the miscalibration is not homogeneous across the field of view. This happens, for example, when a phase-stepping mirror exhibits some amount of parasitic tilting during its piston movement.
- The bias variation is encountered in interferometry when the source is a laser diode, the intensity of which varies with time. It is also common in moiré and grid techniques when the field illumination is nonuniform.
- Vibrations can also cause systematic errors, for example when they make the optical path difference vary sinusoidally during temporal phase-stepping in interferometry. Indeed in that case, the electromagnetic phase variation due to the vibration is correlated with the sampling frequency.

For users, the question is to choose or design the algorithm which will cancel the effects of one or many of the above causes of errors, depending on the particular setup. Recently, several methods have been proposed for designing customized algorithms: the Fourier transform (FT) method [24], the characteristic polynomial (CP) method [25], the data-averaging/data-windowing method [26–28] and the recursion method [29]. It is shown in the following how these different approaches can be related. From a practical point of view, the CP approach requires very simple calculations, and provides a comprehensive set of results.

2.2.1 Characteristic Polynomial and Diagram

The intensity field which is recorded during optical measurements is normally written as

$$I(\phi, A, \gamma) = A\left[1 + \gamma \operatorname{frgn}(\phi)\right], \tag{1}$$

where A is the bias and γ is the contrast. The function $\operatorname{frgn}(\phi)$ is a periodic function, a cosine in the simplest case. In the general case, it contains harmonic terms. The bias and contrast may vary slowly across the field of view: $A = A(x, y)$, $\gamma = \gamma(x, y)$. The bias may also vary with time, if the source intensity is fluctuating: $A = A(t)$. This possibility of bias variation is considered in Sect. 2.2.4. A general example of an intensity pattern is shown in Fig. 1a. In the following, we drop the dependence of I on A and γ, and we simply write the intensity as $I(\phi)$.

Phase-stepping methods require the acquisition of M intensity samples I_k,

$$I_k = I(\phi + k\delta), \tag{2}$$

where $\delta = 2\pi/N$ is the phase shift. In this Section we assume that there is no miscalibration of this phase shift. Phase detection algorithms are most often written as

$$\tilde{\phi} = \arctan\left(\frac{\sum_{k=0}^{M-1} b_k I_k}{\sum_{k=0}^{M-1} a_k I_k}\right), \tag{3}$$

where $\tilde{\phi}$ is the measured phase, lying in the range $[-\pi, \pi]$. Equation (3) can also be interpreted as the computation of the argument of the complex linear combination

$$S(\phi) = \sum_{k=0}^{M-1} c_k I_k, \tag{4}$$

where $c_k = a_k + i\, b_k$. The intensity can be expanded in a Fourier series as

$$I(\phi) = \sum_{m=-\infty}^{+\infty} \alpha_m \exp(i\, m\phi), \tag{5}$$

and consequently the sum $S(\phi)$ can also be expanded as

$$S(\phi) = \sum_{m=-\infty}^{+\infty} s_m \exp(i\, m\phi) = \sum_{m=-\infty}^{+\infty} S_m(\phi). \tag{6}$$

The intensity is real, and so

$$\alpha_m = \alpha^*_{-m}. \tag{7}$$

Without loss of generality, one can suppose that $\alpha_1 = \alpha_{-1}$ is real positive. This simply implies that the phase origin is taken at a bright fringe, and the fundamental signal is then proportional to $\cos\phi$.

For designing algorithms, the basic idea is to choose the set of coefficient $\{c_k, k = 0 \ldots M-1\}$ so that all components S_m cancel except for S_1. Thus, taking the argument of $S(\phi)$ will directly provide ϕ. Of course, this implies an infinite number of equations, and so requires M to be infinite, which is not possible. However, three considerations have to be taken into account. First, the harmonic $m = -1$ must always be removed, as its amplitude is equal to the $m = 1$ one which is sought. Second, the harmonic $m = 0$, which corresponds to the bias has always a large amplitude (as the intensity is always positive) and must also be removed. Third, the amplitudes of higher harmonics usually decrease rapidly, and so canceling harmonics of very high order is usually not a critical issue.

So the first choice the user has to make is to choose the extent up to which the harmonics need to be removed. As an example, the user may want to remove harmonics $m = 0$, $m = -1$, $m = \pm 2$ and $m = \pm 3$. The CP method allows an immediate computation of the corresponding set of coefficients $\{c_k\}$.

The CP is related to the coefficients $\{c_k\}$ by

$$P(x) = \sum_{k=0}^{M-1} c_k x^k. \tag{8}$$

From the above equations, it is easy to write the m-th Fourier component of $S(\phi)$:

$$\begin{aligned} S_m(\phi) &= \alpha_m \sum_{k=0}^{M-1} c_k \exp[im(\phi + k\delta)], \\ &= \alpha_m \exp(im\phi) \sum_{k=0}^{M-1} c_k [\exp(im\delta)]^k, \\ &= \alpha_m \exp(im\phi) P(\zeta^m) , \end{aligned} \tag{9}$$

where $\zeta = \exp(i\delta)$. From this expression, it can be seen that canceling $S_m(\phi)$ requires that $P(\zeta^m) = 0$. This implies that its factorization contains the monomial $x - \zeta^m$. So, if we go back to our example and if the harmonics $m = 0$, $m = -1$ and $m = \pm 2$ are to be removed, the CP should write

$$P(x) = c_{M-1}(x - 1)(x - \zeta^{-1})(x - \zeta^2)(x - \zeta^{-2})(x - \zeta^3)(x - \zeta^{-3}), \tag{10}$$

where c_{M-1} can be arbitrarily chosen. Till now, there is no requirement on the value of the phase shift δ. However, a proper choice will minimize the number of necessary samples. To demonstrate this, it is useful to introduce the characteristic diagram, which is the representation in the complex plane of the roots of the CP on the unit circle. If the phase shift is arbitrary, the characteristic diagram corresponding to the polynomial described by 10 will appear as in Fig. 2a. In that case, the polynomial degree is 6, so $M = 7$ intensities are required. If $\delta = 2\pi/5$, the characteristic diagram will appear as in Fig. 2b where there are two roots less. So this choice requires a total of 5 images instead of 7.

If one wants to eliminate the harmonics up to the order j, the general rule is to take $\delta = 2\pi/N$ with $N = j + 2$. The algorithm which is obtained

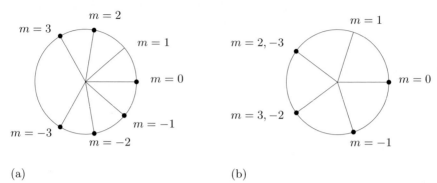

(a) (b)

Fig. 2. Characteristic diagram corresponding to the polynomial described by (10). (a) Arbitrary phase shift (b) Optimum phase shift.

in this case is the classical N-bucket one whose CP is

$$P_N(x) = \zeta \prod_{\substack{k=0 \\ k \neq 1}}^{N-1} (x - \zeta^k)$$

$$= \zeta \frac{x^N - 1}{x - \zeta}$$

$$= 1 + \zeta^{-1} x + \zeta^{-2} x^2 + \ldots + \zeta^{-(N-1)} x^{N-1}. \tag{11}$$

Indeed, this equation shows that $c_k = \zeta^{-k} = \exp(-\mathrm{i}\, 2k\pi/N)$, that is $a_k = \cos(2k\pi/N)$ and $b_k = -\sin(2k\pi/N)$. Thus the final algorithm can be written as

$$\tilde{\phi} = -\arctan \left[\frac{\sum_{k=0}^{N-1} I_k \sin(2k\pi/N)}{\sum_{k=0}^{N-1} I_k \cos(2k\pi/N)} \right]. \tag{12}$$

Surrel [25] calls this algorithm the DFT one, as it corresponds to the computation of the argument of the first coefficient of the discrete Fourier transform of the set of intensities $\{I_k\}$.

2.2.2 Aliased Algorithms

The constant ζ in the product of monomials in (11) has been introduced in order to obtain the N-bucket algorithm in its classical form. The CP is defined up to an arbitrary constant and changing this constant allows one to obtain many aliased versions of the same algorithm. From (9), one gets

$$S_1(\phi) = \alpha_1 \exp(i\phi) P(\zeta). \tag{13}$$

In principle, the algorithm is designed such that $S_1(\phi)$ is the only negligible component of $S(\phi)$, and taking the argument of $S(\phi)$ yields

$$\tilde{\phi} = \arg[S(\phi)] = \phi + \arg(\alpha_1) + \arg[P(\zeta)]. \tag{14}$$

As mentioned in the preceding section, the argument of α_1 depends on the phase origin: if the phase origin is taken so that a null phase corresponds to a bright fringe, the first Fourier component of the intensity is proportional to $\cos\phi$ and α_1 is real positive (it is real negative if the phase origin is chosen at a dark fringe, as in this case the first Fourier component is proportional to $-\cos\phi$). So, with the phase origin taken at a bright fringe, $\arg(\alpha_1) = 0$ and (14) can be rewritten as

$$\tilde{\phi} = \arg[S(\phi)] = \phi + \arg[P(\zeta)]. \tag{15}$$

It can be seen in this equation that multiplying the CP by an arbitrary coefficient β will result in adding the constant $\arg(\beta)$ to the detected phase. So the phases detected by aliased versions of the same algorithm only differ by a constant.

Aliased algorithms can be found in the literature. As an example, the 3-frame algorithm for $\delta = \pi/2$ [12]

$$\tilde{\phi}_1 = \arctan\left(\frac{I_0 - I_1}{I_1 - I_2}\right), \tag{16}$$

corresponds to the CP $P_1(x) = i + (1-i)x - x^2 = -(x-1)(x+i)$. One has in this case $P(\zeta) = P(i) = 2(1+i)$ and following (15), $\tilde{\phi}_1 = \phi + \pi/4$. On the other hand, the algorithm [60]

$$\tilde{\phi}_2 = \arctan\left(\frac{I_0 - 2I_1 + I_2}{I_0 - I_2}\right), \tag{17}$$

corresponds to the CP $P_2(x) = (1+i) - 2ix - (1-i)x^2 = (1-i)P_1(x)$. As $P_2(i) = 4$, the detected phase is $\tilde{\phi}_2 = \phi$. These two algorithms are aliased versions of a single one and correspond to the characteristic diagram presented at line 1 of the table of algorithms in the Appendix. This is not evident from their arctangent form, but is clear from the examination of their respective CPs. The first version given by (16) is more efficient in terms of computing time, as it involves one addition operation less. This is the only reason that can lead one to prefer a particular form. In Sect. 2.2.6 we explain how to choose the multiplying factor of the CP such that a Hermitian symmetry of the coefficients is introduced, allowing us to group the intensity values in pairs, thus reducing the number of multiplication and addition operations.

2.2.3 Self-Calibrating Algorithms

Linear miscalibration. As was mentioned in the introduction, some amount of miscalibration may be present and it is important that the algorithm be designed to be insensitive to this cause of error. A linear miscalibration corresponds to a real phase shift $\delta' = \delta(1 + \epsilon) = \delta + d\delta$. Taking the derivative of $\zeta = \exp(i\delta)$, one obtains $d\zeta = i\zeta d\delta$ and finally:

$$dP(\zeta^m) = P'(\zeta^m)m\zeta^{m-1}i\zeta d\delta,$$
$$= im\mathbf{D}P(\zeta^m)d\delta,$$
$$= im\delta\epsilon \mathbf{D}P(\zeta^m), \tag{18}$$

where $P'(x) = dP(x)/dx$ and the linear operator $\mathbf{D}$ is defined by

$$\mathbf{D}P(x) = xP'(x). \tag{19}$$

Equation (18) gives the first-order variation of $P(\zeta^m)$. If one wants as before to cancel the term proportional to $\exp(im\phi)$ in the sum $S(\phi)$ up to the first order in ϵ, both $P(\zeta^m)$ and $dP(\zeta^m)$ must vanish. This is equivalent to the condition $P(\zeta^m) = \mathbf{D}P(\zeta^m) = 0$, and it is evident from (19) that $\mathbf{D}P(x) = 0$ for $x \neq 0$ if and only if $P'(x) = 0$.

So, we obtain the major result that the m-th harmonic is removed up to the first order in ϵ if ζ^m is a double root of the CP. Clearly, as seen from (18), this condition is not necessary if $m = 0$.

The formula in (18) can be iterated and it is easy to show that an improved insensitivity to miscalibration is obtained for the removal of the harmonic m if ζ^m is a root of increasing order of the CP. More precisely, in the case it is a root of order r, the term in $S(\phi)$ proportional to $\exp(im\phi)$ will contain ϵ^r as the term of lowest power in ϵ [25].

Nonlinear miscalibration. We consider in this subsection the case of a nonlinear phase shift

$$\delta_k = k\delta(1 + \epsilon_1 + k\epsilon_2 + k^2\epsilon_3 + \ldots). \tag{20}$$

In this case, the m-th Fourier component of the sum $S(\phi)$ can be written as

$$S_m(\phi) = \alpha_m \sum_{k=0}^{M-1} c_k \exp\left[im(\phi + \delta_k)\right]$$
$$= \alpha_m \exp(im\phi) \sum_{k=0}^{M-1} c_k \exp(im\delta_k). \tag{21}$$

In the case of the nonlinear phase shift expressed by (20), the sum in (21) can be written as

$$\sum_{k=0}^{M-1} c_k \exp(im\delta_k) = \sum_{k=0}^{M-1} c_k \exp\left[imk\delta(1 + \epsilon_1 + k\epsilon_2 + k^2\epsilon_3 + \ldots)\right]$$

$$= \sum_{k=0}^{M-1} c_k [\exp(i\delta)^m]^k \Big[\, 1 + imk\delta\epsilon_1 + imk^2\delta\epsilon_2 + imk^3\delta\epsilon_3$$

$$+ \ldots$$

$$+ \frac{(imk\delta\epsilon_1)^2}{2!} + \frac{(imk^2\delta\epsilon_2)^2}{2!} + \frac{(imk^3\delta\epsilon_3)^2}{2!} + \ldots$$

$$+ \frac{2(imk\delta\epsilon_1)(imk^2\delta\epsilon_2)}{2!} + \frac{2(imk\delta\epsilon_1)(imk^3\delta\epsilon_3)}{2!} + \ldots$$

$$+ \frac{2(imk^2\delta\epsilon_2)(imk^3\delta\epsilon_3)}{2!} + \ldots \Big]. \tag{22}$$

It can be seen that the preceding expression involves many different polynomials of $\exp(im\delta) = \zeta^m$, with coefficients $\{c_k\}$, $\{kc_k\}$, $\{k^2 c_k\}$ etc. In terms of a dummy variable x, the first one is the characteristic polynomial $P(x)$. The others are deduced from $P(x)$ by the iterative application of the operator $\mathbf{D} = x\frac{d}{dx}$ already introduced in the preceding subsection. So, taking into account (21) and (22), the supplementary term in the sum $S(\phi)$ related to the m-th harmonic becomes in the case of a nonlinear phase shift:

$$S_m(\phi) = \alpha_m \exp(im\phi) \Big[P(\zeta^m)$$

$$+ im\delta\epsilon_1 \mathbf{D} P(\zeta^m) + im\delta\epsilon_2 \mathbf{D}^2 P(\zeta^m) + im\delta\epsilon_3 \mathbf{D}^3 P(\zeta^m) + \ldots$$

$$+ \frac{(im\delta\epsilon_1)^2}{2!} \mathbf{D}^2 P(\zeta^m) + \frac{(im\delta\epsilon_2)^2}{2!} \mathbf{D}^4 P(\zeta^m) + \frac{(im\delta\epsilon_3)^2}{2!} \mathbf{D}^6 P(\zeta^m)$$

$$+ \ldots$$

$$+ \frac{2(im\delta\epsilon_1)(im\delta\epsilon_2)}{2!} \mathbf{D}^3 P(\zeta^m) + \frac{2(im\delta\epsilon_1)(im\delta\epsilon_3)}{2!} \mathbf{D}^4 P(\zeta^m)$$

$$+ \frac{2(im\delta\epsilon_2)(im\delta\epsilon_3)}{2!} \mathbf{D}^5 P(\zeta^m) + \ldots \Big]. \tag{23}$$

The conditions driving the insensitivity to a nonlinear phase shift clearly appear in this equation, and consist in cancelling the adequate terms in

$\epsilon_{r_1}^{s_1}\epsilon_{r_2}^{s_2}\ldots\epsilon_{r_t}^{s_t}$, which requires $\mathbf{D}^{r_1s_1+r_2s_2+\ldots+r_ts_t}P(\zeta^m)$ to cancel. As mentioned before, $\mathbf{D}^n P(\zeta^m)$ cancels if ζ^m is a root of order $n+1$ of the CP. For example, the quadratic nonlinearity has no incidence up to the first order in ϵ_2 if $\mathbf{D}^2 P(\zeta^m) = 0$, that is, if ζ^m is a triple root of the CP. This condition is the same as the one that provides an improved insensitivity to a linear miscalibration, that is, it cancels the term in ϵ_1^2 as well.

2.2.4 Insensitivity to Bias Variation

A bias modulation can be described to a first approximation by a linear variation over the sampling range, as in [30]. So it is assumed that the bias present in intensity I_k includes a term proportional to $k\delta$. This introduces in $S(\phi)$ a supplementary term proportional to $\sum_{k=0}^{M-1} c_k k = P'(1)$. This supplementary term cancels if $P'(1) = 0$, that is, if 1 is a double root of the CP [31]. So, this is the necessary and sufficient condition for an algorithm to be insensitive to a linear bias modulation. It is interesting to see how this condition complements what is presented in the preceding section.

If the bias variation is not linear, the order of the root located at $z = 1$ must be increased [31].

2.2.5 Windowed-DFT Algorithm

A simple generic algorithm can be derived from the results of the two preceding sections. If one wants to keep the properties of the DFT algorithm in the presence of both miscalibration and bias variation, one has to double all the roots of the DFT algorithm. For the special case of $N = 4$, this corresponds to the algorithm presented at line 12 of the table in the Appendix, and the algorithm is written as

$$\phi = \arctan\left(\frac{(I_0 - I_6) - 3(I_2 - I_4)}{2(I_1 + I_5) - 4I_3}\right). \tag{24}$$

As all the roots are doubled, the CP of the (WDFT) algorithm is proportional to $P_N(x)^2$, where $P_N(x)$ is the CP of the DFT algorithm, as before. In order to obtain the Hermitian symmetry of the coefficients (see Sect. 2.2.6), a multiplying factor ζ^{-1} is introduced so that the polynomial finally becomes

$$P(x) = \zeta^{-1} P_N(x)^2 = \sum_{k=0}^{2N-2} t_k \zeta^{-k-1} x^k, \tag{25}$$

where $\{t_k, k = 0, \ldots, 2N - 2\} = \{1, 2, \ldots, N - 1, N, N - 1, \ldots, 2, 1\}$. These coefficients can also be interpreted as the triangular windowing of the signal

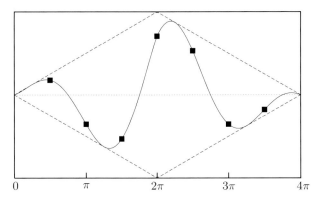

Fig. 3. Triangular windowing performed by the WDFT algorithm

during two periods (Fig. 3). From (25), the WDFT algorithm is written as

$$\phi = \arctan\left[-\frac{\sum_{k=1}^{N-1} k(I_{k-1} - I_{2N-k-1})\sin(2k\pi/N)}{NI_{N-1} + \sum_{k=1}^{N-1} k(I_{k-1} + I_{2N-k-1})\cos(2k\pi/N)}\right]. \tag{26}$$

The total number of samples required by the WDFT algorithm for a phase shift of $2\pi/N$ is $M = 2N - 1$.

It must be recognized that this algorithm will give quite good results in most practical cases, and therefore should be recommended.

2.2.6 Hermitian Symmetry of the Coefficients

The possibility of obtaining aliased algorithms depending on an arbitrary multiplying factor has been presented in Sect. 2.2.2. However, a canonical form can be defined. It is possible to show [32] that the algorithm coefficients can be given the following Hermitian symmetry:

$$c_k^* = c_{M-k-1}, \tag{27}$$

where the star superscript stands for complex conjugation, or equivalently

$$\begin{cases} a_k = a_{M-k-1} \\ b_k = -b_{M-k-1}. \end{cases} \tag{28}$$

This can be achieved when the coefficient c_{M-1} is chosen so that

$$c_0 = c_{M-1}^*, \tag{29}$$

that is, the Hermitian symmetry is obtained for all the coefficients if it is obtained for c_0 and c_{M-1}. Starting from a CP without this symmetry, it is easy to show that it has to be multiplied by

$$\frac{a}{c_0 + c_{M-1}},$$

where a is any real number.

This Hermitian symmetry of the CP coefficients does not confer any specific property to the related algorithm. However, this symmetry is useful for its computer implementation, as it reduces the number of necessary addition and multiplication operations, because it allows us to group intensity terms in pairs, as in the WDFT algorithm presented in (24).

2.2.7 Phasor Representation

It is possible to represent the sum $S(\phi)$ in the complex plane as a sum of vectors, each of which is the representation of a term $c_k I_k$, $k = 0 \ldots M - 1$. That graphical representation of a complex number is often called a phasor.

As expressed by (6), the harmonic m builds up the Fourier component $S_m(\phi)$ of $S(\phi)$. As was explained in Sect. 2.2.1, the set of coefficients $\{c_k\}$ should be chosen so that $S_m(\phi) = 0$ for $m \neq 1$. That is, if the signal is $I(\phi) = \exp(i\phi)$, the individual phasors corresponding to all the terms $c_k I_k$ add constructively so that the resulting phasor has a nonvanishing length. On the other hand, other harmonics should build up the null vector. In this case, the phasor representation of the sum $S_m(\phi)$ is a closed path, starting and ending at the origin.

We present here three examples. The first one corresponds to the DFT (or N-bucket) algorithm with $\delta = \pi/3$. The characteristic diagram for this algorithm is displayed at line 10 of the table in the Appendix. From the examination of the location of the roots, one can deduce that this algorithm is designed to cancel harmonics $m = 0, -1, \pm 2, \pm 3, \pm 4$. The phasor construction of the Fourier components $S_m(\phi)$ is shown in Fig. 4.

The second example corresponds to the WDFT algorithm with $\delta = \pi/3$, that is $N = 6$ and $M = 11$. The characteristic diagram for this algorithm is displayed at line 23 of the table in the Appendix. As in the example above, this algorithm is designed to cancel harmonics $m = 0, -1, \pm 2, \pm 3, \pm 4$, but this time with insensitivity to miscalibration. The phasor construction of the Fourier components $S_m(\phi)$ is shown in Fig. 5.

The last example corresponds to the 9-sample algorithm [33]

$$\phi = \arctan\left[\frac{I_0 - I_8 - 2(I_1 - I_7) - 14(I_2 - I_6) - 18(I_3 - I_5)}{2(I_0 + I_8) + 8(I_1 + I_7) + 8(I_2 + I_6) - 8(I_3 + I_5) - 20I_4}\right], \quad (30)$$

whose diagram is presented at line 21 of the table in the Appendix. The phasor construction of the Fourier components $S_m(\phi)$ is shown in Fig. 6. It

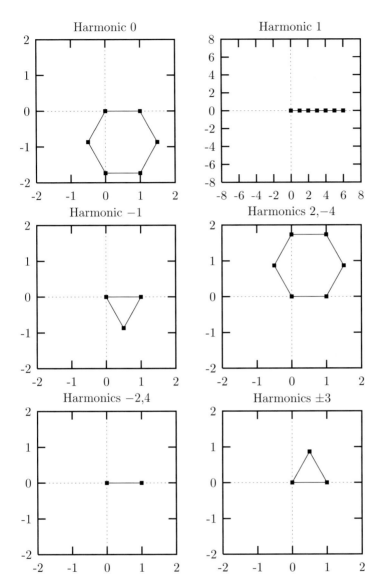

Fig. 4. Phasor representation of the Fourier components of $S(\phi)$ for the DFT algorithm with $\delta = \pi/3$. The characteristic diagram of this algorithm is displayed at line 10 of the table in the Appendix

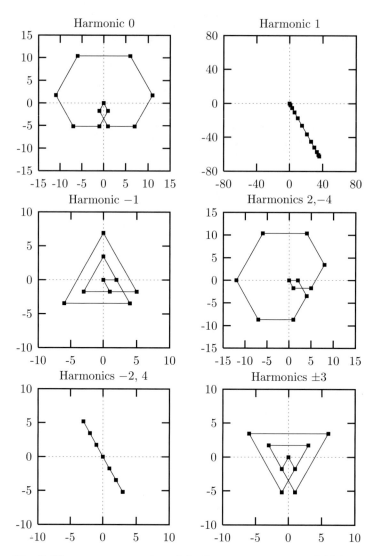

Fig. 5. Phasor representation of the Fourier components of $S(\phi)$ for the windowed-DFT algorithm with $\delta = \pi/3$. The characteristic diagram of this algorithm is displayed at line 23 of the table in the Appendix

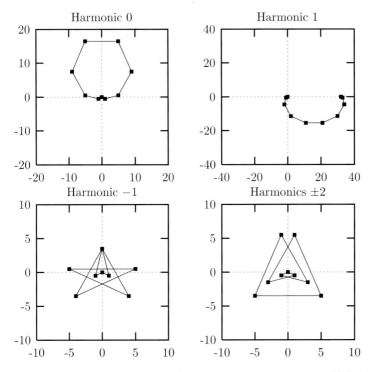

Fig. 6. Phasor representation of the Fourier components of $S(\phi)$ for the 9-sample algorithm described by (30). The characteristic diagram of this algorithm is displayed at line 21 of the table in the Appendix

can be seen that for this algorithm, the construction of $S_1(\phi)$ is not a straight line. That means that the recorded samples are not used in an efficient way. The consequence is that the SNR or "signal strength" will be poor. This effect will be detailed in Sect. 2.5.2.

2.2.8 Relation with the Fourier Transform Approach

The correspondence between our approach and the so-called Fourier transform (FT) approach [24,34] is straightforward. In the FT approach, the sampling functions related to the numerator and denominator of the fraction defining the tangent of the measured phase are introduced:

$$\begin{cases} f_1(\phi) = \sum_{k=0}^{M-1} b_k \delta(\phi + k\delta) \\ f_2(\phi) = \sum_{k=0}^{M-1} a_k \delta(\phi + k\delta) \ . \end{cases} \tag{31}$$

The plus sign is chosen in front of the phase shifts $k\delta$ to be consistent with (2). Actually, in the original work [24], those sampling functions are written as function of time, as it is supposed that the phase shift varies linearly with time. The correspondence is given by

$$\phi = 2\pi\nu_s t, \tag{32}$$

where ν_s is the signal frequency. Then we consider the Fourier transforms $\mathcal{F}$ of the sampling functions f_1 and f_2 and we use ν as the conjugated Fourier variable:

$$\begin{cases} F_1(\nu) = (\mathcal{F}f_1)(\nu) = \sum_{k=0}^{M-1} b_k \exp(\mathrm{i}\, 2k\pi\nu\delta) \\ F_2(\nu) = (\mathcal{F}f_2)(\nu) = \sum_{k=0}^{M-1} a_k \exp(\mathrm{i}\, 2k\pi\nu\delta). \end{cases} \tag{33}$$

The coefficients $\{a_k\}$ and $\{b_k\}$ being real, it can be seen that:

$$\begin{cases} F_1(-\nu) = F_1^*(\nu) \\ F_2(-\nu) = F_2^*(\nu)\,. \end{cases} \tag{34}$$

Still denoting $\zeta = \exp(\mathrm{i}\,\delta)$, we obtain

$$F_2(\nu) + \mathrm{i}\, F_1(\nu) = \sum_{k=0}^{M-1} c_k \exp(\mathrm{i}\, 2k\pi\nu\delta),$$

$$= P(\zeta^m), \tag{35}$$

where $P(x)$ is the CP of the considered algorithm and $m = 2\pi\nu$ is the angular frequency of ϕ. Taking (34) into account, one can write

$$P(\zeta^{-m}) = F_2(-\nu) + \mathrm{i}\, F_1(-\nu) = F_2^*(\nu) + \mathrm{i}\, F_1^*(\nu). \tag{36}$$

Finally, we obtain

$$P(\zeta^m) = F_2(\nu) + \mathrm{i}\, F_1(\nu)$$

$$P^*(\zeta^{-m}) = F_2(\nu) - \mathrm{i}\, F_1(\nu). \tag{37}$$

From this, all the results from the CP theory can be converted in terms of the FT theory, and vice versa. In particular, the elimination of harmonics m and $-m$ corresponds to $P(\zeta^m) = P(\zeta^{-m}) = P^*(\zeta^{-m}) = 0$, that is

$$F_2(\nu) + \mathrm{i}\, F_1(\nu) = 0,$$

$$F_2(\nu) - \mathrm{i}\, F_1(\nu) = 0. \tag{38}$$

This is equivalent to

$$F_1(\nu) = F_2(\nu) = 0, \tag{39}$$

which is the condition given by the FT theory. The detection of the harmonic $+m$ requires us to eliminate the $-m$ component only, that is to cancel $P(\zeta^{-m})$, and so the condition is written as

$$F_1(\nu) = -i\, F_2(\nu), \tag{40}$$

which is the form given by the FT theory, except for the sign. This discrepancy comes from the sign chosen in front of the phase shifts.

2.2.9 Relation with Averaging and Data Windowing

The averaging technique. The averaging technique has been first introduced by *Schwider* et al. [20] and was systematically extended and investigated by *Schmit* and *Creath* [26,27]. The data windowing technique was proposed by *de Groot* [28] (and developed to its ultimate extent by the same author [35]!). In this section we show that the extended averaging and the data windowing techniques can be interpreted using the CP as well.

Schwider et al. [20] noticed that the phase errors in phase-shifting measurements are very often modulated at twice the frequency of the phase ϕ. That is, the measured phase $\tilde{\phi}$ can be written as $\tilde{\phi} = \phi + a\sin(2\phi + \psi)$. So, they proposed to perform two sets of measurements, shifted by a quarter of a period, in order to get two phases $\tilde{\phi}_1 = \phi + a\sin(2\phi + \psi)$ and $\tilde{\phi}_2 = \phi + \pi/2 + a\sin(2\phi + \pi + \psi)$, so that $(\tilde{\phi}_1 + \tilde{\phi}_2 - \pi/2)/2 = \phi$, thus eliminating the phase error. They wrote the arctangent form of an algorithm

$$\tilde{\phi} = \arctan\left(\frac{N}{D}\right). \tag{41}$$

and they showed that, instead of computing two phases $\tilde{\phi}_1$ and $\tilde{\phi}_2$ from (41) and averaging them, it was more efficient to use

$$\tilde{\phi} = \arctan\left(\frac{N_1 + N_2}{D_1 + D_2}\right). \tag{42}$$

This can be applied to any M-sample algorithm but, if the phase shift of the algorithm is $\pi/2$, the two sample sets overlap each other so that the resulting algorithm requires only $M + 1$ samples.

Schmit and *Creath* [26] have extended this idea and proposed to iterate the procedure, that is, to average two $(M+1)$-sample algorithms obtained as before to get an $(M+2)$-sample algorithm:

$$\begin{aligned}\tilde{\phi} &= \arctan\left[\frac{(N_1 + N_2) + (N_2 + N_3)}{(D_1 + D_2) + (D_2 + D_3)}\right], \\ &= \arctan\left(\frac{N_1 + 2N_2 + N_3}{D_1 + 2D_2 + D_3}\right).\end{aligned} \tag{43}$$

In a recent paper, the same authors propose the multiple averaging technique [27], in which not only two but three or more algorithms are averaged. In the latter paper, a large number of new algorithms are derived and analysed in terms of data windowing. As examples, the authors investigated different 8-sample algorithms obtained by averaging many different algorithms in different ways. They showed that the result of this averaging process is the windowing of the intensity data set. However, very similar windowing functions turn out to correspond to algorithms that have very different properties.

In a paper by *de Groot* [28], the reverse approach is used, that is, a data window function is given and then the algorithm is derived. Here too, only an *a posteriori* study is capable of elucidating the properties of the obtained algorithm.

Algebraic interpretation of a shift of the data set. Here, we investigate how a shift of the data set (i.e. using the set $\{I_n, I_{n+1} \ldots I_{n+M-1}\}$ instead of $\{I_0, I_1 \ldots I_{M-1}\}$ as the algorithm input) can be taken into account using the CPs. We will concentrate on algorithms which use a $\pi/2$ phase shift. In any case, results can be straightforwardly extended to the case of a generic phase shift δ.

The N/D notation in (41) introduced by *Schwider* et al. [20] can evidently be related to the sum $S(\phi)$ by (we drop the dependence on ϕ)

$$S = D + i N. \tag{44}$$

When two sets of samples are used as in (42), the sum S is written as

$$S = S_1 + S_2. \tag{45}$$

The first term of the sum has the argument ϕ and the second one the argument $\phi + \phi/2$, because the first sample it involves has already a $\pi/2$ phase lag. Both terms have the same magnitude, so that the argument of the sum S is $\phi + \pi/4$. The complex sum corresponding to the second set of samples is

$$S_2 = \sum_{k=1}^{M} c_k I_k . \tag{46}$$

Hence the corresponding CP is

$$\sum_{k=1}^{M} c_k x^k = x P_1(x) , \tag{47}$$

where $P_1(x)$ is the CP corresponding to the processing of the first data set. In the same way, using the algorithm formula with a third set of samples starting at intensity I_2 to construct a sum S_3 would correspond to the polynomial $x^2 P_1(x)$, and so on.

If the sum S_{n+1}, which has the argument $n\pi/2$, is multiplied by $(-\mathrm{i})^n$, one obtains a complex number of argument ϕ, and a sum of such terms $(-\mathrm{i})^n S_{n+1}$ will also have the argument ϕ. The characteristic polynomial corresponding to $(-\mathrm{i})^n S_{n+1}$ is

$$P_{n+1}(x) = (-\mathrm{i}\,x)^n P_1(x). \tag{48}$$

This result is the algebraic representation of a shift of the input data set.

Example. As an example, let us take an algorithm with $\delta = \pi/2$, whose CP is denoted by $P_1(x)$. Following (48), when two sample sets are averaged the corresponding polynomial will be

$$\begin{aligned} Q_1(x) &= P_1(x) + P_2(x), \\ &= P_1(x)(1 - \mathrm{i}\,x), \\ &= -\mathrm{i}\,P_1(x)(x + \mathrm{i}). \end{aligned} \tag{49}$$

The improvement is evident, as $-\mathrm{i}$ is now a root of higher order, thus increasing the algorithm insensitivity to miscalibration. Now, if three sample sets are averaged, one obtains

$$\begin{aligned} R(x) &= P_1(x) + P_2(x) + P_3(x), \\ &= P_1(x)(1 - \mathrm{i}\,x - x^2), \\ &= -P_1(x)\frac{x^3 + \mathrm{i}}{x - \mathrm{i}}. \end{aligned} \tag{50}$$

The polynomial expressed by the last fraction has as roots the 3rd roots of $-\mathrm{i}$, except for i, that is $\exp(-\mathrm{i}\,\pi/6)$ and $\exp(-\mathrm{i}\,5\pi/6)$. Here, no improvement is obtained with respect either to miscalibration or to harmonics insensitivity. Thus the averaging process cannot ascertain by itself that an interesting algorithm will be obtained. Similarly, the windowing technique (that is, choosing a windowing function using usual ideas of signal processing) will not always provide an efficient algorithm. Moreover, the properties of the obtained algorithms will have to be found by a thourough a posteriori study. So, there is no real advantage of these approaches over the CP method.

A detailed investigation of the various algorithms proposed by *Schmit* and *Creath* [27] is presented in [36].

2.2.10 Relation with the Recursion Method

The recursion method has been introduced recently by *Phillion* [29]. In Sect. 4.F of [29], the author shows some equivalence of his method with the CP one. We extend and complement here his results.

First, starting from the coefficients $\{c_k\}$ of the algorithm (denoted $\{r_j\}$ in the original paper), the weights $\{w_k\}$ are defined by

$$w_k = c_k \exp(\mathrm{i}\,k\delta) = c_k \zeta^k. \tag{51}$$

In Sect. 4.F of [29], the author introduces the w–polynomial $p(x)$ by

$$p(x) = \sum_{k=0}^{M-1} w_k x^k. \qquad (52)$$

It is clear from (51) and (52) that

$$p(x) = P(\zeta x), \qquad (53)$$

where $P(x)$ is still the algorithm's CP. All results in Phillion's paper [29] can be analysed using this relation.

For example, the author introduces recursion relations by

$$\overline{w}_k = \sum_{l=0}^{R} a_k w_{l+k}, \qquad (54)$$

and he notices that this is equivalent to multiplying the CP by

$$Q(x) = \sum_{l=0}^{R} a_l (\zeta^{-1} x)^l. \qquad (55)$$

The recursion relations that are presented correspond to introducing more roots in the CP. For example, the recursion relation for $\delta = \pi/2$, $\overline{w}_k = w_k + w_{k+1}$, is equivalent to multiplying the CP by $Q(x) = 1 - \mathrm{i}\,x = -\mathrm{i}(x+\mathrm{i})$. This increases by one the order of multiplicity of the root located at $-\mathrm{i}$.

It is interesting to compare this with the averaging technique, especially with (49). In general, the recursion and averaging techniques seem to be very closely related.

Also, a "figure of merit" N_F is introduced by Phillion

$$N_\mathrm{F} = \frac{\left|\sum_k w_k\right|}{\left(\sum_k |w_k|^2\right)^{1/2}}. \qquad (56)$$

Owing to (51), the sum in the numerator is nothing else than $P(\zeta)$, and so it can be seen that

$$N_\mathrm{F} = \eta \sqrt{M}, \qquad (57)$$

where η is the efficiency factor defined by (87) in Sect. 2.5.2. However, it should be noticed that $0 \leq \eta \leq 1$ for any algorithm, and therefore the factor η allows more straightforward comparisons. Other results in Phillion's paper concerning the noise evaluation are the same as those presented in Sect. 2.5.

2.3 Global Phase Detection

The global phase detection technique is called the Fourier transform technique [1–8], but it should not be confused with the technique of the same name presented in Sect. 2.2.8. It is called "global" as the whole set of intensity data in the fringe pattern is used to calculate the whole set of phase data.

The requirement on the fringe pattern is that the useful information should appear as the modulation of a spatial carrier of frequency $\boldsymbol{f}_0$, i.e. the intensity signal should appear as

$$I(\boldsymbol{r}) = A\left\{1 + \gamma\cos\left[2\pi \boldsymbol{f}_0 \boldsymbol{r} + \phi(\boldsymbol{r})\right]\right\},$$

$$= A + \frac{A\gamma}{2}\exp(\mathrm{i}\,2\pi \boldsymbol{f}_0 \boldsymbol{r})\exp[\mathrm{i}\,\phi(\boldsymbol{r})]$$

$$+ \frac{A\gamma}{2}\exp(-\mathrm{i}\,2\pi \boldsymbol{f}_0 \boldsymbol{r})\exp[-\mathrm{i}\,\phi(\boldsymbol{r})],$$

$$= A + C(\boldsymbol{r})\exp(\mathrm{i}\,2\pi \boldsymbol{f}_0 \boldsymbol{r}) + C^*(\boldsymbol{r})\exp(-\mathrm{i}\,2\pi \boldsymbol{f}_0 \boldsymbol{r}), \tag{58}$$

where

$$C(\boldsymbol{r}) = \frac{A\gamma}{2}\exp[\mathrm{i}\,\phi(\boldsymbol{r})]. \tag{59}$$

Here we suppose that the fringe intensity profile is sinusoidal. We will mention briefly below how things are modified if the signal contains harmonics.

Taking the Fourier transform of (58), we can write

$$\widehat{I}(\boldsymbol{f}) = A\delta(\boldsymbol{f}) + \widehat{C}(\boldsymbol{f} - \boldsymbol{f}_0) + \widehat{C^*}(\boldsymbol{f} + \boldsymbol{f}_0),$$
$$= A\delta(\boldsymbol{f}) + \widehat{C}(\boldsymbol{f} - \boldsymbol{f}_0) + \widehat{C}^*(-\boldsymbol{f} - \boldsymbol{f}_0). \tag{60}$$

The function $C(\boldsymbol{r})$ is now supposed to have only a low frequency content, i.e. the useful signal $\phi(\boldsymbol{r})$ is supposed to vary slowly with respect to the carrier frequency $\boldsymbol{f}_0$. This means that $\widehat{C}(\boldsymbol{f})$ is concentrated around the origin. This is shown in Fig. 7 in the special case of $\boldsymbol{f}_0 = f_0 \boldsymbol{x}$, where $\boldsymbol{x}$ is the unit vector along direction x. Thus from (60) it can be seen that $\widehat{I}(\boldsymbol{f})$ will consist of three distinct parts: a Dirac delta function corresponding to the mean value of $I(\boldsymbol{r})$ and two side-lobes corresponding to the spectrum of $C(\boldsymbol{r})$ translated by the amount of the carrier frequency, $\boldsymbol{f}_0$ and $-\boldsymbol{f}_0$ (Fig. 8).

Therefore it is possible to band pass filter the intensity spectrum so that only the right-hand side lobe $\widehat{C}(\boldsymbol{f} - \boldsymbol{f}_0)$ is retained (Fig. 9). The resulting function is approximated by

$$g(\boldsymbol{f}) = \widehat{C}(\boldsymbol{f} - \boldsymbol{f}_0)\,. \tag{61}$$

The inverse Fourier transform of this function is $C(\boldsymbol{r})\exp(\mathrm{i}\,2\pi \boldsymbol{f}_0 \boldsymbol{r})$. Following (59), the last step is then to take the argument of that result to obtain

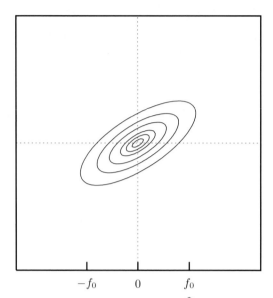

Fig. 7. Low frequency spectrum $|\widehat{C}(f)|$ containing the signal information

$\phi(r) + 2\pi f_0 r$. Then the linear carrier phase variation can be easily subtracted. An equivalent method is to translate the side lobe to the origin, to obtain $\widehat{C}(f)$ before taking its inverse Fourier transform. Practically, the filtering process in the Fourier space should be smoother than a simple stepwise windowing, because it is well known that such a filtering introduces in the original space parasitic ripples that will distort the phase.

It will often be very difficult to separate perfectly the useful signal frequencies. First, the signal profile may not be perfectly sinusoidal and so other side lobes may be present at frequencies $\pm n f_0$, where n is integer. Thus some crosstalk between lobes may be present, mixing fundamental and harmonic information.

The Fourier transform method can be used when two orthogonal carriers encode two different sets of information [37]. A possible example is when a cross-grid is used for measuring displacements [38]. Each grid-line direction is a carrier used for the detection of the displacement component orthogonal to that direction. The carrier frequency vectors can be written f_0 and f_1. In the case of a nonsinusoidal signal, the spectrum will appear as in Fig. 10. The required passing band for the detection of the information encoded by the carrier f_0 appears in Fig. 11.

However, it can be understood that the filtering process cannot be easily automated, as the filtered domain will depend on the shape of the spectrum. The bias itself may vary across the field, thus widening the central Dirac function with also some risk of cross talk with the useful frequencies. The

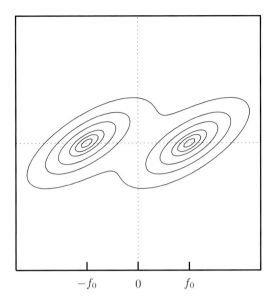

Fig. 8. Spectrum of the intensity, with its two side-lobes centered at $\boldsymbol{f}_0$ and $-\boldsymbol{f}_0$ (the Dirac function at point (0,0) is not represented)

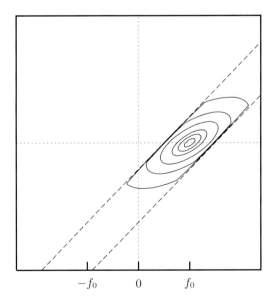

Fig. 9. Bandpass filtered spectrum of the intensity retaining the side-lobe centered at $\boldsymbol{f}_0$

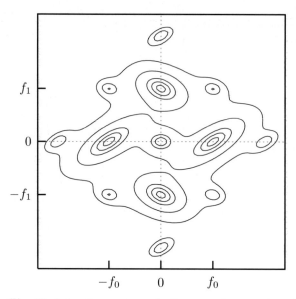

Fig. 10. Intensity spectrum in the case of a non-sinusoidal signal with two orthogonal carriers

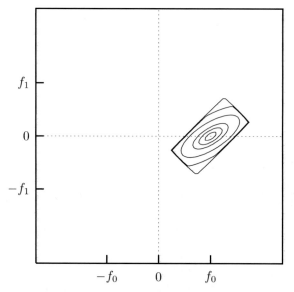

Fig. 11. New definition of the bandpass filter for the spectrum appearing in Fig. 10

fundamental problem is that the resulting errors on the phase are almost impossible to quantify in the general case.

The major advantage of this technique over spatial phase-stepping is that the available range of the fringe frequency is wider. If we consider $[f_0/2, 3f_0/2]$ as a typical band-pass filter width, the local fringe frequency can lie within that range. So the local 'miscalibration', that is the local mismatch between the actual and the carrier frequencies, can be within the range $[-50\%, 33\%]$. However, this range is very theoretical, and in practice the requirements on the fringe frequency variations across the field may not be so different between the Fourier transform and the spatial phase-stepping methods. However, this issue has not yet been precisely addressed.

It is sometimes thought that it is not necessary that the carrier frequency be high. This overlooks a major feature of the phase modulation technique. Indeed, using a low-frequency carrier will decrease the spatial resolution, because it can be considered that the "gauge length" is at least equal to one signal period and so the effective spatial resolution is the carrier period and not the pixel. So even with the Fourier transform method it should be considered desirable that the largest number of fringes appear in the field.

2.4 Residual Phase Errors

A crucial point for real engineering applications is the possibility of evaluating the sensitivity and precision. One requirement is that the phase errors can be evaluated. The CP theory gives analytic formulae for almost any source of systematic errors.

2.4.1 General Evaluation

The residual errors in the phase measurement can easily be evaluated using the CP theory. Normally, the sum $S(\phi)$ should contain $S_1(\phi)$ as the predominant term. From (23) we have

$$S_1 = S_{10} + \Delta S_1 , \tag{62}$$

where

$$S_{10} = \alpha_1 \exp(\mathrm{i}\,\phi) P(\zeta), \tag{63}$$

and where ΔS_1 is a small quantity involving possible miscalibration terms. In the ideal case $\Delta S_1 = 0$ and the phase detected is $\phi + \psi$, where $\psi = \arg[P(\zeta)]$. This was already stated by (15) in Sect. 2.2.2. When harmonics or miscalibration are present, the sum S can be written

$$S = S_{10} + \Delta S ,$$

$$= S_{10}\left(1 + \frac{\Delta S}{S_{10}}\right) ,$$

$$= S_{10}\left[1 + \Re\left(\frac{\Delta S}{S_{10}}\right) + i\Im\left(\frac{\Delta S}{S_{10}}\right)\right] ,$$

$$\approx |S_{10}|\exp[i(\phi + \psi)]\left[1 + i\Im\left(\frac{\Delta S}{S_{10}}\right)\right] , \qquad (64)$$

where $\Re$ and $\Im$ are the real-part and imaginary-part operators. On the other hand,

$$S = |S|\exp[i(\phi + \psi + \Delta\phi)] \approx |S|\exp[i(\phi + \psi)](1 + i\Delta\phi). \qquad (65)$$

So, using $|S| = |S_{10}| + \Delta|S|$, the phase error can be written

$$\Delta\phi = \Im\left(\frac{\Delta S}{S_{10}}\right). \qquad (66)$$

We see from the above equation and from (63) that the phase error formula will always involve a division by $\alpha_1 P(\zeta)$. So, it is a good idea to introduce the *normalized* CP as

$$\tilde{P}(x) = \frac{P(x)}{P(\zeta)} \qquad (67)$$

and the normalized amplitude of the m-th harmonic $\tilde{\alpha}_m = \alpha_m/\alpha_1$. Now, it is very easy to give an analytic formula for the phase error when miscalibration and/or harmonics are present. It can be directly deduced from (23) and (66) and yields

$$\Delta\phi = \sum_m \Im\left\{\tilde{\alpha}_m \exp[i(m-1)\phi]\left[\tilde{P}(\zeta^m) + \right.\right.$$

$$+ im\delta\epsilon_1 \mathbf{D}\tilde{P}(\zeta^m) + im\delta\epsilon_2 \mathbf{D}^2\tilde{P}(\zeta^m) + im\delta\epsilon_3 \mathbf{D}^3\tilde{P}(\zeta^m)\ldots$$

$$+ \frac{(im\delta\epsilon_1)^2}{2!}\mathbf{D}^2\tilde{P}(\zeta^m) + \frac{(im\delta\epsilon_2)^2}{2!}\mathbf{D}^4\tilde{P}(\zeta^m) + \frac{(im\delta\epsilon_3)^2}{2!}\mathbf{D}^6\tilde{P}(\zeta^m)$$

$$+ \ldots + \frac{2(im\delta\epsilon_1)(im\delta\epsilon_2)}{2!}\mathbf{D}^3\tilde{P}(\zeta^m) + \frac{2(im\delta\epsilon_1)(im\delta\epsilon_3)}{2!}\mathbf{D}^4\tilde{P}(\zeta^m)$$

$$\left.\left.+ \frac{2(im\delta\epsilon_2)(im\delta\epsilon_3)}{2!}\mathbf{D}^5\tilde{P}(\zeta^m) + \ldots\right]\right\}. \qquad (68)$$

Notice that in this sum the term $\tilde{\alpha}_1 \tilde{P}(\zeta)$ appears for $m = 1$. Obviously, it is equal to 1 and vanishes when taking the imaginary part, as it should.

Two examples illustrating the application of (68) are presented below.

2.4.2 Effect of the m-th Harmonic

As a first example, let us consider a nonsinusoidal signal and the effects of the m-th harmonic. As stated by (7) of Sect. 2.2.1, both harmonics m and $-m$ are present with conjugated amplitudes, due to the fact that the intensity is a real signal. From (9), we see that the corresponding terms in $S(\phi)$ are $\alpha_m \exp(\mathrm{i}\, m\phi)P(\zeta^m) + \alpha_m^* \exp(-\mathrm{i}\, m\phi)P(\zeta^{-m})$. From (68), the phase error that results is

$$\Delta\phi = \Im\{\tilde{\alpha}_m \exp[\mathrm{i}(m-1)\phi]\tilde{P}(\zeta^m) + \tilde{\alpha}_m^* \exp[-\mathrm{i}(m+1)\phi]\tilde{P}(\zeta^{-m})\}. \quad (69)$$

This error has two components at $(m-1)$ and $(m+1)$ times the fringe frequency.

2.4.3 Effect of the Miscalibration

As a second example, we consider the presence of linear miscalibration, and a purely sinusoidal signal. In that case, only $m = 1$ and $m = -1$ must be considered, with $\epsilon_1 = \epsilon \neq 0$. We take as an example an algorithm that is not self-calibrating. From the results of Sect. 2.2.3, that means $\mathbf{D}P(\zeta^{-1}) \neq 0$. As was already mentioned after (7) in Sect. 2.2.1, $\alpha_1 = \alpha_{-1}$. Taking into account $P(\zeta^{-1}) = 0$, (68) gives for the phase error

$$\Delta\phi = \delta\epsilon\, \Im[\mathrm{i}\, \mathbf{D}\tilde{P}(\zeta) - \mathrm{i}\, \exp(-\mathrm{i}\, 2\phi)\mathbf{D}\tilde{P}(\zeta^{-1})]\,. \quad (70)$$

So there is a static error and a modulation at twice the fringe frequency. Let us consider as an example the DFT algorithm. In this case, one has [25]

$$\begin{cases} \mathbf{D}\tilde{P}(\zeta) = \dfrac{N-1}{2} \\ \mathbf{D}\tilde{P}(\zeta^{-1}) = \dfrac{\mathrm{i}\, \exp(\mathrm{i}\, \delta)}{2\sin\delta}. \end{cases}$$

Thus the phase error can be written for this algorithm as

$$\Delta\phi = \delta\epsilon \left[\frac{N-1}{2} - \frac{\sin(2\phi - \delta)}{2\sin\delta}\right], \quad (71)$$

which corresponds to the formula found in the literature [25,39].

In the case of a self-calibrating algorithm with, say, $\mathbf{D}P(\zeta^{-1}) = 0$ and $\mathbf{D}^2 P(\zeta^{-1}) \neq 0$, (70) would be changed into

$$\Delta\phi = \delta\epsilon\, \Im \left[\mathrm{i}\, \mathbf{D}\tilde{P}(\zeta) - \frac{\delta\epsilon}{2} \exp(-\mathrm{i}\, 2\phi)\mathbf{D}^2\tilde{P}(\zeta^{-1})\right]. \quad (72)$$

It should be clear now how all possible cases can be handled using (68). Other examples of phase error evaluations can be found in [32].

2.5 Effect of Additive Noise

For metrological purposes, it is very important that the sensitivity can be evaluated. A good measurement of the sensitivity is from the standard deviation of the phase noise obtained after detection. A general formula can be given which shows the respective influence of the different parameters at the acquisition and processing levels. Further details of the derivation of the results presented in this section can be found in [40].

2.5.1 Average of the Phase Noise

In this section, the intensity signal is supposed to be a perfect sine:

$$I = A(1 + \gamma \cos \phi).$$

In this case the Fourier coefficient α_1 is

$$\alpha_1 = \frac{A\gamma}{2}. \tag{73}$$

Also, the sampling spacing is assumed to be perfect, that is $\epsilon = 0$, so that the sum $S(\phi)$ can be written as

$$S(\phi) = S_1(\phi) = S_{10}(\phi) = |S_{10}|\exp(i\phi), \tag{74}$$

where $S_{10}(\phi)$ has the same meaning as in Sect. 2.4.1. From (9) and (73) we get

$$|S_1| = \frac{A\gamma}{2}|P(\zeta)|. \tag{75}$$

We now examine the effect on the phase measurement of additive noise present in the intensity signal [40]. Each recorded intensity I_k is supposed to contain an additional centered noise ΔI_k of variance σ^2, supposed not to depend on k:

$$I_k = A\left[1 + \gamma \cos(\phi + k\delta)\right] + \Delta I_k. \tag{76}$$

It is also assumed that there is a statistical independence of the noise corresponding to two different values of intensity, i.e.

$$\langle \Delta I_k \Delta I_m \rangle = \sigma^2 \delta_{km}, \tag{77}$$

where δ_{km} is the Kronecker delta. The signal-to-noise ratio of the signal is

$$\text{SNR} = \frac{A\gamma}{\sigma\sqrt{2}}. \tag{78}$$

In the same way as in Sect. 2.4.1, the phase perturbation can be written as

$$\Delta \phi = \Im\left(\frac{\Delta S}{S_{10}}\right). \qquad (79)$$

Let us define two quantities J and K by

$$\frac{\Delta S}{|S_{10}|} = J + \mathrm{i}\, K, \qquad (80)$$

so that

$$J = \sum_{k=0}^{M-1} a_k \frac{\Delta I_k}{|S_{10}|}, \qquad (81)$$

$$K = \sum_{k=0}^{M-1} b_k \frac{\Delta I_k}{|S_{10}|}. \qquad (82)$$

Then it is easy to evaluate $\Delta \phi$ from (79) and (80) as

$$\Delta \phi = K \cos \phi - J \sin \phi. \qquad (83)$$

Hence as $\langle J \rangle = \langle K \rangle = 0$, the average phase error cancels, i.e.

$$\langle \Delta \phi \rangle = 0. \qquad (84)$$

2.5.2 Variance of the Phase Noise

The variance of the phase error can be evaluated from (83) squared and averaged:

$$\begin{aligned}\langle \Delta \phi^2 \rangle &= \langle K^2 \rangle \cos^2 \phi + \langle J^2 \rangle \sin^2 \phi - 2 \langle JK \rangle \sin \phi \cos \phi, \\ &= \frac{1}{2}\left[\langle J^2 \rangle + \langle K^2 \rangle - (\langle J^2 \rangle - \langle K^2 \rangle) \cos(2\phi) - 2 \langle JK \rangle \sin(2\phi)\right]. \quad (85)\end{aligned}$$

From the algorithm coefficients, three parameters r, β and θ can be defined by

$$\sum_{k=0}^{M-1} |c_k|^2 = \sum_{k=0}^{M-1} (a_k^2 + b_k^2) = r^2$$

$$\sum_{k=0}^{M-1} c_k^2 = \sum_{k=0}^{M-1} (a_k^2 - b_k^2 + 2\mathrm{i}\, a_k b_k) = r^2 \beta \exp(2\mathrm{i}\theta), \qquad (86)$$

as well as the parameter

$$\eta = \frac{|P(\zeta)|}{r\sqrt{M}}, \qquad 0 < \eta \le 1. \qquad (87)$$

This factor η is equal to 1 for the DFT algorithm, and is rarely less than 0.8 for almost all published ones, as can be seen in the table in the Appendix. From all the preceding definitions and results, it is easily shown [40] that the variance of the phase noise is then given by

$$\langle \Delta \phi^2 \rangle = \sigma_\phi^2 = \frac{1}{M\eta^2 \mathrm{SNR}^2} \left\{ 1 - \beta \cos\left[2(\phi - \theta)\right] \right\}. \tag{88}$$

The sinusoidal modulation can usually be neglected, as the parameter β is often less than 0.1, and the following simplified formula can be used:

$$\sigma_\phi^2 = \frac{1}{M\eta^2 \mathrm{SNR}^2}. \tag{89}$$

This formula can be easily analyzed. First, it is normal that the phase variance decreases with the square of the SNR. Second, the factor M is not surprising, as the detection involves the weighted addition of M statistically independent terms. Then the significance of the factor η becomes clear: it characterizes the effect of the weighting terms. In other words, it will measure how efficiently the statistically independent intensities are used to calculate the phase. In some way, and referring to the phasor interpretation presented in Sect. 2.2.7, the factor η characterizes the geometry of the phasor construction of the first Fourier coefficient of the sum $S(\phi)$. This factor is 1 for the DFT algorithm, where collinear phasors of equal length add together (Fig. 4). For the WDFT algorithm with $\delta = \pi/3$, $\eta = 0.898$ the phasor construction is still made of collinear phasors, but of unequal lengths (Fig. 5). The third example given in Sect. 2.2.7 corresponds to $\eta = 0.497$. It can be seen in Fig. 6 that in this case the detection phasor is built in a rather inefficient way. So, this parameter can be called the *efficiency factor* and is a crucial feature of the algorithm.

3 Phase Unwrapping

After the phase is detected, it has to be unwrapped to map the physical quantity (length, displacement, etc.) which is being measured. Unwrapping means removing the 2π-jumps present in the data obtained after the phase detection process (e.g. see Fig. 1b, by adding an adequate multiple of 2π where necessary. The procedure is trivial in the one dimensional case, as it suffices to scan the data from left to right looking for a phase difference between adjacent pixels greater than a prescribed value, typically π, and to remove them by adding (or subtracting, depending on the direction of the jump) 2π to the data set on the right-hand side of the jump. This path-oriented procedure often fails in the two-dimensional case because of the noise present in the data, which causes false jumps to be detected or real jumps to be missed. The essential point is that the error propagates as the phase unwrapping procedure is going on, and a result such as that in Fig. 12 may be obtained.

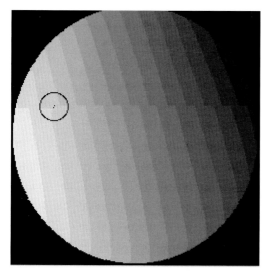

Fig. 12. Propagation of the phase unwrapping error (from the same data set as in Fig. 1). The *circle* is centered at the faulty noisy pixel

Two main strategies exist for phase unwrapping, namely spatial and temporal. However, it should be noted that the temporal one is not always possible.

3.1 Spatial Approach

Phase unwrapping is called spatial when a single wrapped phase map is used. Spatial information and operations are used to carry out the unwrapping. In this case, the problem may be quite odd, as it sometimes happens to have no solution, as in the situation sketched in Fig. 13. In Fig. 13, there are three jumps from P to Q along path A, and only two along path B. Such situations occur in speckle fringe patterns when dislocations are present in the wavefront. Points where the electromagnetic field has zero amplitude have no defined phase, and the wavefront can be skewed around those points [42].

Among the path-dependent methods, a solution is to select from among all possible paths by using amplitude [43] or phase gradient [44,45] information to increase confidence in the unwrapping path. A statistical approach for path selection using genetic algorithms have also been proposed recently [46].

Another possibility is to introduce cuts in the phase map to forbid erroneous paths[1]. It is quite easy to detect points where cuts have to end (*Huntley* and *Buckland* [47] call them "monopoles") by local unwrapping following a 3×3 pixels square around every pixel in the wrapped phase map. When the

---

[1] This is very similar to what is sometimes necessary when integrating a function of a complex variable

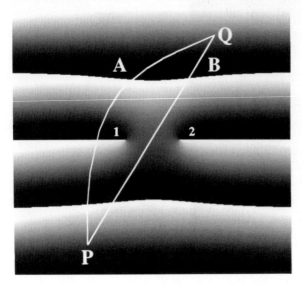

Fig. 13. Example of a phase map that cannot be unwrapped. A cut is necessary between points 1 and 2 to forbid path B (from [41])

phase obtained after looping is the same as the one at the beginning of the closed path, the central pixel is not a monopole. When it differs by $\pm 2\pi$, it is a positive or negative monopole. Once all monopoles have been identified, pairs of opposite signs must be selected to be the ends of the cuts. A simple strategy to pair monopoles is to minimize the total length of the cuts, that is to minimize the total width of the dipoles, with special attention paid to the boundaries [48–50].

It is also possible to segment the phase data into areas between fringe boundaries and then to calculate the phase offset between the disjoint areas [51], or into smaller regions ("tiles") which are then connected according to the confidence attached to the consistency of the phase data at the edges of the tiles [44].

Truly path-independent methods are based on the equivalence between the phase unwrapping problem and the resolution of Poisson's equation with Neumann boundary conditions, which can often be completed using specialized equation solvers. This efficient approach can be found in the literature [52–54], but will not be detailed here because of the level of mathematics involved.

Before closing this topic, it is worth mentioning a very simple approach which may be satisfactory for a large number of practical cases. One must realize that spatial smoothing is a local process that can be "2π-jumps aware", that is, the smoothing algorithm can perform local phase unwrapping before smoothing, and then wrap again the phase. This is illustrated by Fig. 14a,b.

The difference between the raw and smoothed data mostly represents noise (and maybe some signal) and can be retained in a buffer. The smoothed data is then unwrapped using a basic path-dependent unwrapping algorithm, as in Fig. 14c, and the previous phase difference is added back to obtain the unwrapped phase exactly corresponding to the initial wrapped one, as in Fig. 14d.

3.2 Temporal Approach

Temporal phase unwrapping is possible when the frequency of the fringes can be arbitrarily varied. It has been proposed mainly in the domain of profilometry by fringe projection. The principle of profilometry [4] is recalled in Fig. 15. Fringes (or parallel illuminated lines) are obliquely projected onto the object under investigation. The height information z from an arbitrary reference plane surface is encoded into a lateral displacement u_x of the fringe, with

$$z = \frac{u_x}{\tan \theta}, \tag{90}$$

in the simplest case of a collimated illumination shown in Fig. 15a. In (90) it is assumed that the direction of the fringes is aligned with the y-axis. The intensity which can be recorded by a video camera is written as

$$I(x,y) = I_0(x,y)\left\{1 + \gamma(x,y)\cos\left[\frac{2\pi x}{p} + \psi(x,y)\right]\right\}, \tag{91}$$

where $I_0(x,y)$ and $\gamma(x,y)$ are the slowly varying bias and contrast, p is the initial fringe pitch and $\psi(x,y)$ is the phase modulation caused by the height variation:

$$\psi(x,y) = \frac{2\pi u_x}{p} = \frac{2\pi z \tan \theta}{p}. \tag{92}$$

So, the height is known when the unwrapped phase is known. Two specific features of this technique are the possibility of an arbitrary variation of the projected fringe density (usually a slide or video projector is used) and the possible presence of non connected objects or stepwise height variations corresponding to phase steps much greater than 2π, making any spatial phase unwrapping method fail. So the concept of temporal phase unwrapping has emerged. In this approach extra information is needed, thus requiring more images to be recorded.

3.2.1 Gray Code

In this approach, the fringe order is encoded with a succession of 0's and 1's by the use of projected binary masks, the pitch of which varies as successive

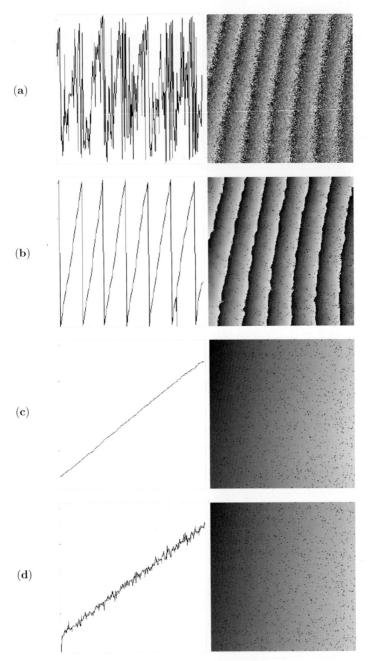

Fig. 14. Phase unwrapping by spatial smoothing: (**a**) noisy wrapped phase, (**b**) spatially smoothed phase, (**c**) smoothed unwrapped phase, and (**d**) noisy unwrapped phase

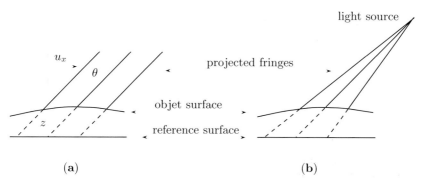

Fig. 15. Principle of projected fringes profilometry: **(a)** collimated projection, **(b)** conical projection

powers of 2. The process is illustrated in Fig. 16. This procedure is used in commercial systems like ATOS from Newport or the products from OMECA Messtechnik. The total number of frames to acquire is the number required by temporal phase-stepping plus the number of masks to project. This number is $\mathrm{INT}(\log_2 F) + 1$, where F is the total number of fringes in the field and INT is the 'integer part' function. For example, if the temporal phase-stepping uses five frames and if $F = 240$, 13 frames $(5 + 8)$ are necessary.

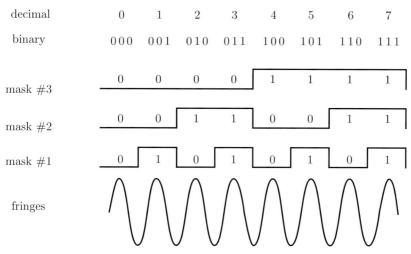

Fig. 16. Encoding of the fringe order by projection of successive binary masks

3.2.2 Multistep Process

In what we call the multistep process, many images with different fringe frequencies are used. In [41,55] the total number of projected fringes is varied over the field of view, one by one from 1 to F, with the central fringe remaining fixed. For each fringe density, a four-frame temporal phase-stepping is done, and the unwrapped phase $\phi(x,y)$ is obtained as the sum of the differences $\varphi_{i+1}(x,y) - \varphi_i(x,y)$ between the phases detected during steps i and $i+1$. The fact that the variation of the fringe number over the field of view is no more than 1 and that the central fringe remains steady ensures that this phase difference lies within the range $[-\pi, \pi]$. The overall procedure is sketched in Fig. 17. As can be seen, the total number of frames required to carry out the processing is $4F$. However, strategies exist to decrease this number [56].

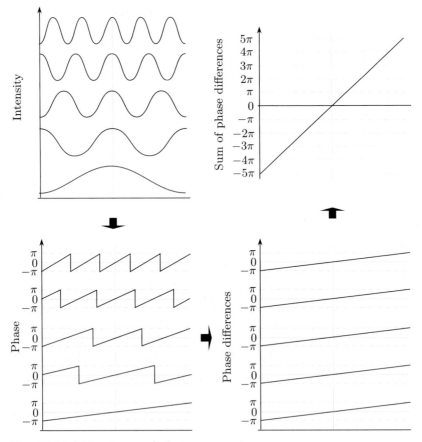

Fig. 17. Multistep temporal phase unwrapping

Recently, *Li* [57] have proposed a method where both a high- and a low-frequency grating are used, with an integer ratio between the frequencies. The high-frequency one provides the sensitivity, and the low-frequency one information for unwrapping. The two gratings are projected simultaneously, and a nice temporal phase-stepping scheme allow to implement an algorithm to sort out the low- and high-frequency phases. The total number of frames required by this method is $2P$, where P is the number of 2π-jumps that can be removed. Actually, the temporal phase-stepping is composed of two sets of P shifts, each set being necessary to deal with one of the frequencies. So, six frames is the minimum number of frames necessary to implement the method, as three frames is the minimum number of steps for the phase detection. With so few frames, only three 2π-jumps can be removed.

3.2.3 Two-Step Process

A two-step method has been proposed in [58], and a very similar method can be found in [59]. To carry out the unwrapping procedure, two fringe patterns with slightly different pitches are projected onto the object, and two images are taken, from which two slightly different phase maps are computed. The difference gives the low-frequency information enabling us to unwrap the phase. This looks like the two-frequency method by *Li* [57] indicated in the preceding section. The difference is twofold: spatial instead of temporal phase-stepping is used, and two high-frequency patterns are used.

More precisely, the slowly varying phase map obtained by subtraction is used to number the fringes. The pitch of the grating is varied between the two images by $(L-1)/L$, where L is an integer. So, the phase difference between the two computed phase maps varies by 2π for every L fringes. This phase difference can be called the "numbering phase". This numbering phase divided by $2\pi/L$ varies by L every L fringes. The integer part of this quotient straightforwardly gives the fringe order. However, the noise present in the data makes this process tail in the vicinity of the 2π-jumps, where the phase and the computed fringe order are "unstable". It is very easy to circumvent this problem. First, the preceding procedure is restricted to the half-fringes where the wrapped phase is in the interval $[-\pi/2, \pi/2]$. Then, π is substracted from the wrapped phase and π/L from the numbering phase, so that the 2π-jumps and the fringe-order transition points move out of the regions concerned. Then, the half fringes where the wrapped phase is in the interval $[-\pi, -\pi/2] \cup [\pi/2, \pi]$ can be processed. This procedure is presented in Fig. 18. However, two points are critical in this procedure: the precision of the pitch variation and the coincidence of the zero phases in the two phase maps at the central point. Obviously, the proposed procedure can unwrap up to L 2π-jumps. As L can be quite large, the dynamic range of phase detection is fairly enlarged.

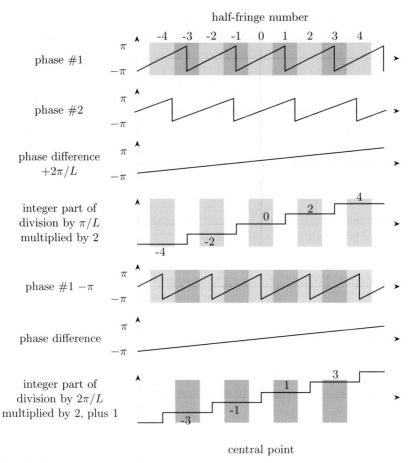

Fig. 18. Two-step temporal phase unwrapping

4 Conclusion

Tremendous activity has been observed in the field of optical metrology during the last decade. The possibility of easy and cheap image processing has completely modified the available optical tools. Methods that were most by more than one century old have been revised and their use is now spreading into industry.

Actually, it is necessary to assess the metrological performance that is required by industrial applications. Issues such as spatial resolution, sensibility and accuracy have to be addressed by following a well-defined methodology. The first step is the assessment of the phase detection from intensity fields. For the phase-stepping process, the influence of the noise and the amplitude of the residual errors can be evaluated. The same level of achievement has not yet obtained for the Fourier transform method.

The CP theory allows us to readily analyze the properties of the numerous algorithms that have been published over the years. For practical applications, only two algorithms need be considered in the first instance, namely the DFT (or N-bucket) and the WDFT ones, the parameters of which have to be determined as a function of harmonic content, amount of miscalibration and possible number of frames.

Phase unwrapping remains a difficult problem to handle in the most general case, and so remains difficult to fully automate. However, a number of strategies exist that will allow us to successfully unwrap the phase in most practical situations.

Appendix: Table of Algorithms

In Table 1, we give information about a large number of algorithms that have been published. We indicate the total number M of frames in the algorithm, the nominal phase shift δ and the characteristic diagram. The coefficients of the algorithm

$$\tilde{\phi} = \arctan\left(\frac{\sum_{k=0}^{M-1} b_k I_k}{\sum_{k=0}^{M-1} a_k I_k}\right),$$

are written as

$$\frac{b_0, b_1, \ldots, b_{M-1}}{a_0, a_1, \ldots, a_{M-1}}.$$

Following (15) in Sect. 2.2.2, each algorithm detects $\phi + \psi$, where $\psi = \arg[P(\zeta)]$ is a constant that is given in the Table. The sensitivity to noise propagation depends on the efficiency parameter η and on the parameter β that appear in (88) giving the standard deviation of the phase noise in Sect. 2.5.2.

Table 1. Algorithms that can be found in the literature

#	M	δ	Diagram	Coefficients	ψ	η	β	Ref.
1	3	$\dfrac{\pi}{2}$		$\dfrac{1,-1,0}{0,1,-1}$ $\dfrac{1,-2,1}{1,0,-1}$	$\dfrac{\pi}{4}$ 0	0.816	0.500	[12,60]
2	3	$\dfrac{2\pi}{3}$		$\dfrac{\sqrt{3}[0,-1,1]}{[2,-1,-1]}$	0	1	0	[12]
								continued next page

Table 1. *(continued)*

#	M	δ	Diagram	Coefficients	ψ	η	β	Ref.
3	4	$\dfrac{\pi}{2}$	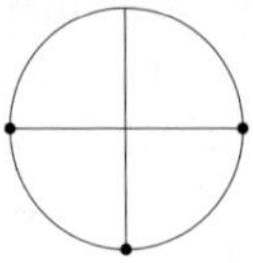	$\dfrac{1,1,-1,-1}{-1,1,1,-1}$ $\dfrac{0,-1,0,1}{1,0,-1,0}$	$\dfrac{3\pi}{4}$ 0	1	0	[61]
4	4	$\dfrac{\pi}{2}$	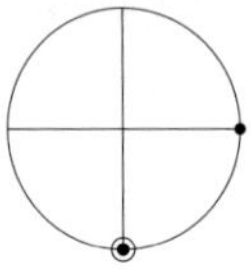	$\dfrac{0,-2,2,0}{1,-1,-1,1}$	0	0.816	0.333	[62,26]
5	5	$\dfrac{\pi}{2}$	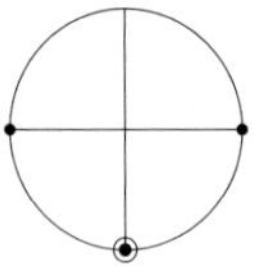	$\dfrac{0,-2,0,2,0}{1,0,-2,0,1}$	0	0.956	0.143	[63,26]
6	5	$\dfrac{\pi}{2}$	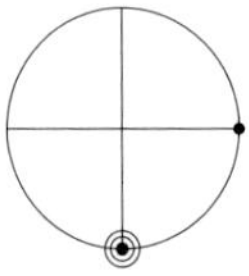	$\dfrac{1,-4,0,4,-1}{1,2,-6,2,1}$	0	0.800	0.150	[26]
7	6	$\dfrac{\pi}{2}$	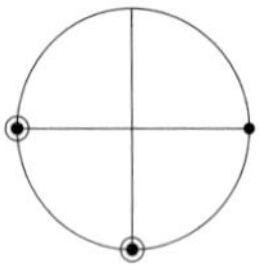	$\dfrac{0,-2,-2,2,2,0}{1,1,-2,-2,1,1}$	$\dfrac{\pi}{4}$	0.873	0.143	[64]
8	6	$\dfrac{\pi}{2}$	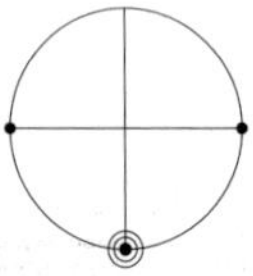	$\dfrac{0,-3,0,4,0,-1}{1,0,-4,0,3,0}$	0	0.906	0	[26]
9	6	$\dfrac{\pi}{2}$	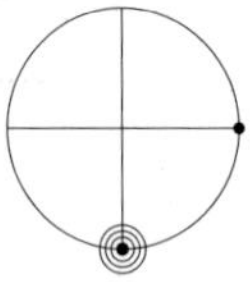	$\dfrac{0,-4,4,4,-4,0}{1,-1,-6,6,1,-1}$	$-\dfrac{\pi}{4}$	0.781	0.086	[26]
10	6	$\dfrac{\pi}{3}$	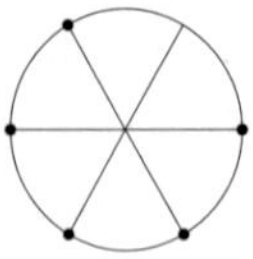	$\dfrac{\sqrt{3}[-1,-2,-1,1,2,1]}{3,0,-3,-3,0,3}$	$-\dfrac{\pi}{6}$	1	0	

continued next page

Table 1. (continued)

#	M	δ	Diagram	Coefficients	ψ	η	β	Ref.
11	6	$\frac{\pi}{3}$	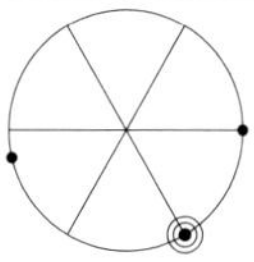	$\frac{\sqrt{3}[-5,6,17,-17,-6,5]}{1,-26,25,25,-26,1}$	$\frac{5\pi}{6}$	0.495	0.107	[33]
12	7	$\frac{\pi}{2}$	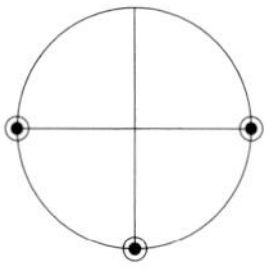	$\frac{1,0,-3,0,3,0,-1}{0,2,0,-4,0,2,0}$	$\frac{\pi}{2}$	0.912	0.091	[65,25]
13	7	$\frac{\pi}{2}$	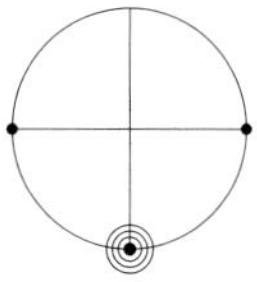	$\frac{-1,0,7,0,-7,0,1}{0,-4,0,8,0,-4,0}$	π	0.864	0.020	[28]
14	7	$\frac{\pi}{3}$	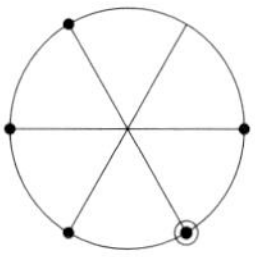	$\frac{\sqrt{3}[-1,3,3,0,-3,-3,1]}{3[-1,-1,1,2,1,-1,-1]}$	π	0.953	0.118	[34]
15	7	$\frac{\pi}{3}$	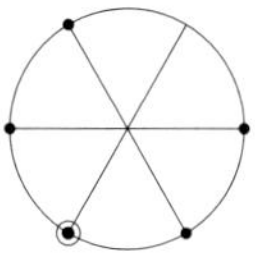	$\frac{\sqrt{3}[0,1,1,0,-1,-1,0]}{-1,-1,1,2,1,-1,-1}$	π	0.967	0.091	[34]
16	8	$\frac{\pi}{2}$	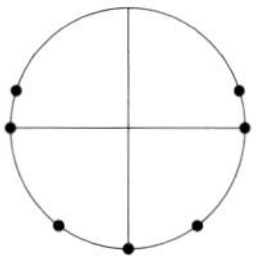	$\frac{-1,-2,3,4,-4,-3,2,1}{1,-2,-3,4,4,-3,-2,1}$	$-\frac{\pi}{4}$	0.913	0	[26]
17	8	$\frac{\pi}{2}$	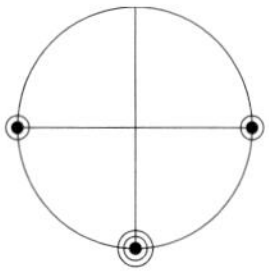	$\frac{-1,-3,5,7,-7,-5,3,1}{1,-3,-5,7,7,-5,-3,1}$	$-\frac{\pi}{4}$	0.873	0	[26]
18	8	$\frac{\pi}{2}$	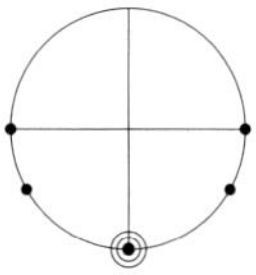	$\frac{-1,-4,8,11,-11,-8,4,1}{1,-4,-8,11,11,-8,-4,1}$	$-\frac{\pi}{4}$	0.844	0	[26]

continued next page

Table 1. (continued)

#	M	δ	Diagram	Coefficients	ψ	η	β	Ref.
19	8	$\dfrac{\pi}{2}$		$\dfrac{-1,-5,11,15,-15,-11,5,1}{1,-5,-11,15,15,-11,-5,1}$	$-\dfrac{\pi}{4}$	0.830	0	[26]
20	9	$\dfrac{\pi}{4}$		$\dfrac{0,1,2,1,0,-1,-2,-1,0}{-1,-1,0,1,2,1,0,-1,-1}$	π	0.970	0.091	[28]
21	9	$\dfrac{\pi}{2}$		$\dfrac{1,-2,-14,-18,0,18,14,2,-1}{2,8,8,-8,-20,-8,8,8,2}$	π	0.497	0.140	[33]
22	10	$\dfrac{\pi}{3}$		$\dfrac{\sqrt{3}[1,-1,-7,-11,-6,6,11,7,1,-1]}{3[1,3,3,-1,-6,-6,-1,3,3,1]}$	$\dfrac{\pi}{2}$	0.830	0.106	[25]
23	11	$\dfrac{\pi}{3}$		$\dfrac{\sqrt{3}[-1,-2,0,4,5,0,-5,-4,0,2,1]}{1,-2,-6,-4,5,12,5,-4,-6,-2,1}$	$-\dfrac{\pi}{3}$	0.898	0.055	[40]
24	11	$\dfrac{\pi}{3}$		$\dfrac{\sqrt{3}[0,1,4,7,6,0,-6,-7,-4,-1,0]}{-2,-5,-6,-1,8,12,8,-1,-6,-5,-2}$	$-\dfrac{\pi}{3}$	0.681	0.205	[65]
25	11	$\dfrac{\pi}{2}$		$\dfrac{1,0,-8,0,15,0,-15,0,8,0,-1}{0,4,0,-12,0,16,0,-12,0,4,0}$	$\dfrac{\pi}{2}$	0.851	0.003	[28]

References

1. F. Roddier, C. Roddier: Imaging with a Multi-Mirror Telescope, in F. Pacini, W. Richter, R. N. Wilson (Ed.): *ESO Conference on Optical Telescopes of the Future, Geneva*, ESO, CERN, Geneva (1978)
2. C. Roddier, F. Roddier: Imaging with a Coherence Interferometer in Optical Astronomy, in C. van Schooneveld (Ed.): *IAU Colloquium 49 on Formation of Images from Spatial Coherence Functions in Astronomy, Groningen, 1978*. Reidel, Norwell, MA (1979)
3. M. Takeda, H. Ina, S. Kobayashi: Fourier-Transform Method of Fringe-Pattern Analysis for Computer-Based Topography and Interferometry, J. Opt. Soc. Am. A. **72**, 156–160 (1982)
4. M. Takeda, K. Mutoh: Fourier Transform Profilometry for the Automatic Measurement of 3-D Object Shapes, Appl. Opt. **22**, 3977–3982 (1983)
5. W. W. Macy, Jr.: Two-Dimensional Fringe-Pattern Analysis, Appl. Opt. **22**, 3898–3901 (1983)
6. K. H. Womack: Frequency Domain Description of Interferogram Analysis, Opt. Eng. **23**, 396–400 (1984)
7. M. Takeda: Spatial Carrier Fringe Pattern Analysis and its Applications to Precision Interferometry and Profilometry: An Overview, Ind. Metrol. **1**, 79–99 (1990)
8. J. B. Liu, P. D. Ronney: Modified Fourier Transform Method for Interferogram Fringe Pattern Analysis, Appl. Opt. **36**, 6231–6241 (1997)
9. P. Carré: Installation et Utilisation du Comparateur Photoélectrique et Interférentiel du Bureau International des Poids et Mesures, Metrologia **2**, 13–23 (1966)
10. J. H. Bruning, D. R. Herriott, J. E. Gallagher, D. P. Rosenfeld, A. D. White, D. J. Brangaccio: Digital Wavefront Measuring Interferometer for Testing Optical Surfaces and Lenses, Appl. Opt. **13**, 2693–2703 (1974)
11. J. C. Wyant: Use of an AC Heterodyne Lateral Shear Interferometer with Real Time Wavefront Correction Systems, Appl. Opt. **14**, 2622–2626 (1975)
12. C. L. Koliopoulos: Interferometric Optical Phase Measurement Techniques, PhD thesis, University of Arizona, Tucson, AZ (1981)
13. K. A. Stetson, W. R. Brohinsky: Electrooptic Holography and its Application to Hologram Interferometry, Appl. Opt. **24**, 3631–3637 (1985)
14. J. H. Bruning: Fringe Scanning Interferometers, in D. Malacara (Ed.): *Optical Shop Testing*, Wiley, New York (1987), 414
15. K. H. Womack: Interferometric Phase Measurement Using Spatial Synchronous Detection, Opt. Eng. **23**, 391–395 (1984)
16. S. Toyooka, M. Tominaga: Spatial Fringe Scanning for Optical Phase Measurement, Opt. Commun. **51**, 68–70 (1984)
17. M. Kujawińska, J. Wójciak: Spatial-Carrier Phase-Shifting Technique of Fringe Pattern Analysis, in W. P. Jüptner (Ed.): *Industrial Applications of Holographic and Speckle Measuring Techniques*, Proc. SPIE **1508**, 61–67 (1991)
18. X. Colonna de Lega: Processing of Non-stationary Interference Patterns: Adapted Phase-Shifting Algorithms and Wavelet Analysis. Application to Dynamic Deformation Measurements by Holographic and Speckle Interferometry, PhD thesis, EPFL, Lausanne, Switzerland (1997)

19. J. L. Maroquin, M. Servin, R. Rodriguez-Vera: Adaptive Quadrature Filters and the Recovery of Phase from Fringe Pattern Images, Appl. Opt. **14**, 1742–1753 (1997)
20. J. Schwider, R. Burow, K.-E. Elssner, J. Grzanna, R. Spolaczyk, K. Merkel: Digital Wave-Front Measuring Interferometry: Some Systematic Error Sources, Appl. Opt. **22**, 3421–3432 (1983)
21. C. Ai, J. C. Wyant: Effect of Piezoelectric Transducer Nonlinearity on Phase Shift Interferometry, Appl. Opt. **26**, 1112–1116 (1987)
22. J. van Wingerden, H. J. Frankena, C. Smorenburg: Linear Approximation for Measurement Errors in Phase-Shifting Interferometry, Appl. Opt. **30**, 2718–2729 (1991)
23. K. Creath: Phase Measurement Interferometry: beware these Errors, in R. J. Pryputniewicz (Ed.): *Laser Interferometry IV: Computer-Aided Interferometry*, SPIE **1553**, 213–220 (1992)
24. K. Freischlad, C. L. Koliopoulos: Fourier Description of Digital Phase-Measuring Interferometry, J. Opt. Soc. Am. A. **7**, 542–551 (1990)
25. Y. Surrel: Design of Algorithms for Phase Measurements by the Use of Phase-Stepping, Appl. Opt. **35**, 51–60 (1996)
26. J. Schmit, K. Creath: Extended Averaging Technique for Derivation of Error-Compensating Algorithms in Phase-Shifting Interferometry, Appl. Opt. **34**, 3610–3619 (1995)
27. J. Schmit, K. Creath: Window Function Influence on Phase Error in Phase-Shifting Algorithms, Appl. Opt. **35**, 5642–5649 (1996)
28. P. de Groot: Derivation of Algorithms for Phase-Shifting Interferometry Using the Concept of Data-Sampling Window, Appl. Opt. **34**, 4723–4730 (1995)
29. D. W. Phillion: General Methods for Generating Phase-Shifting Interferometry Algorithms, Appl. Opt. **36**, 8098–8115 (1997)
30. R. Onodera, Y. Ishii: Phase-Extraction Analysis of Laser-Diode Phase-Shifting Interferometry that is Insensitive to Changes in Laser Power, J. Opt. Soc. Am. A. **13**, 139–146 (1996)
31. Y. Surrel: Design of Phase-Detection Algorithms Insensitive to Bias Modulation, Appl. Opt. **36**, 805–807 (1997)
32. Y. Surrel: Phase-Shifting Algorithms for Nonlinear and Spatially Nonuniform Phase Shifts: Comment, J. Opt. Soc. Am. A. **15**, 1227–1233 (1998)
33. K. Hibino, B. F. Oreb, D. I. Farrant, K. G. Larkin: Phase-Shifting Algorithms for Nonlinear and Spatially Nonuniform Phase Shifts, J. Opt. Soc. Am. A. **14**, 918–930 (1997)
34. K. G. Larkin, B. F. Oreb: Design and Assessment of Symmetrical Phase-Shifting Algorithms, J. Opt. Soc. Am. A. **9**, 1740–1748 (1992)
35. P. de Groot: 101-Frame Algorithm for Phase-Shifting Interferometry, in: *Optical Inspection and Micromeasurements II*, Proc. SPIE **3098**, 283–292 (1997)
36. Y. Surrel: Extended Averaging and Data Windowing Techniques in Phase-Stepping Measurements: An Approach Using the Characteristic Polynomial Theory, Opt. Eng. **37**, 2314–2319 (1998)
37. D. J. Bone, H. A. Bachor, R. John Sandeman: Fringe-pattern Analysis Using a 2-D Fourier Transform, Appl. Opt. **25**, 1653–1660 (1986)
38. H. T. Goldrein, S. J. P. Palmer, J. M. Huntley: Automated Fine Grid Technique for Measurement of Large-Strain Deformation Map, Opt. Lasers Eng. **23**, 305–318 (1995)

39. Y. Surrel: Phase Stepping: A New Self-Calibrating Algorithm, Appl. Opt. **32**, 3598–3600 (1993)
40. Y. Surrel: Additive Noise Effect in Digital Phase Detection, Appl. Opt. **36**, 271–276 (1997)
41. J. M. Huntley, H. O. Saldner: Temporal Phase-Unwrapping Algorithm for Automated Interferogram Analysis, Appl. Opt. **32**, 3047–3052 (1993)
42. N. B. Baranova, B. Ya. Zel'dovich: Dislocations of the Wave-front Surface and Zeros of the Amplitude, Sov. Phys.-JETP **30**, 925–929 (1981)
43. M. Takeda, T. Abe: Phase Unwrapping by a Maximum Cross-Amplitude Spanning Tree Algorithm: A Comparative Study, Opt. Eng. **35**, 2345–2351 (1996)
44. D. P. Towers, T. R. Judge, P. J. Bryanston-Cross: Automatic Interferogram Analysis Techniques Applied to Quasi-heterodyne Holography and ESPI, Opt. Lasers Eng. **14**, 239–282 (1991)
45. T. R. Judge, C. Quan, P. J. Bryanston-Cross: Holographic Deformation Measurements by Fourier Transform Technique with Automatic Phase Unwrapping, Opt. Eng. **31**, 533–543 (1992)
46. A. Collaro, G. Franceschetti, F. Palmieri, M. S. Ferreiro: Phase Unwrapping by Means of Genetic Algorithms, J. Opt. Soc. Am. A **15**, 407–418 (1998)
47. J. M. Huntley, J. R. Buckland: Characterization of Sources of 2π Phase Discontinuity in Speckle Interferograms, J. Opt. Soc. Am. A. **12**, 1990–1996 (1995)
48. R. Cusack, J. M. Huntley, H. T. Goldrein: Improved Noise-Immune Phase-Unwrapping Algorithm, Appl. Opt. **35**, 781–789 (1995)
49. J. R. Buckland, J. M. Huntley, S. R. E. Turner: Unwrapping Noisy Phase Maps by Use of a Minimum-Cost-Matching Algorithm, Appl. Opt. **34**, 5100–5108 (1995)
50. J. A. Quiroga, A. González-Cano, E. Bernabeu: Stable-Marriages Algorithm for Preprocessing Phase Maps with Discontinuity Sources, Appl. Opt. **34**, 5029–5038 (1995)
51. P. G. Charette, I. W. Hunter: Robust Phase-unwrapping Method for Phase Images with High Noise Content, Appl. Opt. **35**, 3506–3513 (1996)
52. D. C. Ghiglia, L. A. Romero: Robust Two-Dimensional Weighted and Unweighted Phase Unwrapping that Uses Fast Transforms and Iterative Methods, J. Opt. Soc. Am. A. **11**, 107–117 (1994)
53. D. C. Ghiglia, L. A. Romero: Minimum L^p-Norm Two-Dimensional Phase Unwrapping, J. Opt. Soc. Am. A. **13**, 1999–2013 (1996)
54. G. H. Kaufmann, G. E. Galizzi: Unwrapping of Electronic Speckle Pattern Interferometry Phase Maps: Evaluation of an Iterative Weighted Algorithm, Opt. Eng. **37**, 622–628 (1998)
55. H. O. Saldner, J. M. Huntley: Profilometry Using Temporal Phase Unwrapping and a Spatial Light Modulator-Based Fringe Projector, Opt. Eng. **36**, 610–615 (1997)
56. J. M. Huntley, H. O. Saldner: Error Reduction Methods for Shape Measurement by Temporal Phase Unwrapping, J. Opt. Soc. Am. A. **14**, 3188–3196 (1997)
57. J.-L. Li, H.-J. Su, X.-Y. Su: Two-Frequency Grating Used in Phase-Measuring Profilometry, Appl. Opt. **36**, 277–280 (1997)
58. Y. Surrel: Two-Step Temporal Phase Unwrapping in Profilometry, in C. Gorecki (Ed.): *Optical Inspection and Micromeasurements II, 16–19 June 1997, Munich*, Proc. SPIE **3098**, 271–282 (1997)

59. M. Fujigaki, Y. Morimoto: Automated Shape Analysis for Multiple Phase Fringes by Phase-Shifting Method Using Fourier Transform, in I. M. Allison (Ed.): *Experimental Mechanics - Advances in Design, Testing and Analysis*, Balkema, Rotterdam (1998), 711–714
60. J. B. Hayes: Linear Methods of Computer Controlled Optical Figuring, PhD thesis, University of Arizona, Tucson, AZ (1984)
61. W. R. C. Rowley, J. Hamon: Quelques Mesures de Dissymétrie de Profils Spectraux, R. Opt. Théor. Instrum. **42**, 519–523 (1963)
62. J. Schwider, O. Falkenstörfer, H. Schreiber, A. Zöller, N. Streibl: New Compensating Four-Phase Algorithm for Phase-Shift Interferometry, Opt. Eng. **32**, 1883–1885 (1993)
63. P. Hariharan, B. F. Oreb, T. Eiju: Digital Phase-Shifting Interferometry: A Simple Error-Compensating Phase Calculation Algorithm, Appl. Opt. **26**, 2504–2506 (1987)
64. B. Zhao, Y. Surrel: Phase Shifting: A Six-Step Self-Calibrating Algorithm Insensitive to the Second Harmonic in the Fringe Signal, Opt. Eng. **34**, 2821–2822 (1995)
65. K. Hibino, B. F. Oreb, D. I. Farrant, K. G. Larkin: Phase-Shifting for Nonsinusoidal Waveforms with Phase-Shift Errors, J. Opt. Soc. Am. A. **12**, 761–768 (1995)

Principles of Holographic Interferometry and Speckle Metrology

Pramod K. Rastogi

Stress Analysis Laboratory, Swiss Federal Institute of Technology Lausanne,
CH-1015 Lausanne, Switzerland
pramod.rastogi@epfl.ch

Abstract. Because of their nonintrusive, real-time and full-field features, holographic interferometry [1,2,3,4,5,6,7,8,9,10] and speckle metrology [11,12,13,14,15] [16,17,18,19] have received considerable attention in experimental mechanics. As a result, a broad range of measurement possibilities have emerged that are capable of handling challenges that applications pose to test methods. The active interest in these fields is testified by a proliferation of research papers that these fields have produced over the last three decades. It would thus be illusive to attempt to cover the subject in depth and in its entirety in the present review. The intent of this chapter is limited to describing a sample of approaches and techniques applied to deformation analysis and shape measurements, and to presenting examples drawn from these areas. The main emphasis is on giving the reader some appreciation of the variety of basic procedures that have been developed over the years for deformation measurement.

1 Holographic Interferometry

Conceived by *Denis Gabor* [20] in 1947 and constrained to sit out the next fifteen years of its existence in the antechamber of discoveries, the concept of holography was propelled into prominence with the advent of lasers in the early sixties [21,22]. The discovery led to the award of the Nobel Prize in Physics to its inventor in 1971. The concept provided the rare opportunity of recording and later reconstructing the amplitude and phase of a light wave. The technique is accomplished by combining on a photographic film or a plate a known "reference" wave with another wave scattered off an opaque object surface Fig. 1a. The interference of two waves produces a dense fringe pattern on the photographic plate. The developed plate is known as a hologram. The reconstruction of the original object wave from the hologram is done by illuminating the hologram with a beam of light that is identical to the original reference wave Fig. 1b. The interference grating recorded on the hologram diffracts the light into a wavefront that is an exact replica of the object wave that had originally served to create the interference grating when the hologram was recorded. An observer looking through the hologram window sees the reconstructed three-dimensional image of the object.

In essence a hologram can be considered as a device to store a complicated wave pattern. That the stored wave pattern can be retrieved for reconstruc-

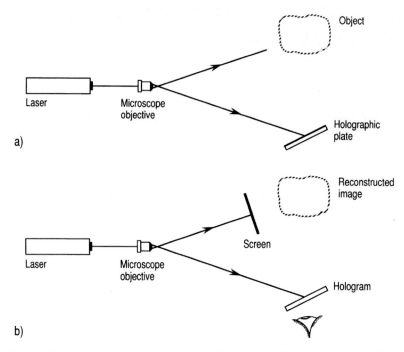

Fig. 1. Typical schematics for (**a**) recording, and (**b**) reconstructing a hologram

tion at a later time offers the unique possibility to compare two complicated wave patterns that have albeit existed at different times. This statement is fundamental to the understanding of holographic interferometry as it makes it feasible to compare the shape of a diffusely reflecting object with the shape of the same object after it has been deformed. Assuming the surface deformations to be very small, the interference of the two complicated wave patterns forms a set of interference fringes that are indicative of the amount of displacement and deformation undergone by the diffuse object.

1.1 Types of Holographic Interferometry

There are several schemes that have been implemented to obtain the interferometric comparison of the wavefronts. The most important of these are briefly described below.

In the real-time form of holographic interferometry, a hologram is made of an arbitrarily shaped rough surface Fig. 2. After development the hologram is placed back in exactly the same position. This is feasible as the holographic plate is mounted on a kinematic mount. Upon reconstruction, the hologram produces the original wavefront. A person looking through the hologram sees a superposition of the original object and its reconstructed image. The object wave interferes with the reconstructed wave to produce a dark field due to

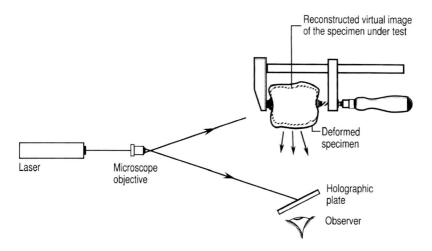

Fig. 2. Typical schematic for performing real-time holographic interferometry

destructive interference. If the object is now slightly deformed, interference fringes are produced which are related to the change in shape of the object. The interference process occurs in real-time and changes in the object are continuously monitored onto a TV monitor, which can be connected to a video-recording tape to store the time-varying fringe patterns. A dark fringe is produced whenever

$$\varphi(x,y) - \varphi'(x,y) = 2n\pi \qquad n = 0, 1, 2, \ldots, \qquad (1)$$

where n is the fringe order and φ and φ' are the phases of the waves from the object in its original and deformed states, respectively. The method is very useful for determining the direction of the object displacement, and for compensating on the interferogram the influence of the overall motion of the object when subjected to a stress field.

Double exposure is another way to implement holographic interferometry. In this scheme Fig. 3, two exposures are made on the holographic plate: one exposure with the object in its original state and a second exposure with the object in a deformed state. On readout, two reconstruction fields are produced, one corresponding to the object in its original state and the other corresponding to the object in its stressed state. The reconstructed image is overlaid by a set of interference fringes that contain information about how and how much the object was distorted between the two exposures. The fringe pattern is permanently stored on the holographic plate. The interference fringes denote the loci of points that have undergone an equal change of optical path between the light source and the observer. A bright fringe is produced whenever

$$\varphi(x,y) - \varphi'(x,y) = 2n\pi \qquad n = 0, 1, 2, \ldots . \qquad (2)$$

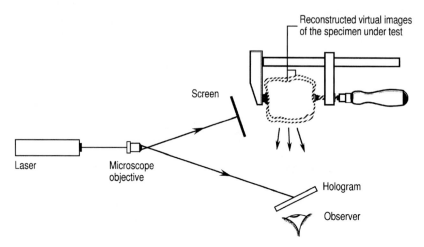

Fig. 3. Typical schematic for performing double exposure holographic interferometry

In the third form of holographic interferometry, two holographic plates placed in the same plate-holder with their emulsions facing the object are exposed simultaneously Fig. 4a. Then the object is deformed. A new pair of plates is similarly exposed to the object in its deformed state Fig. 4b. After development, a sandwich [23] is formed by superposing the back plate from the first pair onto the front plate from the second pair Fig. 4c. Illuminated by the original reference wave, the reconstructed image of the object is covered by an interference pattern indicative of the surface deformation. A tilt of the sandwich then allows us to compensate for the rigid body movement of the object Fig. 4d. In sandwich holography the information on the direction of displacement is not lost so long as one remembers the plate corresponding to the first exposure. In double exposure this information is lost, as the plate does not remember the time sequence of recording. Fringe control, to remove high spatial frequency fringe patterns arising as a result of the overall motion of the object, can also be achieved in a holographic set-up containing two reference beams [24,25]. The technique works by moving one object reconstructed image in relation to the other.

An interesting configuration of holographic interferometry has been developed to obtain interference fringes even in the presence of extraneous large object motions. In this approach the holographic plate is rigidly attached to the object surface [26]. The beam that illuminates the object-plate tandem serves both as object and reference waves. The reference wave is the one that falls directly on the back of the holographic plate, while the object illumination beam is the one that is transmitted by the plate. A feature of the method is that the reconstructed interference image can also be viewed in white light [27]. Fastening a hologram onto an object surface and the

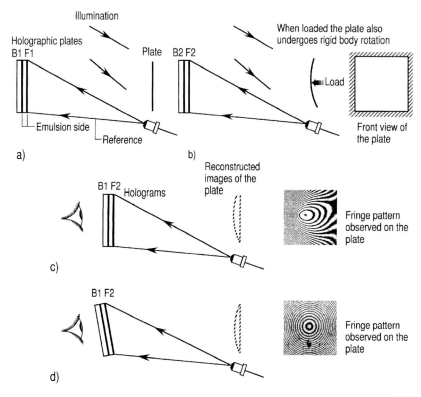

Fig. 4. Illustration of the principle of sandwich holographic interferometry: (**a**) first exposure, (**b**) second exposure, (**c**) fringe reconstruction, and (**d**) object tilt compensation

limitations imposed by the plate on the size of the object are two major inconveniences of the method.

Recently, the concept of digital holographic interferometry has been theoretically proposed and experimentally demonstrated [28,29]. It consists of recording two holograms corresponding to two different states of an object using a CCD camera. These images are stored in a frame grabber. The two holograms are then reconstructed separately in the computer. The contours of constant optical phase change between the two exposures are obtained by comparing the two reconstructed wave fields. The applications of the method are limited mainly by the pixel size in CCD arrays.

1.2 Thermoplastic Camera

In recent years, thermoplastic recording material has become increasingly important because it can be processed rapidly using electrical and thermal processes, and is erasable and reusable at least several hundred times. It has

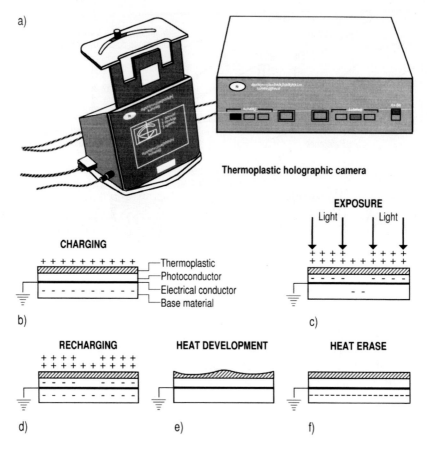

Fig. 5. (a) schematic of the commercially available Newport thermoplastic holographic camera (**b–f**) principle of operation

thus brought to a virtual halt the use of silver halide photographic plates and the associated wet processing in holographic interferometry. An example of such a system commercially available from the Newport Corp. is sketched in Fig. 5a. In thermoplastic recording the information is recorded as a thickness variation corresponding to a charge intensity pattern deposited on the thermoplastic layer. The recording process is illustrated in Fig. 5b-f. First, the film is evenly charged by corona charging Fig. 5b. The film surface is then exposed to spatially modulated light. The exposure selectively discharges the illuminated areas Fig. 5c. The film is then recharged to enhance the charge variation Fig. 5d. Finally, the thermoplastic is heated, which results in the deformation of the film as a function of the charge pattern Fig. 5e. Cooling of the thermoplastic freezes the relief pattern on the film surface. Being a phase hologram, its diffraction efficiency can be as high as 30%. To erase the recorded relief, the thermoplastic is heated to a higher temperature Fig. 5f. Surface tension smooths out the thickness variation.

1.3 Mapping of the Resolved Part of Displacement

In holographic interferometry the two wavefronts related to the states of a surface before and after deformation are reconstructed simultaneously and compared. Suppose that an object is deformed such that a point P on its surface moves to P' as shown in Fig. 6. The surface is illuminated by light incident in the direction SP and is viewed along the direction PT. The change of optical phase due to displacement d is

$$\Delta \varphi = \frac{2\pi d}{\lambda}(\cos \psi_1 - \cos \psi_2) , \qquad (3)$$

where λ is the wavelength of the light used and ψ_1 and ψ_2 are the angles which the illumination and observation waves make with respect to the direction of displacement. A more useful representation of (3) is obtained by rewriting it in the form

$$\Delta \varphi = \frac{4\pi d \cos \eta \cos \psi}{\lambda} , \qquad (4)$$

where ψ is the angle of bisection of the illumination and viewing directions, and η is the angle which the bisector makes with the direction of displacement. The term $d \cos \eta$ is the resolved part of the displacement PP' in the direction of the bisector. Hence, (4) shows that the fringe pattern contours the component of displacement along the bisector of the angle between the incident and viewing directions. Since a single observation yields information only about the resolved part of the surface displacement in one particular direction, the approach to measuring a three-dimensional vector displacement would be to record holograms from three different directions in order to obtain

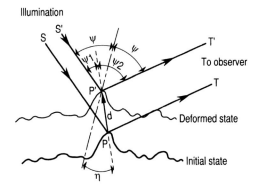

Fig. 6. Geometry showing changes of optical path length due to displacement d of a surface point P

three independent resolved components of displacement. Hence, in principle, it is possible by way of computation to obtain from a single hologram the complete vector displacement field, $\boldsymbol{d}(x,y)$, for all points on a surface.

1.4 Determination of the Wavefront Phase Using Phase-shifting Interferometry

The last 15 years have seen many significant developments in phase-shifting interferometry [30,31,32,33,34]. Driven by the need for automatic fringe analysis, interferometers based on holography are routinely incorporating phase-shifting techniques into their respective instrumentation. Specific phase-detection procedures are discussed in the contribution by Yves Surrrel.

The image intensity of a holographic interferogram is

$$I(x,y) = I_0(x,y)[1 + V(x,y)\cos\Delta\,\varphi(x,y)]\;, \qquad (5)$$

where $I_0(x,y)$ is the background intensity and $V(x,y)$ is the fringe contrast. A series of holograms is recorded by introducing artificially known steps of phase differences in the interference pattern

$$I_i(x,y) = I_0(x,y)\{1 + V(x,y)\cos[\Delta\,\varphi(x,y) + \Delta\,\psi_i]\}\;, \qquad (6)$$

where $\Delta\,\psi_i$ is the phase shift which is produced by shifting the phase of the reference wave. A minimum of three intensity patterns is required to calculate the phase at each point on the object surface. One of the common methods of inducing a phase shift in an interferometer is by moving a mirror mounted on a piezoelectric transducer (PZT). The role of the PZT is to shorten or elongate the reference beam path by a fraction of a wavelength.

The use and advantages of phase shifting in the quantitative determination of object deformations, vibration modes, surface shapes, and flow patterns are well known. The method provides accurate phase measurements and eliminates the phase sign ambiguity of the interference fringes. The independence of the calculated phase term, $\Delta\,\varphi(x,y)$, in relation to $I_0(x,y)$ and $V(x,y)$ reduces considerably the dependence of the measurement accuracy on the fringe quality.

An important advance in holographic interferometry is electronic or electro-optic holographic interferometry [35,36,37,38] in which the phase information is stored electronically. The method consists of recording the relative phase distributions of the object wave before and after the object state is changed by applying a load. The phase maps in the two states being stored as image files, the phase difference relative to deformation is found by subtracting the phase maps one from the other. Although electronic holography may involve rather lengthy calculations, it reduces significantly the amount of speckle noise in the resulting phase map and opens up the possibility of obtaining modulo-2π phase in real-time at video rates.

1.5 Out-of-plane Displacement Measurement

First we will see how to measure an out-of-plane rigid body rotation. The surface, initially supposed to lie in the $x-y$ plane, is rotated around the y-axis by a small angle β (Fig. 7). The phase change between the scattered waves is given by

$$\Delta\varphi = \frac{2\pi x \beta(\cos\vartheta_e + \cos\vartheta_0)}{\lambda}, \tag{7}$$

where ϑ_e and ϑ_0 are the angles which the illumination beam and the observation direction, respectively, make with the surface normal. Equation (7) represents a set of equispaced straight fringes running parallel to the y-axis. The magnitude of rotation is obtained as

$$\beta = \frac{\lambda}{p_f(\cos\vartheta_e + \cos\vartheta_0)}, \tag{8}$$

where p_f is the fringe spacing.

Interference fringes are localized in the region where the variation in phase difference is a minimum over the range of viewing directions [39,40]. This requirement leads to the relation

$$z_0 = \frac{y \sin\vartheta_0 \cos^2\vartheta_0}{\cos\vartheta_e + \cos\vartheta_0}. \tag{9}$$

Equation (9) shows that the surface of localization always intersects the object, $z_0 = 0$, at the axis of rotation $y = 0$. The fringes are localized on the object when the observation is made in a direction perpendicular to the object surface, i.e. $\theta_0 = 0$. For the case of $\theta_0 \neq 0$, the fringes are localized either in front ($z_0 > 0$) or behind ($z_0 < 0$) the object surface.

The schematic of an optical set-up to observe fringes corresponding to out-of-plane displacements is shown in Fig. 8. The holographic plate is placed

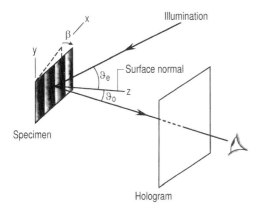

Fig. 7. Observation of fringes on a surface rotated around the y-axis

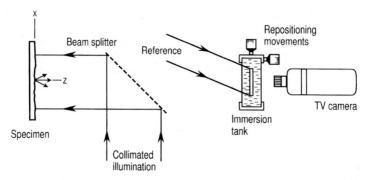

Fig. 8. Optical configuration for measuring the out-of-plane component of displacement

inside an immersion tank mounted on a universal translation and rotation stage. The reconstructed interference image is observed in real-time on the TV monitor. The fringe equation is given by

$$w = \frac{n\lambda}{2}, \tag{10}$$

where w is the out-of-plane component of displacement in the z direction. An example of a fringe pattern depicting the out-of-plane displacements in a rectangular aluminum plate-clamped along the edges and drawn out at the center by means of a bolt-subjected to three-point loading is shown in Fig. 9. A simple portable holographic system used to observe (a) the deformation behaviour in real time of concrete due to shrinkage and (b) the prenatal development of motility in a chick embryo is shown in Fig. 10. The entire system is mounted on a granite slab placed over inflated inner tubes. The system is assembled around a dual-purpose microprocessor-controlled cabinet, which serves both as a freezing box and as an incubator. The interferogram in Fig. 11a shows the damage induced in a concrete specimen due to shrinkage. Cracking is seen at the mortar-aggregate interfaces and in the matrix, extending in a direction normal to the contours of the aggregate. Figure 11b refers to an interferogram taken from the study of the embryonic behaviour of embryos during incubation.

1.6 In-plane Displacement Measurement

The measurement of in-plane components of displacement using holographic interferometry has been a topic of active investigation [41,42,43,44,45]. The first whole-field interferometer to be described for in-plane displacement measurement is shown in Fig. 12. Two collimated beams illuminate the surface symmetrically with respect to the z-axis. The in-plane displacements, in a direction containing the two beams, are mapped as a moiré pattern between

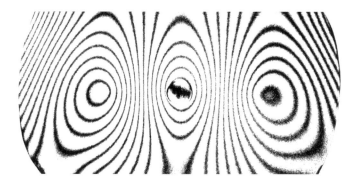

Fig. 9. Fringe pattern formed in real-time holographic interferometry on a rectangular aluminum plate - clamped along the edges and drawn out at the center by means of a bolt - subjected to three-point loading

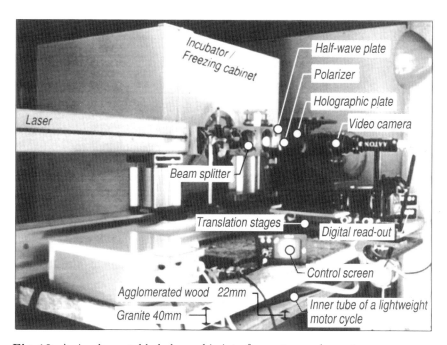

Fig. 10. A simple portable holographic interferometry equipment

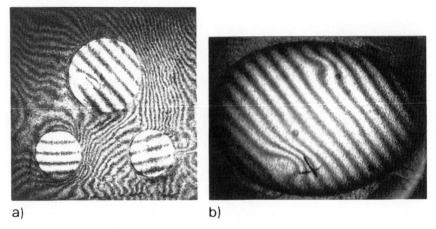

Fig. 11. Examples of real-time fringe contours corresponding to out-of-plane displacements. Interferogram in (**a**) shows the damage induced in a concrete specimen due to shrinkage and in (**b**) illustrates a specific limb movement of a duck embryo during incubation

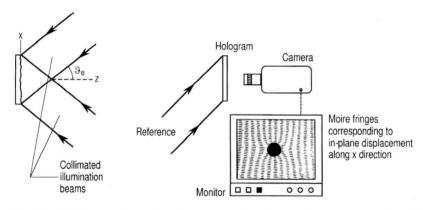

Fig. 12. Optical configuration for measuring the in-plane component of displacement

the interferograms due to each illumination beam. However, the fringe patterns in the two interferograms may not be of sufficient density to produce a reasonably good quality moiré pattern. This difficulty is overcome by introducing artificially an additional phase difference much larger than that introduced by the object deformation. The additional phase difference gives rise to a system of parallel equispaced fringes of high density localized on the object surface. The moiré pattern appears as a family of fringes

$$u = \frac{n_m \lambda}{2 \sin \vartheta_e} , \qquad (11)$$

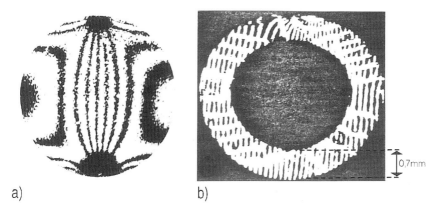

Fig. 13. Examples of in-plane displacement fringes obtained (**a**) on a disk under diametral compression, and (**b**) on the end surface of a reed stem (holomicroscopic moiré, image magnification ×15)

where u is the in-plane component of displacement in the x direction and n_m is the moiré fringe number. The TV screen in Fig. 12 displays an interferogram obtained on a thin aluminum sheet with a hole in it and loaded in uniaxial tension. Figure 13a shows the u pattern corresponding to a disk under diametral compression. The moiré pattern is obtained after filtering the interference image. Figure 13b shows an example of a u pattern, on a reed cross-section, obtained by a holo-microscopic method to study the mechanical response of reed stems growing in eutrophic and in healthy water bodies. Two examples of phase-shifted holographic moiré patterns corresponding to in-plane displacements are shown in Fig. 14. The phase distributions refer to the measurement of in-plane displacements on: (a) a notched concrete specimen subjected to a wedge-splitting test and (b) a thin aluminum sheet loaded in uniaxial tension. The microcracks on the concrete surface in Fig. 14a are detected by the presence of discontinuities in the phase distribution.

1.7 Holographic Shearing Interferometry

This method provides directly the patterns of slope-change contours by laterally shearing wavefronts diffracted from the object surface [46,47]. There are numerous ways to achieve shearing. The role of the shearing device is to permit the observation of a point on the object along two distinct neighbouring directions. Lateral wavefront shearing ensues and as a result a point in the image plane receives contributions from two different points on the object. The fringe-contour relative to slope-changes of the deformed object appear as a moiré pattern between two systems of laterally shifted displacement fringes. Assuming that the illumination beam lies in the $x-z$ plane and makes an angle ϑ_e with the z-axis, and the observation is carried out along the direction normal to the object surface, the optical phase change due to

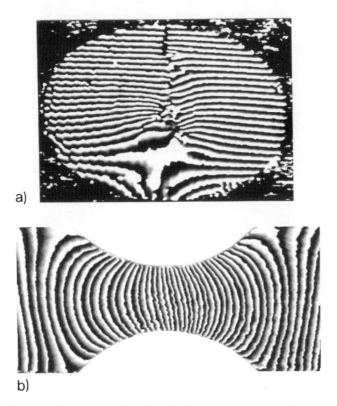

Fig. 14. Examples of modulo-2π holographic moiré phase maps corresponding to in-plane displacements produced on (**a**) an aluminum sheet under tension, and (**b**) a notched concrete specimen subjected to a wedge splitting test

deformation is

$$\Delta\varphi = \frac{2\pi\Delta x}{\lambda}\left[\frac{\partial u}{\partial x}\sin\vartheta_e + \frac{\partial w}{\partial x}(1+\cos\vartheta_e)\right], \tag{12}$$

where Δx is the object shear along the x direction. For $\vartheta_e = 0$, the fringe equation becomes

$$\frac{\partial w}{\partial x} = \frac{n_m \lambda}{2\Delta x}. \tag{13}$$

Equation (13) displays the family of fringes corresponding to the contours of constant slope change, $\partial w/\partial x$.

A drawback of the method is that, since the shearing mechanism is incorporated in the recording process, the sensitivity and direction of the slopes cannot be changed during an experiment. This makes any a posteriori manipulation of the shear impossible. A schematic of the procedure, which delinks the wavefront shearing from the recording stage but still incorporates it in the

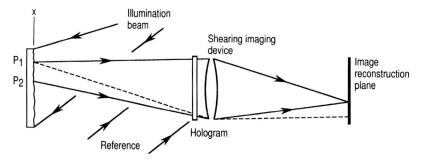

Fig. 15. Schematic of a holographic shearing interferometer

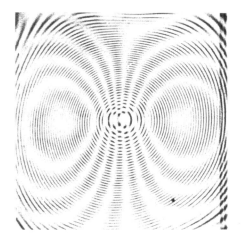

Fig. 16. Holographic moiré fringes corresponding to slope change distribution $\partial w/\partial x$ produced on a centrally loaded square aluminum plate clamped along its boundary

image reconstruction is shown in Fig. 15. This configuration gives the holographic slope measurement the flexibility to change at will both the sensitivity and the direction of slopes. An example of a fringe contour corresponding to the slope distribution of a centrally-loaded square aluminum plate clamped along its boundary is shown in Fig. 16. A dual image-shearing device [47] has been used to obtain whole-field phase maps corresponding to curvature and twist distributions.

1.8 Comparative Holographic Interferometry

The concept of comparative holography is of great interest in non-destructive testing. The method provides the contours of path variations related to the difference in displacements or shapes of two objects. The possibility of instantaneously comparing two objects considerably broadens the scope of holographic interferometry in non-destructive inspection. The main difficulty associated with the development of the method has been inherent in the task itself, which aims at comparing two macroscopically similar but physically different surfaces. A number of approaches have been developed to address the problem [48,49,50,51,52]. The technique provides a tool for comparing the mechanical responses and also the shapes of two nominally identical specimens subjected to the same loading, and also the shapes of two nominally identical specimens. This feature offers a versatile and practical way of detecting anomalies in the test specimen with respect to the flaw-free master specimen.

An example of a fringe pattern depicting the out-of-plane difference displacement component Δw is shown in Fig. 17a. The interferogram obtained by a holographic moiré technique compares deflections of two square aluminum plates clamped along the edges and submitted to centrally concentrated loads. Figure 17b shows an application to flaw detection. Two thick rubber plates of the same size and shape are machined from the same rubber sheet. In one rubber plate a defect is introduced by drilling a small blind hole on its rear surface. The other rubber plate is considered free of defects. The two plates are held in their respective loading frames, in such a way that the in-plane displacements at the edges are constrained. The closed set of moiré fringes clearly shows the presence and location of the flaw. Auxiliary displacement fringes are added to obtain a good formation of the moiré pattern.

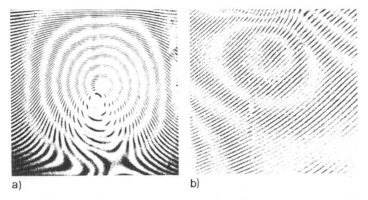

Fig. 17. (a) Example of a holographic moiré pattern corresponding to difference in displacements of two square plates clamped along the edges and subjected to centrally concentrated loads; (b) An application to nondestructive inspection: the moiré pattern clearly displays the presence and location of the flaw in a test piece

1.9 Vibration Analysis

Another important application of holographic interferometry is in the study of the resonant modes of vibration of an object [53,54,55,56,57,58,59,60], such as a turbine blade. Consider a hologram recording of an object which is vibrating sinusoidally in a direction normal to its surface Fig. 18a. The exposure time is supposed to be much longer than the period of vibration. The intensity distribution of the image reconstructed by this hologram is

$$I(x,y) = I_0(x,y) J_0^2 \left[\frac{2\pi}{\lambda} d(x,y)(\cos\theta_e + \cos\theta_o) \right] , \tag{14}$$

where J_0 is the zero-order Bessel function of the first kind. The virtual image is modulated by the $J_0^2(\xi)$ function. The dark fringes correspond to the zeros of the function $J_0^2(\xi)$. The plot of the function, shown in Fig. 18b, is characterized by a comparatively brighter zero-order fringe which corresponds to the nodes, a decreasing intensity and an unequal spacing between the successive zeros. The photographs in Fig. 19 show the mode mapping of a vibrating helicopter component. Figures 19a,b show holographic reconstructions corresponding to two different modes of vibrations at 1349 Hz and 3320 Hz, respectively. The nodes, which represent zero motion, are clearly seen as the brightest areas in the time-average holograms.

The possibility of studying the response of a vibrating object in real-time extends the usefulness of the technique so as to identify the resonance states (resonant-vibration mode shapes and resonant frequencies) of the object. A single exposure is made of the object in its state of rest. The plate is processed, returned to its original condition and reconstructed. The observer looking through the hologram at the sinusoidally vibrating object sees the

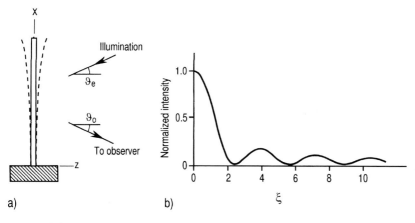

Fig. 18. (a) Schematic of the optical configuration of a vibrating cantilever beam. (b) Plot of fringe functions for time-average holography

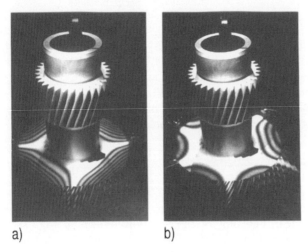

Fig. 19. Vibration-mode mapping of a helicopter component at (**a**) 1349 Hz, and (**b**) 3320 Hz. (Coutresy of L.M. Rizzi, ENEL Research)

time-averaged intensity

$$I(x,y) = I_0(x,y)\left\{1 - J_0\left[\frac{2\pi}{\lambda}d(x,y)(\cos\theta_e + \cos\theta_o)\right]\right\}, \qquad (15)$$

where it is assumed that vibration frequencies higher than the time constant of the human eye (≈ 0.04 s) are being observed. The method has half the sensitivity of time-average holography. The fringe contrast is much lower than that of time-average fringes.

Stroboscopic holography is another interesting variant in the study of vibrations. The hologram is recorded by exposing the photographic plate twice for short time intervals during a vibration cycle. The pulsed exposures are synchronized with the vibrating surface, which is equivalent to making the surface virtually stationary during the recording. Reconstruction of the hologram yields cosinusoidal fringes which are characteristic of double exposure holography. In real-time stroboscopic holography a hologram of a non-vibrating object is first recorded. If the vibrating object is illuminated stroboscopically and viewed through the hologram, the reconstructed image from the hologram interferes directly with the light scattered from the object to generate live fringes.

Double-pulsed holography has become a routine technique for vibration analysis of non-rotating objects. The double pulse freezes the object at two points in the vibration cycle. On the other hand, the study of the vibration of rotating structures would require some form of rotation compensation to ensure correlation between the wave fields scattered from the two states of the vibrating and rotating object. Several approaches are proposed in the literature to prevent image decorrelation due to object rotation [60,61,62].

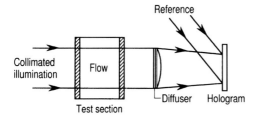

Fig. 20. Schematic of a holographic configuration for studying flow

1.10 Flow Visualization

Applications of holographic interferometry to flow visualization [63,64,65,66,67] and the measurement of spatial refractive index, density or temperature distributions have led to an advancement of understanding in areas such as aerodynamics, plasma diagnostics and heat transfer. In the holographic interferometer shown in Fig. 20 two consecutive exposures are made, usually the first exposure without flow and the second in the presence of a flow field. Double-pulsed holography is used if the flow field is changing rapidly. The optical phase change due to flow between the exposures is

$$\Delta \varphi(x,y) = \frac{2\pi}{\lambda} \int_0^t [n'(x,y,z) - n'_0] \, dz \, , \qquad (16)$$

where $n'(x,y,z)$ is the refractive index distribution during the second exposure, n'_0 is the uniform refractive index during the first exposure and t is the length of the test section. Assuming that the properties of the flow are constant in the z direction, the expression for phase change can be expressed as

$$\Delta \varphi(x,y) = \frac{2\pi K t[\rho(x,y) - \rho_0]}{\lambda} \, , \qquad (17)$$

where K is the Gladstone-Dale constant, ρ is the density of the gas, and ρ_0 is the density in a no-flow situation The interference pattern contours the change in the density field of the flow. The change in density per fringe is given by λ/Kt. An advantage of holographic interferometry is that high-quality interferograms can be obtained through glass windows of relatively poor quality. An example of temperature measurements in an axisymmetric, laminar, premixed fuel-lean propane-air flame is illustrated in Fig. 21.

1.11 Holographic Contouring

A practical way to display the shape of an object is to obtain contour maps showing the intersection of the object with a set of equidistant planes orthogonal to the line of sight. A number of methods exist to measure relief variations of an arbitrarily shaped object [68,69,70,71,72]. The optical set-up

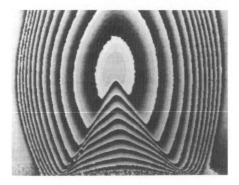

Fig. 21. Modulo 2π phase distribution of a test flame. (Courtesy of S.M. Tieng)

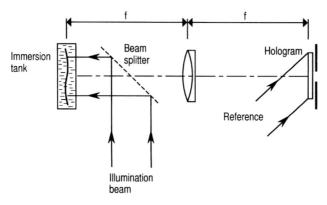

Fig. 22. Schematic of the refractive-index-change method for shape measurement

of one of the earlier described methods is shown schematically in Fig. 22. In this method the model is placed in a glass tank filled with a liquid (or gas) of refractive index n'_1. A hologram of the object is recorded, and the liquid contained in the tank is replaced by one of refractive index n'_2. An observer looking through the hologram sees the object surface modulated by a set of interference fringes arising from the change of optical phase in the light rays traversing the two liquids. The contour interval is given by

$$\Delta z = \frac{\lambda}{2(n'_1 - n'_2)} . \tag{18}$$

In another approach, the surface to be contoured is illuminated by means of two coherent collimated beams aligned symmetrically with respect to the optical axis (Fig. 23). The two beams are supposed to be in the $x - z$ plane. The method is based on the principle of real-time four-wavefront mixing. Two of these wavefronts, are recorded on the holographic plate. These are played back during the reconstruction of the hologram. The two wavefronts are due to one illumination each. The other two wavefronts, also due to one

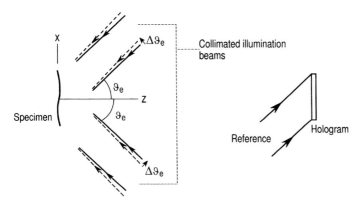

Fig. 23. Holographic configuration of the two-beam multiple source method for shape measurement

illumination each, correspond to the real-time wavefronts issuing from the object surface and observed through the holographic plate. The illumination beams are then tilted by the same amount, $\Delta\vartheta_e$, and in the same direction around an axis perpendicular to the plane of the paper [70]. An observer looking through the hologram sees a family of moiré fringes. The increment of height between two consecutive contour planes is given by

$$\Delta z = \frac{\lambda}{2\sin\vartheta_e \sin\Delta\vartheta_e} \,. \tag{19}$$

A fairly wide range of contour intervals has been obtained by this method. An example of the results obtained by applying this approach is shown in Fig. 24. The distances between the two adjacent coutouring planes in the pat-

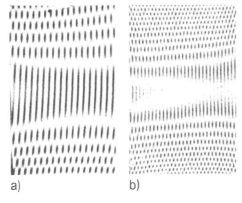

Fig. 24. Illustration of the continuous generation of sensitivities obtained by two-beam multiple source holography. The sensitivities generated in (**a**, **b**) are given by 380 and 172 μm, respectively

terns displayed in (a) and (b) are given by 380 and 172 μm, respectively. It is interesting to note that relief sensitivity can also be obtained by slightly rotating the object surface in a direction perpendicular to the plane containing the two beams.

Another interesting method in this class of interferometers requires illuminating an object obliquely by a collimated beam Fig. 25. The scattered waves are recorded on a holographic plate, and the object and its reconstructed image are viewed through the hologram. A set of parallel equidistant fringes is projected onto the object surface by introducing a small tilt, $\Delta \vartheta_e$, to the illuminating beam. Contou surfaces normal to the line of sight are generated by introducing a rotation to the reference beam combined with an appropriate translation to the holographic plate. The contour sensitivity per fringe is given by

$$\Delta z = \frac{\lambda}{\sin \vartheta_e \sin \Delta \vartheta_e} . \tag{20}$$

The sensitivity of the method can be tuned over a wide range [71]. It offers the possibility of generating contour planes in real time at any angle to the line of sight. An example of topographic contour patterns is shown in Fig. 26b. The distance between two adjacent coutouring planes in the patterns displayed in (a) and (b) are, respectively, given by 1231 and 216 μm. This approach has also been employed to obtain measurement of large out-of-plane deformations undergone by a deformed object [73].

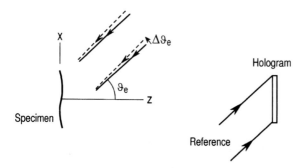

Fig. 25. Schematic of multiple source holographic contouring

2 Speckle Metrology

Denis Gabor once said, "The novelty in holography is speckle noise.". It was soon to be shown that this apparent source of noise which holographers once aspired to eliminate, could also be a very useful engineering tool. When an optically rough surface is illuminated with coherent light, the rays are

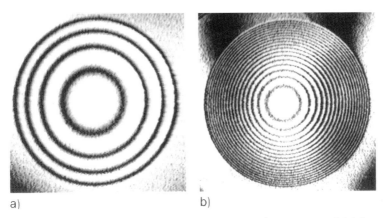

Fig. 26. Illustration of the real-time generation of contour sensitivities of (**a**) 1231 and (**b**) 216 μm obtained by multiple source holography

scattered in all directions and at all distances from the surface. The scattered wave distribution displays a random spatial variation of intensity called a speckle pattern. The speckled appearance is caused by the interference of the elementary waves emanating from the surface. The interference pattern looks like a random distribution of dark and bright spots of different shapes and sizes.

Intensive developments led to the development of a wide range of procedures in speckle metrology, which are classified into three broad categories: speckle photography, speckle interferometry and speckle shearing interferometry. Speckle metrology is now firmly established as a major tool in nondestructive testing and inspection. These techniques are distinguished by their practical approach, which offers problem-solving directly on field models. A significant feature of these methods lies in their extreme flexibility in handling large deformation problems.

2.1 Focused Speckle Photography

The method [74,75,76] consists of recording in the image plane the intensity distributions of a coherently illuminated object surface in its two states of deformation (Fig. 27). At an arbitrary point on the object surface, the components of speckle shifts $\Omega_x(x,y)$ and $\Omega_y(x,y)$ along x and y directions are, respectively, given by

$$\Omega_x(x,y) = u(x,y) + \frac{x}{d_0}w(x,y) \;, \tag{21}$$

$$\Omega_y(x,y) = \nu(x,y) + \frac{y}{d_0}w(x,y) \;, \tag{22}$$

where v is the in-plane component of displacement in the y direction. Equations (21,22) show that the influence of the out-of-plane displacement com-

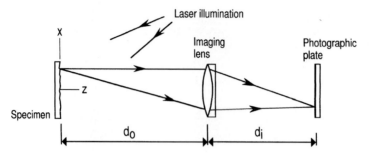

Fig. 27. Focused speckle photography configuration

ponent, including tilts on the in-plane component of speckle shift, becomes progressively important as the observation point on the object surface shifts away from the optical axis.

The photographic plate that records two laterally displaced speckle patterns could then be analysed in two ways to yield the in-plane displacement undergone by the object surface. The point-by-point approach of analyzing the specklegram consists of illuminating the processed plate with a narrow laser beam (Fig. 28a). The laser beam diffracted by the speckles lying within the beam area gives rise to a diffraction halo. The halo is modulated by an equidistant system of fringes arising from the interference of two identical but displaced speckles. The direction of these fringes is perpendicular to the direction of displacement. The magnitude of displacement, inversely proportional to the fringe spacing p, is given by

$$d_\xi = m\sqrt{\Omega_x^2 + \Omega_y^2} = \frac{\lambda L}{p}, \tag{23}$$

where d_ξ is the displacement in the image plane, m is the magnification factor of the speckle recording imaging system, and L is the distance between the specklegram and the observation screen.

On the other hand, the reconstruction of a whole-field interference image requires placing the recorded specklegram in an optical filtering arrangement, as in Fig. 28b. An observer, looking through a small offset aperture placed in the focal plane of the Fourier transform lens, sees contours of speckle displacement along the direction defined by the position of the aperture. The fringe contours are described as the loci of points where

$$d_\xi = \frac{n\lambda f}{\zeta_\xi}, \tag{24}$$

where f is the focal length of the Fourier transform lens; ξ denotes the azimuth and ζ_ξ is the radial distance from the position of the zero diffraction order. An example of a fringe pattern depicting the in-plane displacements along the x direction in a wooden beam structure subjected to deformation is shown in Fig. 29.

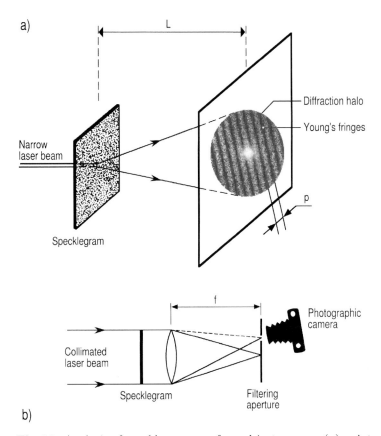

Fig. 28. Analysis of specklegrams performed in two ways (**a**) point-wise filtering, and (**b**) whole-field filtering

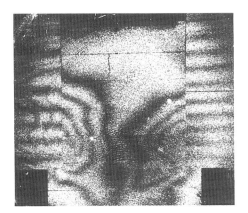

Fig. 29. Illustration of in-plane displacement contours obtained by speckle photography on a wooden beam structure subjected to deformation

A limitation on the measurable range in speckle photography arises from the presence of secondary speckles in the diffraction halo. To successfully carry out the measurements, the Young's interfringe, p, must satisfy the condition

$$\langle s_{\text{halo}} \rangle < p > D_{\text{halo}}, \tag{25}$$

where the size of the secondary speckles in the halo is s_{halo} and the diameter of the halo is D_{halo}.

In speckle photography we essentially measure the lateral shifts in the speckle pattern to obtain the in-plane displacement fields. This led to the development of a method which substituted laser by white light to form speckle patterns. The method [77] requires a camera to resolve an artificial granular structure in the image of an object illuminated by white light. Each resolved grain thus serves as a reference point, specifying the initial and final position of that many points on the object surface, before and after the object deformation. During the analysis, the same correlation techniques apply as in laser speckle photography.

Simple to implement and relatively tolerant on stability requirements, the technique of focused speckle photography is well suited for use under factory floor conditions. However, there are some important drawbacks which hold back the widespread use of speckle photography in nondestructive testing. These are: nonconcommitant time mode of operation, tedious and time-consuming processing of the Young's fringes, and relatively poor quality constant-displacement fringes obtained in whole-field analysis. These difficulties have prompted work towards developing means to make the technique user-friendly. A quasi real time realization of the speckle photography technique has been obtained by using the photorefractive crystals to record the specklegrams [78,79,80]. On the other hand, the applications of heterodyne [81,82] and phase shifting [83] to speckle photography have helped in considerably reducing the last two drawbacks. Finally, the development of automatic microcomputer-based image-processing systems has greatly facilitated the evaluation of Young's fringes [84,85,86,87].

2.2 Defocused Speckle Photography

A broad range of methods in coherent speckle photography do not necessitate the focusing of the object surface on the film plane [88,89,90]. The coherent speckles are present all over the space surrounding the illuminated object surface. The schematic of an optical system for the measurement of out-of-plane derivatives [88] is shown in Fig. 30. The method is based on photographing the speckles contained in a plane, in front of or behind the object. In the study of a thin-plate bending problem for which this system was initially conceived, the in-plane displacement field is much smaller than the out-of-plane displacement field. Moreover, rigid body rotations and translations can

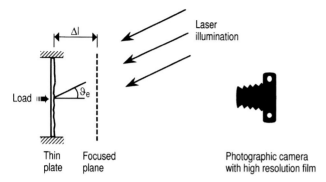

Fig. 30. A defocused speckle photography arrangement for slope measurement

be eliminated by means of a well-built loading device. Under these assumptions, the speckle shifts for points lying on the object in the neighbourhood of the optical axis are given by

$$\Omega_x = \Delta l \left[\frac{\partial u}{\partial x} \sin \vartheta_e + \frac{\partial w}{\partial x} (1 + \cos \vartheta_e) \right] , \tag{26}$$

$$\Omega_y = \Delta l \left[\frac{\partial v}{\partial y} \sin \vartheta_e + \frac{\partial w}{\partial y} (1 + \cos \vartheta_e) \right] . \tag{27}$$

It is possible to vary the measurement sensitivity of the method by modifying the amount of defocusing Δl. Using this method, one can rely on the measurement of slopes, along the x and y-axes, only as long as the lateral and axial displacements are deemed to be negligible.

The concept of focusing the imaging lens on a plane in front of or behind the specimen surface has been developed and widely applied to separate out and measure different displacement and deformation components. Aiming at speckles lying in a defocused plane, the domain of application of speckle photography has been extended to the measurement of out-of-plane rotations [89,90]. Also based on the principle of defocused speckle photography is the conceptually elegant technique of tandem speckle photography [91]. The technique requires recording a focused plane and two tandem specklegrams simultaneously for at least three different directions of illumination. Subsequent solving of the linear system of equations allows one to determine displacements, slopes, and surface strains undergone by the object.

2.3 Speckle Shearing Photography

The concept of speckle shearing photography [92,93] represents a simple and direct solution to the problem of measuring the derivatives of in-plane displacements. The concept is based on the property that the wave field scattered from a coherently lit object point and passing through two different aperture portions, on its way to the photographic plate, gives rise to two distinct speckle patterns in the image plane when each pattern is viewed independently. The successive exposure of the photographic plate to each of the two speckle patterns enables us to safeguard their identity on the recording medium.

A rough surface placed in the $x - y$ plane is imaged on a photographic emulsion by means of a lateral shearing system. The z-axis is taken to be along the optical axis of the shearing system. The shearing device is constituted of two apertures, each of which focuses an image of the object point on the film plane. The two images are laterally sheared with respect to one another. The method requires exposing the photographic plate to the speckle patterns arising independently from the two apertures of the shearing device. The first exposure is made with the scattered light passing through one aperture and the second exposure is made with the light made to pass only through the other aperture. The object is deformed and a series of two more exposures are made under conditions identical to the above. The dual double-exposure specklegram thus obtained has the strain-field information frozen on the plate. Point-wise read-out of the specklegram uses the diffraction properties of the recorded fields. A narrow beam of laser light is directed on the photographic plate at normal incidence. Two perfectly overlapping diffraction halos related to the lens apertures are generated in the far field. The halos are each modulated by a system of so-called Young's fringes. The two systems of Young's fringes beat together to form a Young's moiré pattern. An example of a diffraction halo containing Young's moiré fringes is shown in Fig. 31.

Suppose that the shearing system is adjusted to give a shift in the x direction. In this case, the relative incremental changes in speckle displacement components along the x and y-axes follow as

$$\delta\Omega_x = \Delta\, x \left(\frac{\partial u}{\partial x} + \frac{x}{d_0} \frac{\partial w}{\partial x} \right), \tag{28}$$

$$\delta\Omega_y = \Delta\, x \left(\frac{\partial v}{\partial x} + \frac{y}{d_0} \frac{\partial w}{\partial x} \right). \tag{29}$$

In the neighbourhood of the optic axis ($x/d_0; y/d_0 \approx 0$) relative increments in the speckle shift due to the slopes can be neglected. With this approximation

$$\delta\Omega_x = \Delta\, x \frac{\partial u}{\partial x},$$
$$\delta\Omega_y = \Delta\, x \frac{\partial v}{\partial x}, \tag{30}$$

which leads to the Young's-moiré-fringe equation

$$\sqrt{\left(\frac{\partial u}{\partial x}\right)^2 + \left(\frac{\partial v}{\partial x}\right)^2} = \frac{\lambda L}{\Delta\, x p_m}. \tag{31}$$

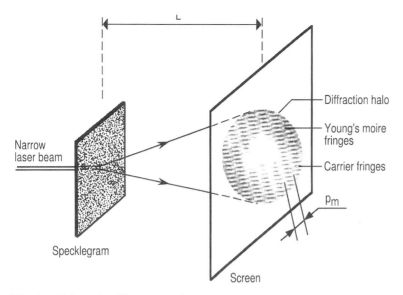

Fig. 31. Point-wise filtering to obtain Young's moiré fringes

It follows that the in-plane normal strain component $\partial u/\partial x$ and the shearing strain component $\partial \nu/\partial x$ in the observation plane are given by

$$\frac{\partial u}{\partial x} = \frac{\lambda L}{\Delta\, x p_{\mathrm{m}}} \cos \phi \;,$$
$$\frac{\partial \nu}{\partial x} = \frac{\lambda L}{\Delta\, x p_{\mathrm{m}}} \sin \phi \;, \qquad (32)$$

where ϕ is the angle which the perpendicular to the moiré fringes make with the x_{f} axis in the Fourier plane. Measurement of strains by manual and point-wise reading of the speckle-shearogram could be a tedious experience requiring considerable time to complete the data reduction. For these reasons and to reduce readout errors, it is advisable to first digitise the Young's moiré fringes and later analyze the data for strain information. Approaches similar to those applied in laser speckle photography for automatic digital readout can be used and adapted for the reconstruction of the strain fields from the Young's moiré fringes.

2.4 Speckle Interferometry

Speckle interferometry [94,95,96,97,98,99,100,101,102,103,104,105,106,107] [108,109,110] has proven to be a valuable tool in speckle metrology. Of sensitivity comparable to that of holographic interferometry, the method is based on the coherent addition of speckle fields diffracted by object and reference fields. The reference field can either be a plane wave or another speckle field not necessarily originating from the object surface under study.

The introduction of a reference field brings about a significant change in the behaviour of the speckle pattern when the object is deformed. Since the observed speckle pattern is formed by the interference of two coherent fields, the intensity in the resultant pattern depends on the relative phase distributions of the added fields. The fringes in speckle interferometry arise from the correlation of two speckle pattern intensity distributions obtained before and after object deformation.

2.4.1 Schemes to Obtain Visualization of Speckle Correlation Fringes

Several methods have been developed to obtain the display of the correlation fringes. The first, though not the most employed, of these methods is the mask technique. The technique consists of recording the initial speckle intensity, I, on a photographic plate. The plate after processing is replaced in its original position. The speckle distribution after object deformation, I', is observed through the processed photographic plate. In the regions where I and I' are identical, the images are complementary. The flux transmitted in these regions is much weaker than in the regions where I and I' are uncorrelated. As a result correlation fringes are seen to appear, covering the image in the form of light and dark bands arising from the variation of light transmitted through the mask. These fringes are obtained in real-time.

A method quite akin to the mask technique relies on photographic subtraction to detect the positions of correlation. The speckle distribution I' is recorded on a separate plate and its positive contact print is made on another plate. The two plates, one containing the negative image of the distribution I, and the other containing the positive image of the distribution I', are placed in register and adjusted to yield correlation fringes to an observer looking through the pair of plates. This mode of generating correlation fringes is very interesting as it allows for compensating rigid-body movements with the result that the fringe contours are of relatively good quality.

Double-exposure methods have been used to detect the positions of correlation. Speckle patterns are recorded on a photographic plate before and after object deformation. The regions where the phase difference between the two speckle patterns remains the same or changes by $2n\pi$ gives rise to a coherent addition of the two speckle fields. The regions where the phase changes by $(2n+1)\pi$ between the two exposures gives rise to an incoherent addition of the two speckle fields. Although these regions have same average intensities, they do differ in their textures. The detection can be improved by making use of the non-linearity of the photographic process. In the regions with speckle texture (coherent combination) the flux transmitted is much higher than in the regions of incoherent combination of the speckle fields.

A useful method of generating correlation fringes uses a filtering technique that consists of recording on a photographic plate two slightly shifted images of the object. The recorded film when illuminated by a plane wave gives rise to a diffraction pattern which is modulated by fringes of spacing inversely proportional to the given shift. The light intensity at the minima of the diffraction fringes receives contributions only from the uncorrelated regions on the plate. A slit centered on a dark fringe in the diffraction halo passes light from only these regions. The dark fringes correspond to phase variations of $2n\pi$, while the bright fringes correspond to phase variations of $(2n+1)\pi$.

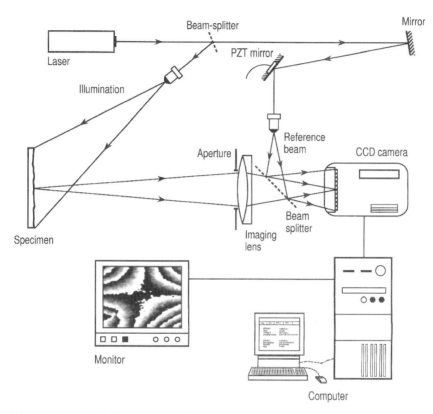

Fig. 32. A typical electronic speckle-pattern interferometer

Another attractive method of observing correlation fringes is based on the electronic subtraction of intensities, before and after object deformation. Unlike traditional photographic film based methods, this system allows quantitative phase data to be obtained quickly and in electronic form. The speckle intensities are recorded directly onto a charge-coupled device (CCD) detector and stored and processed in a computer. A schematic of an electronic speckle pattern interferometer (ESPI) system is shown in Fig. 32. Supposing that the image intensity displayed on the TV monitor is proportional to $|I - I'|$ the ESPI presents an analogy, at least qualitatively, in the imaging aspects to the mask technique. The speed of ESPI is best illustrated by its ability to produce an interferogram every 1/25 of a second. ESPI has been combined with the phase-shifting technique to yield a quantitative display of the phase of the object's displacement corresponding to the interference fringes. Phase stepping is introduced in one of the two arms of the interferometer, irrespective of the arm's dependence on object deformation. Digital image processing techniques are being routinely applied to obtain improved quality correlation fringes.

2.4.2 Speckle Interferometers Sensitive to Out-of-plane Displacements

The schematic of an optical system to measure out-of-plane displacements is shown in Fig. 33. It basically uses a set-up analogous to a Michelson interferometer in which both mirrors are replaced by scatter surfaces [94]. The resultant speckle pattern in the image plane is formed by the interference of speckle fields issuing independently from the two scatter surfaces. Of the two surfaces, one is subjected to deformation and the other is used as the source of reference speckles.

Under the small-angle approximation and in the configuration of collimated illumination at zero incidence, the equation of the observed fringe contours is given by (10). An example of a fringe pattern obtained by this

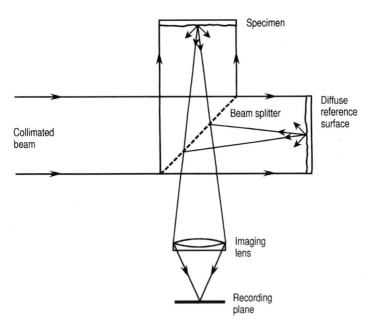

Fig. 33. Speckle interferometer for measuring the out-of-plane component of displacement

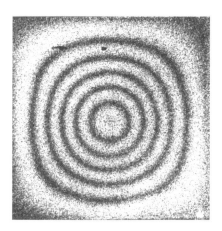

Fig. 34. An example of speckle correlation fringes representing contours of constant out-of-plane displacement, w

method is shown in Fig. 34. The fringe pattern obtained by the double-exposure procedure described in the last Subsection corresponds to the deformation of a square plate clamped along its edges and subjected to a centrally concentrated load. The method has been used to determine the difference in deformations of two macroscopically similar but different object surfaces [101,102]. This difference is observed in the form of fringes that contour the change in the surface shapes of two specimens, namely, master and test, caused by deformations. The measurement of the difference in deformations has a considerable application in nondestructive testing. The technique has been used for fringe compensation and flaw detection in applications related to nondestructive testing.

In another method in this class of interferometers, a smooth reference wave travelling along the optical axis is added to the object speckle field in the image plane. The resultant speckle pattern is due to the interference of the speckle field with the smooth reference wave. The reference speckle field is independent of the object deformation. This interferometer has found extensive applications in out-of-plane displacement measurements and vibration analysis.

2.4.3 Speckle Interferometers Sensitive to In-plane Displacements

Figure 35 illustrates the basic optical system used for the measurement of in-plane displacements undergone by a diffuse object surface. The object is illuminated by two collimated waves aligned symmetrically and making an angle with respect to the surface normal [94,95]. The resulting speckle pattern in the image plane is generated by the coherent superposition of the speckle fields corresponding to each of the two illumination beams. When

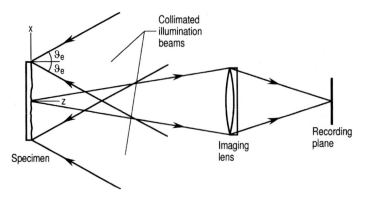

Fig. 35. Speckle interferometer for measuring the in-plane component of displacement

the object is deformed, the phase difference between the illuminating waves at any point will change. The relative phase change between the two speckle patterns is the same as in (11). The corresponding fringes represent contours of constant in-plane displacement u. Equation (11) shows that the system is independent of the out-of-plane and as well as in-plane displacements in a direction perpendicular to the plane containing the two illumination beams. This configuration is also one of those rare speckle metrology methods which escape from off-axis error terms. Figure 36 shows the u displacement fringes obtained on a ring subjected to diametrical compression. The fringe pattern was obtained with the mask technique. Figure 37 displays phase distributions obtained in an ESPI configuration. The photograph corresponds to phase

Fig. 36. An example of speckle correlation fringes representing contours of constant in-plane displacement, u

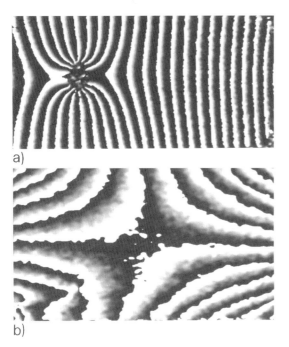

Fig. 37. Modulo-2π phase distribution corresponding to in-plane displacements on (**a**) a thin aluminum sheet with a hole in it and loaded in uniaxial tension, and (**b**) a wooden beam subjected to four-point bending

maps obtained on (a) a thin aluminum sheet with a hole in it and loaded in uniaxial tension, and (b) a wooden beam subjected to four-point bending.

Another method for determining in-plane displacements consists of imaging the object surface through a lens fitted with a double aperture located symmetrically about the lens center. The imaging system is illustrated in Fig. 38a. The lens focused on the surface sees an arbitrary point on the object surface along two different directions. The two viewing apertures are separated by distance h. Large speckles of characteristic size dependent on the aperture diameters are formed in the image plane. As a result of the coherent superposition of the beams emerging from the two apertures, each speckle is modulated by a grid structure running perpendicular to the line joining the apertures. The pitch of the fringe system embedded in each speckle depends on the separation of the two apertures.

For movements which are out-of-plane, or in-plane in a direction perpendicular to the line joining the apertures, the two beams diffused by a point on the object undergo equal phase changes and the speckle grid remains unchanged. In reality this statement is rigorously true only for those object points which lie on the optical axis. On the other hand, the movements in a direction parallel to the line joining the two apertures cause the speckle grid

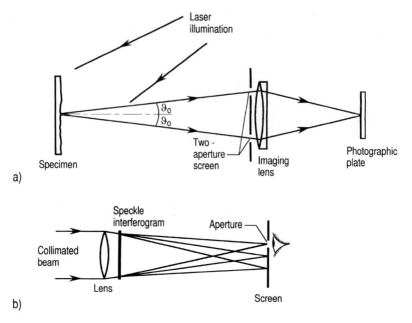

Fig. 38. (a) Two-aperture speckle interferometer for measuring the in-plane component of displacement; (b) observation of fringe contours

to shift parallel to itself by a grating interval for every successive surface displacement equivalent to the grid spacing. For those areas that have undergone integral values of this displacement, the grid pattern becomes out-of-phase with the original pattern, resulting in a loss of the periodic structure of the grid within the speckle in the case of a double-exposure recording.

Neglecting the second-order terms, the equation of the fringe pattern can be expressed as [96]

$$\frac{h}{d_0}\left(u + \frac{x}{d_0}w\right) = n\lambda. \tag{33}$$

Fringes can be observed and photographed in the Fourier-filtering set-up of Fig. 38b. The technique was later extended to measure simultaneously the in-plane components u and v by placing a mask containing two pairs of apertures, one each along the x- and y-axes, in front of the lens.

Unlike the previous two methods, which either use a single observation direction and two illumination beams or a single illumination beam and two observation directions, an interferometer has recently been developed which uses two illumination beams and two observation directions [111]. A salient feature of this interferometer is that it improves upon the sensitivity of the first interferometer by a factor of two. The experimental arrangement is shown in Fig. 39. Two collimated beams are incident at equal angles on opposite sides of the surface normal illuminating the object surface. The observation

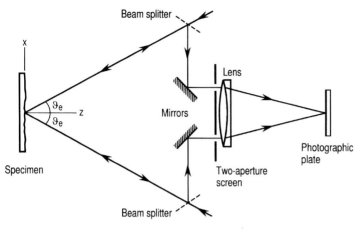

Fig. 39. A speckle interferometer for obtaining increased sensitivity for the in-plane component of displacement

is also made along two directions. The phase change due to deformation can be expressed as

$$\Delta\varphi = \frac{8\pi u \sin\vartheta_e}{\lambda} \ . \tag{34}$$

A five-aperture speckle interferometer is displayed in Fig. 40a. An advantage of this type of configuration is that it is sensitive to the three displacements and enables each of these component fields to be visualized separately. A laser beam illuminates the diffuser placed in front of the central aperture to produce a diffuse reference beam. The schematic of the diffraction halo generated by the specklegram in the Fourier plane is shown in Fig. 40b.

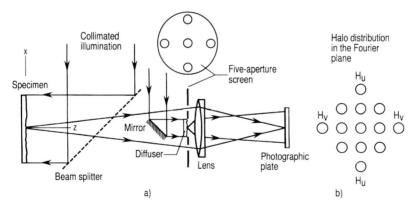

Fig. 40. (a) A speckle interferometer for simultaneously measuring the out-of-plane and in-plane components of displacement; (b) halo distribution at the Fourier plane

Filtering through the H_u and H_v halos yields a family of fringes corresponding to the u and v fields, respectively. The out-of-plane fringes are reconstructed by filtering through any one of the four halos situated next to the zero-order and along the horizontal and vertical directions in the Fourier plane.

2.4.4 Shape Contouring

The schematic of an optical system for contouring three-dimensional objects is shown in Fig. 41. The specimen is illuminated by two beams incident at equal angles, ϑ_e, on both sides of the optical axis [112]. The combined speckle field in the image plane results from the superposition of the individual fields generated by each of the two illumination beams. The two illumination beams are tilted by the same amount, $\Delta\vartheta_e$, and in the same direction around an axis perpendicular to the plane of the paper. The relative phase change introduced in the interferometer as a result of this rotation is given by

$$\Delta\varphi = \frac{4\pi \sin\vartheta \sin\Delta\vartheta_e}{\lambda} . \tag{35}$$

The increment of height between two contour planes is given by (19). The equation is independent of the direction of observation. An example of a contouring pattern obtained on pyramidical surface is shown in Fig. 42. The contour planes lie parallel to the $x - y$ plane.

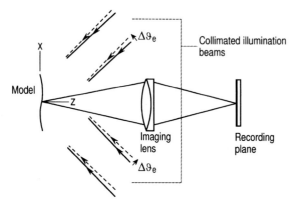

Fig. 41. A two-beam speckle interferometer for determining object shapes

2.5 Speckle Shearing Interferometry

Speckle shearing interferometry [113,114,115,116,117,118,119,120] is a well established and essential technique for the measurement of the derivatives of surface displacements. The principle of this class of methods is the same as

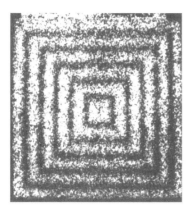

Fig. 42. Contouring pattern obtained for a pyramid shape object. (courtesy of C. Joenathan)

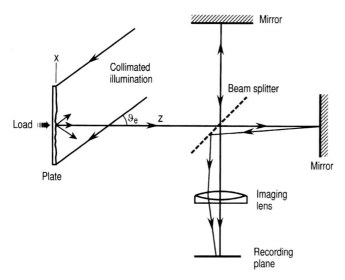

Fig. 43. A speckle shearing interferometer using the Michelson-type mirror arrangement

that of speckle interferometry, except that the two interfering speckles now originate from the same object and are sheared laterally with respect to each other.

A schematic of the first method in this class of interferometers is shown in Fig. 43. The object, illuminated by a collimated beam lying in the $x-z$ plane and making an angle with the surface normal, is imaged by means of a lens through a Michelson interferometer [113,114]. A small tilt of one of the two mirrors introduces a lateral shift Δx between the two speckle patterns. When the object is deformed, an arbitrary point (x,y) on the object surface is displaced to $(x+u, y+v, w)$ and a neighbouring point $(x+\Delta x, y)$ is displaced to $(x + \Delta x + u + \delta u, y + v + \delta v, w + \delta w)$. In an optical set-up designed to make the angle subtended by the imaging lens at the object surface small, the relative phase change due to the displacement between the two contributing points is given by (12). The equation for the fringe contour is

$$\Delta x \left[\frac{\partial u}{\partial x} \sin \vartheta_e + \frac{\partial w}{\partial x} (1 + \cos \vartheta_e) \right] = n\lambda . \tag{36}$$

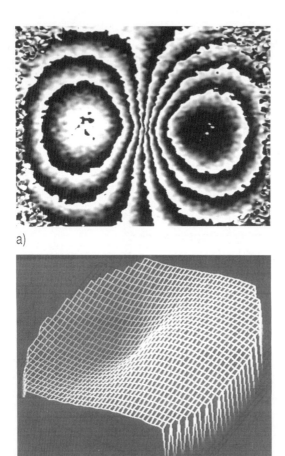

Fig. 44. (a) Modulo-2π phase distribution corresponding to slope change for a centrally loaded edge-clamped square aluminum plate; (b) three-dimensional plot of the phase distribution in (a)

This interferometer is used to observe fringes depicting pure slope change contours in the configuration of $\vartheta_e = 0$. Figure 44a shows the phase map depicting the horizontal slope component for a centrally loaded edge-clamped square aluminum plate. Figure 44b gives a three dimensional plot of the phase map corresponding to slopes.

Another method whose optical system is quite similar to the two-aperture speckle interferometer is shown in Fig. 45a. A shearing mechanism, in the form of two symmetrically oriented glass wedges, is inserted in front of the two apertures contained in the mask covering the imaging lens [115]. The lens is focused on the object surface. This wavefront-splitting arrangement brings to focus at two different points on the image plane the two split diffused wavefronts emerging from a point on the object. In other words, the interferometer causes two laterally shifted speckle fields to interfere in the image plane. The magnitude and orientation of shear can be varied by appropriate control of the wedge plates mounted in front of the two apertures.

The object is illuminated by a plane parallel wave lying in the $x-z$ plane. When the object is deformed, the two shifted points on the object surface undergo different deformations. The mechanism of fringe formation is exactly

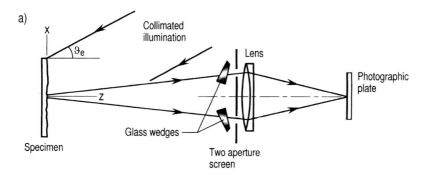

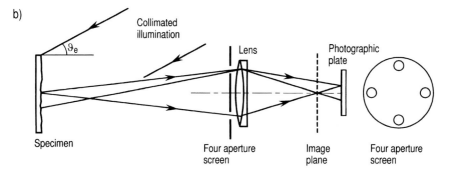

Fig. 45. Schematic of (a) two-aperture, and (b) defocused, speckle shearing interferometers for measuring slopes

similar to that described for the two-aperture speckle interferometer, with the sole exception that it is the relative difference in optical phase between the lights scattered by the two neighbouring points which now causes a shift of the speckle grid. Under the small-angle approximation the relative phase change between the two points is given by (12). This equation shows that contrary to the two-aperture speckle interferometer, the sensitivity of the method is independent of the separation between the slots. If the object is illuminated at normal incidence, the fringe equation corresponding to slope contours simply becomes

$$\frac{\partial w}{\partial x} = \frac{n\lambda}{2\Delta x} \, . \tag{37}$$

These fringes represent contours of constant slope change. A variant of the method [116] consists of eliminating the glass plates from the optical system, and producing the shear by introducing a slight defocusing in the imaging system, as shown in Fig. 45b. The advantages of this variant are that its sensitivity can be varied during fringe readout and one can display fringes depicting the contours of the derivatives of out-of-plane displacements along any desired direction in the $x - y$ plane. The limitations of these methods have been described in the literature [96,117].

Several methods for measuring lateral, radial and rotational contour slopes of an object surface have been reported [120,121]. An example of a phase map corresponding to slope change variations of a curved surface using a two-beam illumination technique is shown in Fig. 46.

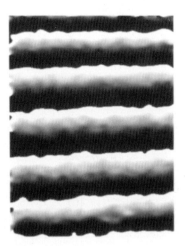

Fig. 46. Modulo-2π phase map corresponding to slope change variations of a curved surface

3 Conclusions

Several holograpic interferometers and speckle metrology techniques for use in experimental mechanics have been briefly described. The techniques presented here are only a sample of the procedures available in these fields. The use of hologram interferometery and speckle metrology has become important and even indispensable in some applications of nondestructive testing and inspection. The development leading to the integration of phase-stepping techniques and advanced computer-aided evaluation concepts to these methods promises to have a major impact on fringe processing in both static and dynamic environments. These developments augur well for the widespread use of these methods in diverse fields of engineering sciences.

References

1. R. K. Erf: *Holographic Non-Destructive Testing*, Academic, New York (1974)
2. C. M. Vest: *Holographic Interferometry*, Interscience, New York (1979)
3. Y. I. Ostrovsky, M. M. Butusov, G. V. Ostrovskaya: *Interferometry by Holography*, Springer, Berlin, Heidelberg (1980)
4. W. Schumann, M. Dubas: *Holographic Interferometry*, Springer, Berlin, Heidelberg (1979)
5. W. Schumann, J. P. Zürcher, D. Cuche: *Holography and Deformation Analysis*, Springer, Berlin, Heidelberg (1985)
6. Y. I. Ostrovsky, V. P. Schepinov, V. V. Yakovlev: *Holographic Interferometry in Experimental Analysis*, Springer, Berlin, Heidelberg (1991)
7. P. K. Rastogi: *Holographic Interferometry - Principles and Methods*, Springer, Berlin, Heidelberg (1994)
8. T. Kreis: *Holographic Interferometry - Principles and Methods*, Akademie, Berlin (1996)
9. N. Abramson: *The Making and Evaluation of Holograms*, Academic, New York (1981)
10. R. S. Sirohi, K. D. Hinsch: Selected Papers on Holographic Interferometry - Principles and Techniques, SPIE Milestone Ser. **144**, SPIE Optical Eng., Washington (1998)
11. A. E. Ennos: Speckle interferometry, in J. C. Dainty: Topics in Applied Physics **9**, Springer, Berlin, Heidelberg (1975), pp. 203–253
12. K. A. Stetson: A Review of Speckle Photography and Speckle Interferometry, Opt. Eng. **14**, 482–489 (1975)
13. A. E. Ennos: Speckle interferometry, in E. Wolf: Progress in Optics **16**, North-Holland, Amsterdam (1978) pp. 233–288
14. R. K. Erf: *Speckle Metrology*, Academic, New York (1978)
15. K. A. Stetson: Miscellaneous topics in speckle metrology, in R. K. Erf: *Speckle Metrology*, Academic, New York, (1978) pp. 295–320
16. R.Jones, C. Wykes: Holographic and speckle interferometry, Cambridge Univ. Press, London (1983)
17. R. S. Sirohi: Selected Papers on Speckle Metrology, SPIE Milestone Ser. **35**, SPIE Optical Eng., Washington, DC (1991)

18. R. S. Sirohi: Speckle Metrology, Dekker, New York (1993)
19. P. Meinlschmidt, K. D. Hinsch, R. S. Sirohi: Selected Papers on Speckle Pattern Interferometry – Principles and Practice, SPIE Milestone Ser. **132**, SPIE Optical Eng., Washington, DC (1996)
20. D. Gabor: A New Microscopic Principle, Nature **161**, 777–778 (1948)
21. E. N. Leith, J. Upatnieks: Wavefront Reconstruction with Diffused Illumination and Three-Dimensional Objects, J. Opt. Soc. Am. **54**, 1295–1301 (1964)
22. P. Hariharan: *Optical Holography*, Cambridge Univ. Press., London (1985)
23. N. Abramson: Sandwich Hologram Interferometry: A New Dimension in Holographic Comparison, Appl. Opt. **13**, 2019–2025 (1974)
24. J. D. Trolinger: Application of Generalized Phase Control During Reconstruction of Flow Visualization Holography, Appl. Opt. **18**, 766–774 (1979)
25. R. Dändliker: Two-Reference-Beam Holographic Interferometry, in P. K. Rastogi (Ed.): *Holographic Interferometry - Principles and Methods*, Springer, Berlin, Heidelberg (1994) pp. 75–108
26. D. B. Neumann, R. C. Penn: Object Motion Compensation Using Reflection Holography, J. Opt. Soc. Am. **62**, 1373 (1972)
27. J. P. Waters: Holography, in R. K. Erf (Ed.): *Holographic Nondestructive Testing*, Academic, New York (1974) pp. 5–59
28. U. Schnars, W. Jüptner: Direct Recording of Holograms by a CCD Target and Numerical Reconstruction, Appl. Opt. **33**, 179–181 (1994)
29. G. Pedrini, Y. L. Zou, H. J. Tiziani: Digital Double Pulse Holographic Interferometry for Vibration Analysis, J. Mod. Opt. **42**, 367–374 (1995)
30. P. Hariharan, B. F. Oreb, N. Brown: A Digital Phase-Measurement System for Realtime Holographic Interferometry, Opt. Commun. **51**, 393–396 (1982)
31. K. Creath: Phase-Measurement Interferometry Techniques, in E. Wolf, (Ed.): Prog. Opt. **26**, Elsevier, Amsterdam (1988) pp. 349–393
32. D. W. Robinson, G. T. Reid: *Interferogram Analysis - Digital Fringe Pattern Measurement Techniques*, Institute of Physics Publishing, Bristol (1993)
33. K. Creath: Phase-Shifting Holographic Interferometry, in P. K. Rastogi, (Ed.): *Holographic Interferometry - Principles and Methods*, Springer, Berlin, Heidelberg (1994) pp. 109–150
34. J. E. Greivenkamp, J. H. Bruning: Phase Shifting Interferometry, in D. Malacara (Ed.): *Optical Shop Testing,* Wiley, New York (1992) pp. 501–598
35. K. A. Stetson, W. R. Brohinsky: Electro-Optic Holography and its Application to Hologram Interferometry, Appl. Opt. **24**, 3631–3637 (1985)
36. K. A. Stetson, W. R. Brohinsky: Electro-Optic Holography System for Vibration Analysis and Nondestructive Testing, Opt. Eng. **26**, 1234–1239 (1987)
37. R. J. Pryputniewicz, K. A. Stetson: Measurement of Vibration Patterns Using Electro-Optic Holography, SPIE Proc. **1162**, 456–467 (1989)
38. K. A. Stetson: Phase-Step Interferometry of Irregular Shapes by Using an Edge-Following Algorithm, Appl. Opt. **31**, 5320–5325 (1992)
39. K. A. Stetson: A Rigorous Theory of the Fringes of Hologram Interferometry, Optik **29**, 386–400 (1969)
40. M. Dubas, W. Schumann: On Direct Measurement of Strain and Rotation in Holographic Interferometry Using the Line of Complete Localization, Opt. Acta **22**, 807–819 (1975)
41. C. A. Sciamarella, J. A. Gilbert: A Holographic-Moiré Technique to Obtain Separate Patterns for Components of Displacement, Exp. Mech. **16**, 215–220 (1976)

42. P. K. Rastogi, M. Spajer, J. Monneret: In-Plane Deformation Measurement Using Holographic Moiré, Opt. Lasers Eng. **2**, 79–103 (1981)
43. P. K. Rastogi, E. Denarić: Visualization of In-Plane Displacement Fields by Using Phase-Shifting Holographic Moiré: Application to Crack Detection and Propagation, Appl. Opt. **31**, 2402–2404 (1992)
44. P. K. Rastogi: Holographic In-Plane Measurement Using Reference-Wave Reconstruction: Phase Stepping and Application to a Deformation Problem, Appl. Opt. **34**, 7194–7196 (1995)
45. L. Pirodda: Conjugate Wave Holographic Interferometry for the Measurement of In-Plane Deformations, Appl. Opt. **26**, 1842–1844 (1989)
46. P. K. Rastogi: A Real-Time Holographic Moiré Technique for the Measurement of Slope Change, Opt. Acta **31**, 159–167 (1984)
47. P. K. Rastogi: Visualization and Measurement of Slope and Curvature Fields Using Holographic Interferometry: An Application to Flaw Detection, J. Mod. Opt. **38**, 1251–1263 (1991)
48. D. B. Neumann: Comparative Holography: A Technique for Eliminating Background Fringes in Holographic Interferometry, Opt. Eng. **24**, 625–627 (1985)
49. Z. Füzessy, F. Gyimesi: Difference Holographic Interferometry : Technique for Optical Comparison, Opt. Eng. **32**, 2548–2556 (1993)
50. P. K. Rastogi: Comparative Holographic Interferometry: A Nondestructive Inspection System for Detection of Flaws, Exp. Mech. **25**, 325–337 (1985)
51. P. K. Rastogi: Comparative Phase Shifting Holographic Interferometry, Appl. Opt. **30**, 722–728 (1991)
52. P. K. Rastogi: Direct and Real-Time Holographic Monitoring of Relative Changes in Two Random Rough Surfaces, Phys. Rev. A **50**, 1906–1908 (1994)
53. K. A. Stetson, R. L. Powell: Interferometric Hologram Evaluation and Real-Time Vibration Analysis of Diffuse Objects, J. Opt. Soc. Am. **55**, 1694–1695 (1965)
54. R. J. Pryputniewicz: Time-Average Holography in Vibration Analysis, Opt. Eng. **24**, 843–848 (1985)
55. G. C. Brown, R. J. Pryputniewicz: Holographic Microscope for Measuring Displacements of Vibrating Microbeams using Time-Averaged, Electro-optic Holography, Opt. Eng. **37**, 1398–1405 (1998)
56. P. Hariharan, B. F. Oreb: Stroboscopic Holographic Interferometry: Applications of Digital Techniques, Opt. Commun. **59**, 83–86 (1986)
57. S. Nakadate: Vibration Measurement Using Phase-Shifting Time-Average Holographic Interferometry, Appl. Opt. **25**, 4155–4161 (1986)
58. K. A. Stetson, J. Wahid: Real-Time Phase Imaging for Nondestructive Testing, Exp. Tech. **22**, 15–17 (1998)
59. C. S. Vikram: Study of Vibrations, in P. K. Rastogi (Ed.): *Holographic Interferometry - Principles and Methods* Springer, Berlin, Heidelberg (1994) pp. 293–318
60. S. Urgela: Grading of Wooden Plates for Musical Instrument making by means of Holographic Interferometry, Opt. Eng. **37**, 2108–2118 (1998)
61. K. A. Stetson: The Use of an Image Derotator in Hologram Interferometry and Speckle Photography of Rotating Objects, Exp. Mech. **18**, 67–73 (1978)
62. M. A. Beek: Pulsed Holographic Vibration Analysis on High-Speed Rotating Objects: Fringe Formation, Recording Techniques, and Practical Applications, Opt. Eng. **31**, 553–561 (1992)

63. J. D. Trolinger, J. C. Hsu: Flowfield Diagnostics by Holographic Interferometry and Tomography, in W. Jüptner and W. Osten (Ed.): *Fringe'93*, Akademie, Berlin (1993) pp. 423–439
64. R. J. Parker, D. G. Jones: The Use of Holographic Interferometry for Turbomachinery Fan Evaluation During Rotating Tests, J. Turbomach. **110**, 393–399 (1988)
65. S. M. Tieng, W. Z. Lai: Temperature Measurement of Reacting Flowfield by Phase-Shifting Holographic Interferometry, J. Thermophys. Heat Trans. **6**, 445–451 (1992)
66. S. P. Sharma, S. M. Ruffin: Density Measurements in an Expanding Flow Using Holographic Interferometry, J. Thermophys. Heat Trans. **7**, 261–268 (1993)
67. T. A.W.M. Lanen: Digital Holographic Interferometry in Flow Research, Opt. Commun. **79**, 386–396 (1990)
68. J. R. Varner: Holographic Contouring Methods, in H. J. Caulfield (Ed.): *Handbook of Optical Holography*, Academic, New York (1979) pp. 595–600
69. N. Abramson: Sandwich Hologram Interferometry 3: Contouring, Appl. Opt. **15**, 200–205 (1976)
70. P. K. Rastogi, L. Pflug: A Fresh Approach of Phase Management to Obtain Customized Contouring of Diffuse Object Surfaces of Broadly Varying Depths Using Real-Time Holographic Interferometry, J. Mod. Opt. **37**, 1233–1246 (1990)
71. P. K. Rastogi, L. Pflug: A Holographic Technique Featuring Broad Range Sensitivity to Contour Diffuse Objects, J. Mod. Opt. **38**, 1673–1683 (1991)
72. P. Carelli, D. Paoletti, G. S. Spagnolo: Holographic Contouring Method: Application to Automatic Measurements of Surface Defects in Artwork, Opt. Eng. **30**, 1294–1298 (1991)
73. P. K. Rastogi, L. Pflug: Measurement of Large Out-of-Plane Displacements Using Two Source Holographic Interferometry, J. Mod. Opt. **41**, 589–594 (1994)
74. E. Archbold, A. E. Ennos: Displacement Measurement from Double Exposure Laser Photographs, Opt. Acta **19**, 253–271 (1972)
75. R. P. Khetan, F. P. Chiang: Strain Analysis by One-Beam Laser Speckle Interferometry 1: Single Aperture Method, Appl. Opt. **15**, 2205–2215 (1976)
76. P. K. Rastogi (Ed.): Special Issue on Speckle Photography, Opt. Lasers Eng. **29**, 81–225 (1998)
77. P. Jacquot, P. K. Rastogi: Influence of Out-of-Plane Deformation and its Elimination in White Light Speckle photography, Opt. Lasers Eng. **2**, 33–55 (1981)
78. H. J. Tiziani, K. Leonhardt, J. Klenk: Real Time Displacement and Tilt Analysis by a Speckle Technique Using $Bi_{12}SiO_{20}$ Crystals, Opt. Commun. **34**, 327–331 (1980)
79. K. Nakagawa, T. Takatsuji, T. Minemoto: Measurement of the Displacement Distribution by Speckle Photography Using BSO Crystal, Opt. Commun. **76**, 206–212 (1990)
80. N. Krishna Mohan, J. S. Darlin, M. H. Majles Ara, M. P. Kothiyal, R. S. Sirohi: Speckle Photography with $BaTiO_3$ Crystal for the Measurement of In-Plane Displacement Field distribution of distant Objects, Opt. Lasers Eng. **29**, 211–216 (1998)
81. G. B. Smith, K. A. Stetson: Heterodyne Read-Out of Specklegram Halo Interference Fringes, Appl. Opt. **19**, 3031–3033 (1980)

82. J. S. Kim, S. Musazzi, V. Perini, M. Giglio: Heterodyne Read-Out System for Dual Plate Speckle Photography: Analysis of Error Sources and Performance Evaluation, Appl. Opt. **28**, 1862–1868 (1989)
83. G. H. Kaufmann, P. Jacquot: Phase Shifting of Whole Field Speckle Photography Fringes, Appl. Opt. **29**, 3570–3572 (1990)
84. D. W. Robinson: Automatic Fringe Analysis With a Computer Image-Processing System, Appl. Opt. **22**, 2169–2176 (1983)
85. R. Erbeck: Fast Image Processing With a Microcomputer Applied to Speckle Photography, Appl. Opt. **24**, 3838–3841 (1985)
86. J. M. Huntley: An Image Processing System for the Analysis of Speckle Photographs, J. Phys. E. **19**, 43–49 (1986)
87. F. Ansari, G. Ciurpita: Automated Fringe Measurement in Speckle Photography, Appl. Opt. **26**, 1688–1692 (1987)
88. F. P. Chiang, R. M. Juang: Laser Speckle Interferometry for Plate Bending Problems, Appl. Opt. **15**, 2199–2204 (1976)
89. H. J. Tiziani: Vibration Analysis and Deformation Measurement, in R. K. Erf (Ed.): *Speckle Metrology*, Academic, New York (1978) pp. 73–110
90. D. A. Gregory: Topological Speckle and Structural Inspection, in R. K. Erf (Ed.): *Speckle Metrology*, Academic, New York (1978) pp. 183–223
91. K. A. Stetson, I. R. Harrison: Determination of the Principle Surface Strains on Arbitrary Deformed Objects Via Tandem Speckle Photography, in: *Proc. 6th Int. Conf. on Exp. Stress Analysis*, Düsseldorf, pp. 149–154 (1978)
92. P. K. Rastogi: Speckle Shearing Photography: A Tool for Direct Measurement of Surface Strains, Appl. Opt. **37**, 1292–1298 (1998)
93. P. K. Rastogi: Determination of Surface Strains by Speckle Shear Photography, Opt. Lasers Eng. **29**, 103–116 (1998)
94. J. A. Leendertz: Interferometric Displacement Measurement on Scattering Surfaces Utilizing Speckle Effect, J. Phys. E. **3**, 214–218 (1970)
95. K. A. Stetson: Analysis of Double Exposure Photography with Two Beam Illumination, J. Opt. Soc. Am. **64**, 857–861 (1974)
96. P. K. Rastogi, P. Jacquot: Speckle Metrology Techniques: A Parametric Examination of the Observed Fringes, Opt. Eng. **21**, 411–426 (1982)
97. O. J. Lokberg: Recent Developments in Video Speckle Interferometry, in R. S. Sirohi (Ed.): *Speckle Metrology*, Dekker, New York (1993) pp. 157–194
98. P. K. Rastogi (Ed.): Special Issue on Speckle and Speckle Shearing Interferometry 1, Opt. Lasers Eng. **26**, 83–278 (1997)
99. P. K. Rastogi (Ed.): Special Issue on Speckle and Speckle Shearing Interferometry 2, Opt. Lasers Eng. **26**, 279–460 (1997)
100. D. E. Duffy: Measurement of Surface Displacement Normal to the Line of Sight, Exp. Mech. **14**, 378–384 (1974)
101. P. K. Rastogi, P. Jacquot: Measurement of Difference Deformation Using Speckle Interferometry, Opt. Lett. **12**, 596–598 (1987)
102. A. R. Ganesan, C. Joenathan, R. S. Sirohi: Real-time Comparative Digital Speckle Pattern Interferometry, Opt. Commun. **64**, 501–506 (1987)
103. S. Nakadate, T. Yatagai, H. Saito: Electronic Speckle Pattern Interferometry Using Digital Image Processing Techniques, Appl. Opt. **19**, 1879–1883 (1980)
104. C. Wykes: Use of Electronic Speckle Pattern Interferometry (ESPI) in the Measurement of Static and Dynamic Surface Displacements, Opt. Eng. **21**, 400–406 (1982)

105. K. Creath: Phase-Shifting Speckle Interferometry, Appl. Opt. **24**, 3053–3058 (1985)
106. D. W. Robinson, D. C. Williams: Digital Phase-Stepping Speckle Interferometry, Opt. Commun. **57**, 26–30 (1986)
107. F. M. Santoyo, M. C. Shellabear, J. R. Tyrer: Whole Field In-Plane Vibration Analysis Using Pulsed Phase-Stepped ESPI, Appl. Opt. **30**, 717–721 (1991)
108. K. Creath: Averaging Double-Exposure Speckle Interferograms, Opt. Lett. **10**, 582–584 (1985)
109. R. Höfling, W. Osten: Displacement Measurement by Image-Processed Speckle Patterns, J. Mod. Opt. **34**, 607–617 (1987)
110. Y. Y. Hung: Displacement and Strain Measurement, in R. K. Erf (Ed.): *Speckle Metrology*, Academic, New York (1978) pp. 51–71
111. R. S. Sirohi: Speckle Methods in Experimental Mechanics, in R. S. Sirohi (Ed.): *Speckle Metrology*, Dekker, New York (1993) pp. 99–155
112. C. Joenathan, B. Pfister, H. J. Tiziani: Contouring by Electronic Speckle Pattern Interferometry Employing Dual Beam Illumination, Appl. Opt. **29**, 1905–1911 (1990)
113. J. A. Leendertz , J. N. Butters: An Image-Shearing Speckle Pattern Interferometer for Measuring Bending Moments, J. Phys. E. **6**, 1107–1110 (1973)
114. Y. Y. Hung, C. E. Taylor: Speckle-Shearing Interferometric Camera - A Tool for Measurement of Derivatives of Surface Displacements, SPIE Proc. **41**, 169–175 (1973)
115. Y. Y. Hung, C. Y. Liang: Image-Shearing Camera for Direct Measurement of Surface Strains, Appl. Opt. **18**, 1046–1051 (1979)
116. Y. Y. Hung, I. M. Daniel, R. E. Rowlands: Full-Field Optical Strain Measurement having Post-Recording Sensitivity and Direction Selectivity, Exp. Mech. **18**, 56–60 (1978)
117. M. Keinan, J. Politch: Second Order Phenomena in Strain Measurement by Speckle Techniques, Opt. Lasers Eng. **15**, 149–182 (1991)
118. O. M. Peterson: Digital Speckle Pattern Shearing Interferometry: Limitations and Prospects, Appl. Opt. **30**, 2730–2738 (1991)
119. D. K. Sharma, R. S. Sirohi, M. P. Kothiyal: Simultaneous Measurement of Slope and Curvature with a Three-Aperture Speckle Shearing Interferometer, Appl. Opt. **23**, 1542–1546 (1984)
120. C. Joenathan, C. S. Narayanmurthy, R. S. Sirohi: Radial and Rotational Slope Contours in Speckle Shear Interferometry, Opt. Commun. **56**, 309–312 (1986)
121. P. K. Rastogi: An Electronic Pattern Speckle Shearing Interferometer for the Measurement of Surface Slope Variations of Three-Dimensional objects, Opt. Lasers Eng. **26**, 93–100 (1997)

Moiré Methods for Engineering and Science – Moiré Interferometry and Shadow Moiré

Daniel Post[1], Bongtae Han[2], and Peter G. Ifju[3]

[1] Professor Emeritus, Virginia Polytechnic Institute and State University
Department of Engineering Science and Mechanics
1007 Horseshoe Lane, Blacksburg, VA 24060 USA
danpost@vt.edu
[2] University of Maryland, Department of Mechanical Engineering
College Park, MD 20742 USA bthan@eng.umd.edu
[3] University of Florida
Aerospace Engineering, Mechanics, and Engineering Science
Gainesville, FL 32611 USA
pgi@aemes.aero.ufl.edu

Abstract. Moiré interferometry and shadow moiré are extraordinarily versatile and effective methods for determining in-plane and out-of-plane displacement fields, respectively. The basic concepts are reviewed for both methods, topics on practice and analysis are addressed, and numerous examples of important applications are presented.

The moiré data are received as whole-field fringe patterns, or contour maps, of displacements. For moiré interferometry with the typical reference grating frequency of 2400 lines/mm, the contour interval is 0.417 µm per fringe order; the sensitivity is 2.4 fringes per µm displacement. Orthogonal U and V displacements are measured, and normal and shear strains are determined from these in-plane displacement fields. For microscopic moiré interferometry, sensitivity corresponding to 17 nm per fringe contour has been achieved by means of the optical/digital fringe multiplication algorithm.

The patterns of moiré interferometry are characterized by excellent fringe contrast and spatial resolution, including patterns from complex applications. The applications reviewed here address laminated composites, including the study of free-edge effects along the cylindrical surface of holes in laminated plates; thermal deformation of microelectronics devices; the damage wake along a crack path; and a micromechanics study of grain deformations in titanium.

The examples of shadow moiré show the out-of-plane displacements W for pre-buckling and post-buckling of columns; and W displacements of electronic packages subjected to temperature changes. Phase-stepping analyses were used for the electronic packages to increase sensitivity, providing 12.54 µm per fringe contour. Since W is typically much larger than U and V, the sensitivity of shadow moiré can be adjusted to serve broad categories of engineering applications.

1 Introduction

Moiré fringes are formed by the superposition of two gratings, i.e. two arrays of (nearly) uniformly spaced lines. The moiré fringes are contour maps of the difference between the gratings, and as such, they have been attractive in experimental solid mechanics for the determination of surface deformations. For in-plane displacement measurements, the technology has evolved beautifully from low-sensitivity geometric moiré to the powerful capabilities of *moiré interferometry*. Moiré measurements are performed routinely in the interferometric domain with fringes representing subwavelength displacements per contour. Since moiré responds only to geometric changes, it is equally effective for elastic, viscoelastic, and plastic deformations, for isotropic, orthotropic and anisotropic materials, and for mechanical, thermal, and dynamic loadings.

The technique called *shadow moiré* has evolved as the moiré method most widely chosen for out-of-plane measurements. Although its sensitivity is lower, engineering applications seldom require the subwavelength sensitivity of interferometric methods for out-of-plane measurements. Instead, the low sensitivity, plus the recently achieved intermediate sensitivity of shadow moiré, satisfies nearly all the practical applications.

Consequently, this review on moiré methods in engineering and science will be devoted to the two moiré methods: moiré interferometry and shadow moiré. These techniques are treated in greater detail in [1]. Additional moiré techniques are covered in the recent literature by *Patorski* [2], *Kafri* and *Glatt* [3], *Cloud* [4], and others.

Moiré interferometry and shadow moiré have become extremely important tools in the electronics industry for electronic packaging studies. Their application–mostly to evaluate thermal strains, but also for mechanical loading and material characterization–is introduced at design and development, evaluation, and process control stages. Applications for the study of composite materials and components are extensive and increasing, but not yet rivaling the electronics packaging activity. Other areas of strong and growing interest include fracture mechanics, biomechanics, rheology of plastics, metallurgy and ceramic science. Both the scope and volume of analyses by these two techniques are growing together with the increasingly complex developments in engineering and science, and in juxtaposition and support of the increasingly complex analyses by numerical computer studies.

2 Moiré Interferometry

2.1 Basic Concepts

The general scheme of moiré interferometry is illustrated in Fig. 1. A high-frequency cross-line grating on the specimen, initially of frequency f_s, deforms together with the specimen. A parallel (collimated) beam, B_1, of laser light strikes the specimen and a portion is diffracted back, nominally perpendicular to the specimen, in the +1 diffraction order of the specimen grating. Light from the mutually coherent collimated beam B_2 is diffracted back in its −1 order. Since the specimen grating is deformed as a result of the applied loads,

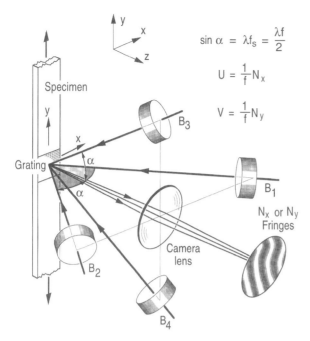

Fig. 1. Schematic illustration of four-beam moiré interferometry to record the N_x and N_y fringe patterns, which depict the U and V displacement fields. (Figure partly or wholly extracted from [1] with permission of Springer-Verlag; subsequent acknowledgments indicated by [1].)

these diffracted beams are no longer collimated. Instead, they are beams with warped wavefronts, where the warpages are related to the deformation of the grating. These two coherent beams interfere in the image plane of the camera lens, producing an interference pattern of dark and light bands, which is the N_x moiré pattern.

Similarly, mutually coherent collimated beams B_3 and B_4, centered in the vertical plane, are diffracted in $+1$ and -1 diffraction orders by the nominally horizontal lines of the deformed specimen grating. These two diffracted beams interfere to produce the N_y moiré pattern. In practice, beams B_1 and B_2 are blocked, so the N_y fringes are viewed alone. Alternatively, B_3 and B_4 are blocked to view the N_x fringes.

These moiré patterns are contour maps of the U and V displacement fields, i.e. the displacements in the x and y directions, respectively, of each point in the specimen grating. The relationships, for every x, y point in the field of view, are

$$U(x,y) = \frac{1}{2f_s} N_x(x,y),$$
$$V(x,y) = \frac{1}{2f_s} N_y(x,y). \tag{1}$$

In the moiré interferometry applications shown in this review, $f_s = 1200$ lines/mm (30480 lines/in.). In the fringe patterns, the contour interval is $1/2f_s$, which is 0.417 µm displacement per fringe order. The sensitivity is its reciprocal, 2.4 fringes per micrometer displacement. For microscopic moiré interferometry, described later, a sensitivity of 57.6 fringe contours per micrometer displacement has been achieved.

The basis of moiré interferometry is described above by this sequence: (1) two beams are incident upon the specimen grating, (2) diffraction of each beam by the deformed grating produces two warped wavefronts emerging from the specimen, and (3) the coherent addition of these two beams produces the moiré pattern by constructive and destructive interference. This is an excellent physical explanation. It is consistent with the mathematical derivation, which defines the intensity distribution in the moiré pattern and its relationship (1) to the fringe orders [1].

However, another physical explanation is equally compelling-perhaps it is more closely related to our experience and intuition, and perhaps it is more helpful for understanding the relationship between the deformation and the moiré pattern. It is very simple.

Figure 2a illustrates unobstructed beams B_1 and B_2 in a two-dimensional cross-sectional view. Assuming the two beams have propagated equal distances from the source to point a, constructive interference occurs along the plane of symmetry, ab. The path lengths between the source and any point in this plane are equal for the two beams. Along an adjacent plane, the path lengths of B_1 and B_2 differ by $\lambda/2$ at every point, so destructive interference – or the absence of light – occurs in that plane. Along another adjacent plane, the path difference is λ, creating a plane of constructive interference, etc. Accordingly, the volume of space where beams B_1 and B_2 coexist is filled with a regular array of planes of constructive interference, alternating with planes of destructive interference. These represent bright regions of light separated by regions of darkness. The volume of space is said to be filled with *walls of interference*.

In Fig. 2b, the walls are cut by plane AA, which becomes illuminated by an array of bright bands separated by dark bands. The illumination resembles a bar-and-space grating, and it is fully analogous to the reference grating of geometrical moiré. In moiré interferometry, it functions as a reference grating. There is no physical reference grating in moiré interferometry, but the two incident beams create a *virtual reference grating*. The volume of space where the two beams coexist is called a *virtual grating* and the array of bright and dark bands that illuminates the specimen is called a virtual reference grating.

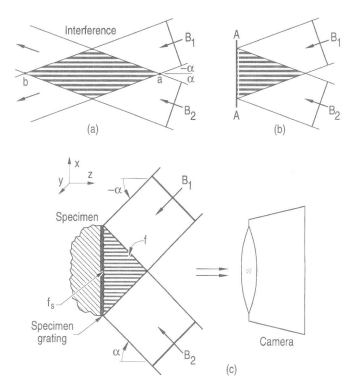

Fig. 2. Schematic diagrams illustrating (**a**) constructive and destructive interference at the intersection of two coherent beams; (**b**) a virtual reference grating on plane AA; and (**c**) the interaction of a virtual reference grating and a specimen grating, which creates a moiré pattern [1]

In the second physical description of moiré interferometry, the virtual reference grating and the deformed specimen grating interact to form the moiré pattern, which is recorded by a camera. The system is illustrated in Fig. 2c. The effect is analogous to geometrical moiré, but since it has been proved that the equations of moiré interferometry and geometrical moiré are identical [5], the description is fully justified.

The frequency, f, of the virtual reference grating is determined by the angle α and wavelength λ by

$$f = \frac{2}{\lambda} \sin \alpha \ . \tag{2}$$

For the macromechanics applications shown in this review, $f = 2400$ lines/mm (60960 lines/in.). Moiré interferometry utilizes a virtual reference grating of twice the initial frequency of the specimen grating, i.e.

$$f = 2f_s \ . \tag{3}$$

This corresponds to moiré fringe multiplication by a factor of two [6]. The resulting sensitivity is determined by the reference grating, corresponding to geometrical moiré with gratings of frequency f.

Again, the second physical explanation is consistent with the mathematical derivation, in which the intensity distribution in the moiré pattern, and also (1), are derived [1]. Fringe order N_x determines the in-plane displacements U at each point in the field, uniquely; although the specimen and specimen grating experiences V and W displacements (where W is the out-of-plane displacement), the N_x pattern is a simple function of U, alone. Similarly, the mathematical derivation proves that the N_y pattern is a function of V, independent of U and W.

2.2 Equipment

2.2.1 Moiré Interferometers

Any configuration of optical and mechanical components that produces the four beams illustrated in Fig. 1 can be used. Of course, each pair of beams must be mutually coherent. Several configurations are illustrated and described in [1]. Fine adjustments of beam direction are required to precisely control the initial null field and, when desired, the carrier fringes. An extremely important feature is robust construction to minimize sensitivity to external disturbances – primarily vibrations from mechanical and acoustical sources, and secondarily, air currents and thermal disturbances. Compact construction is an asset in this regard.

2.2.2 Load Application

The deformation to be measured might be induced by mechanical loading, temperature changes, water or chemical absorption, etc. For mechanical loading, it is often most effective to construct a compact loading fixture for each class of tests under consideration. Examples for tension, compression and shear are given in [1]. Special care should be taken to avoid unwanted in-plane rotation of the specimen during load application. Numerous fringes can be induced by this rigid-body rotation because of the high sensitivity to all in-plane displacements. Consequently, loading fixtures should be designed to minimize unwanted lateral motion when axial forces are applied.

Moiré interferometers have been used successfully on universal testing machines. Special fixtures on the columns or cross-heads must be constructed to attach the interferometer to the testing machine. On some relatively quiet machines, the fringe patterns can be recorded during load application. On others, it has been necessary to stop the machine at different load levels to record the progress of the deformation. In general, screw-type testing machines have exhibited less mechanical noise than hydraulic machines.

In real-time thermal loading, the specimen is supported in an oven or environmental chamber, which is positioned directly in front of the interferometer. Again, if the apparatus incorporates a forced-air blower (or fan), it might be necessary to stop the blower prior to recording the fringe patterns. Usually, the specimen is placed close to the observation window, so that the optical path lengths within the chamber are small. Use of a vacuum chamber to eliminate movement of the air has not proved to be necessary for temperatures below 150° C.

A very simple technique for measuring thermal deformation induced by a single temperature increment is described below under the heading *Bithermal Loading*.

2.3 Specimen Gratings

The bar-and-space gratings of geometrical moiré cannot be printed with very high frequencies. Instead, phase gratings are used, which means that the grating surface consists of a regular array of hills and valleys. The specimen grating is usually applied by a replication process, illustrated by cross-sectional views in Fig. 3. A special mold is used, which is a plate with a cross-line phase grating on its surface. The grating is overcoated with a highly reflective metallic film, usually evaporated aluminum. A small pool of liquid adhesive is

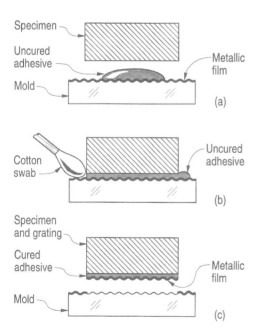

Fig. 3a–c. Steps in producing the specimen grating by a casting or replication process; the reflective metallic film is transferred to the specimen grating [1]

poured on the mold, and the specimen is pressed into the pool to spread the adhesive into a thin film. Excess adhesive is cleaned off repeatedly as it flows out. The mold is prized off after the adhesive has hardened. The weakest interface is between the metallic film and the cross-line grating, so the film is transferred to the specimen. Thus, a thin, highly reflective cross-line grating is firmly attached to the specimen surface, such that it deforms together with the specimen surface. Details can be found in Chap. 4 and Appendix C of [1].

The adhesive thickness is typically about 25 μm (0.001 in.) for larger specimens – of area greater than 300 mm^2 (0.5 in.^{2}) – and about 2 μm (80 μin.) for small specimens. For most analyses the thickness and stiffness of the grating is negligible. Various room temperature curing adhesives can be used, including epoxies, acrylics, urethanes and silicone rubbers. Recent reports of success with instant cyanoacrylate cements have been circulated. Adhesives that cure by exposure to ultraviolet light have been used successfully.

Special techniques have been developed for replicating specimen gratings on electronic packages, to cope with the small size and tiny openings [7].

2.4 Bithermal Loading (Isothermal Loading)

Thermal deformations can be analyzed by room temperature observations. In this technique, the specimen grating is applied at an elevated temperature and allowed to cool to room temperature before being observed in the moiré interferometer. Thus, the deformation incurred by the temperature increment is locked into the grating and recorded at room temperature [1,8].

A typical temperature increment is 100° C, in which the grating is applied at about 120° C and observed at about 20° C. An adhesive that cures at elevated temperature is used, usually an epoxy. The specimen and mold are preheated to the application temperature, and the adhesive is applied and allowed to cure at the elevated temperature. The mold is a grating on a zero-expansion substrate, so its frequency is the same at elevated and room temperatures. Otherwise, a correction is required for the thermal expansion of the mold.

These measurements can also be performed at cryogenic temperatures. In one test, the specimen grating was applied at −40° C using an adhesive that cured in ultraviolet light [1].

This technique has come to be known as *isothermal loading* in recent years, but the name implies constant temperature and therefore is not descriptive. We propose to call the technique *bithermal loading*, implying two discrete temperatures.

2.5 Fringe Counting

The assignment of fringe orders is treated in more detail in [1]. Because deformations are determined by the relative displacements of each point in a body, the zero-order fringe is arbitrary – the zeroth order can be assigned

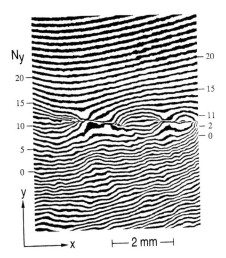

Fig. 4. Weld defects in a stainless steel tension specimen. $f = 2400$ lines/mm. (Courtesy of *S. A. Chavez*) [1]

to any fringe in the moiré pattern. In Fig. 4, it is assigned in the lower left region of the specimen. Of course, at every point along this continuous fringe, $N_y = 0$.

Figure 4 shows the deformation of a tensile coupon cut from a stainless steel plate in the region of a weld. Cracks appear in the weld, but along the left edge of the specimen the material is continuous. Since the strain is tensile, or positive, the fringe orders along the left edge increase monotonically in the $+y$ direction, as shown. The fringe orders can therefore be assigned at every point in the field by following continuous fringes across the field. Fringe orders along the right edges were assigned accordingly.

Where a crack is present, the V displacements are different along the upper and lower lips of the crack. This accounts for the crack opening that results from the tensile load. For example, at the right edge the fringe order changes from $N_y = 3$ to $N_y = 10$, indicating a crack-opening displacement of 7 fringe orders, or 2.9 μm (115 μin.).

Clues derived from known loading conditions and known specimen geometry are often sufficient to establish the sign of the fringe gradient at every point, i.e. to establish whether the fringe order is increasing or decreasing as the count progresses from one fringe to the next neighboring fringe. Occasionally the clues might not be sufficient, but there is always a simple experimental way to determine the sign of the fringe gradient [1].

2.6 Strain Analysis

Strains can be determined from the two displacement fields by the relationships for engineering strain

$$\varepsilon_x = \frac{\partial U}{\partial x} = \frac{1}{f}\left(\frac{\partial N_x}{\partial x}\right),$$

$$\varepsilon_y = \frac{\partial V}{\partial y} = \frac{1}{f}\left(\frac{\partial N_y}{\partial y}\right), \tag{4}$$

$$\gamma_{xy} = \frac{\partial U}{\partial y} + \frac{\partial V}{\partial x} = \frac{1}{f}\left(\frac{\partial N_x}{\partial y} + \frac{\partial N_y}{\partial x}\right), \tag{5}$$

where ε is the normal strain and γ is the shear strain at the surface of the specimen. Although it is not indicated here by the (x, y) suffix [shown in (1)], these equations apply for every point in the field. Thus, it is the fringe gradients that determine the strains, both normal and shear.

In principle, the exact differential can be extracted at any point by plotting a curve of fringe orders along a line through the point and measuring the slope of the curve at the point. Often, however, the finite increment approximation is sufficient, whereby (as an example) $\partial N_x/\partial x$ is taken to be equal to $\Delta N_x/\Delta x$. In that case, strain is determined by measuring Δx, the distance between neighboring fringes, in the immediate neighborhood of the point of interest.

Shear strains are determined as readily as normal strains. Numerous examples of fringe counting and strain analysis are given in [1]. In nearly all cases of strain analysis, the strains are sought at specific points (e.g. where the fringes are most closely packed, indicating strain maxima) or along specific lines. Manual methods and computer-assisted methods are of practical use for such cases.

2.7 Carrier Fringes

An array of uniformly spaced fringes can be produced by adjustment of the moiré interferometer, and these are called *carrier fringes*. With the specimen in the no-load condition, a carrier of extension is produced by changing angle α (in Fig. 1). These fringes are parallel to the lines of the reference grating, just like the fringes of a pure tensile (or compression) specimen. Carrier fringes of rotation are produced by a rigid-body rotation of the specimen or the interferometer, and these are perpendicular to the lines of the reference grating.

It is frequently valuable to modify the load-induced fringe patterns with carrier fringes. Figure 5 is an example, where the load-induced fringes are shown in (a). The specimen is cut from a graphite/epoxy composite material and is loaded in a compact shear fixture [1,9]. In (b), the fringes in the test section between notches are subtracted off by carrier fringes of rotation.

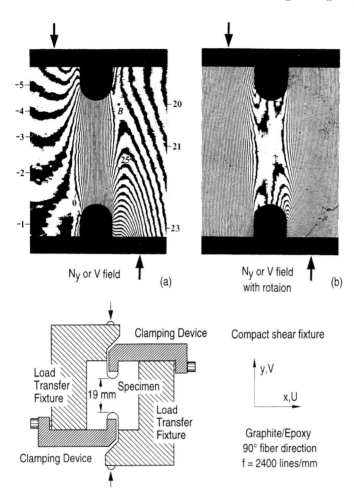

Fig. 5. N_y fringe patterns for a composite shear specimen (**a**) without rotation and (**b**) with rotation. Local irregularities in the fringes are caused by local variations in the composite material [1]

The carrier fringes are vertical and uniformly spaced, and their gradient is opposite to that of the shear-induced fringes. What is the benefit? It shows that the normal strain ε_y is zero along the vertical center line of the specimen, that ε_y is extremely small throughout the test zone, and that the shear deformation is nearly uniform over a wide test region.

Another compelling example is shown in Fig. 6 [1,10]. Here carrier fringes of extension are used to eliminate potential ambiguities of fringe ordering, and to enhance the ease and accuracy of strain analysis.

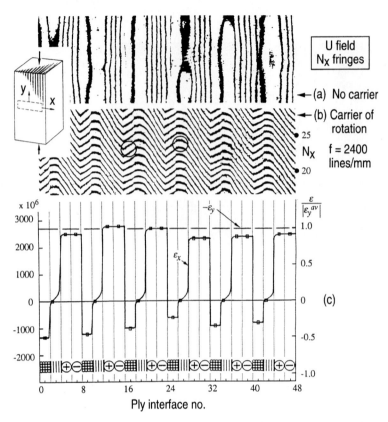

Fig. 6. The N_x moiré pattern in the region of the dashed box for a laminated composite in compression. Carrier fringes were applied to evaluate ε_x. The graph shows the transverse normal strains on a ply-by-ply basis [1]

2.8 Out-of-Plane Deformation

This is largely a pedagogical issue, inasmuch as the literature may not be sufficiently clear. Various publications show sensitivity to out-of-plane rotations about an axis parallel to the lines of the reference grating. However, the rigorous mathematical analysis of [1] proves that moiré interferometry is insensitive to all out-of-plane motions, including rotations. Where do the publications deviate?

Figure 7 should help clarify the issue. It represents the *walls of interference* of the virtual reference grating, where the plane of the undeformed specimen grating is abcd. Let P be a point fixed on the specimen grating, and let P move to another point in space, P′, when the specimen is deformed. The deformed surface surrounding P′ might have any slope, as indicated by the dashed box. The component of displacement in the x direction, i.e. U, cuts through walls of interference and therefore causes a change of fringe order in

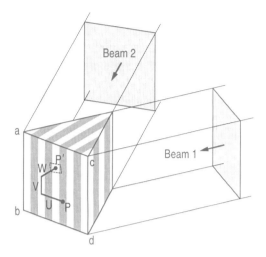

Fig. 7. Walls of interference of the virtual reference grating

the moiré pattern. For displacement components V and W, the point remains precisely in the same plane (within the wall of interference) as the end of the U component, and therefore these components have no effect on the fringe order. The moiré pattern remains insensitive to V and W regardless of the surface slope. This corroborates the proof that moiré interferometry senses only the component of displacement perpendicular to the walls of the virtual reference grating.

Engineering strain measured by moiré is the ratio of the local change of length to the original length, when both are projected onto the original specimen surface abcd. Thus, engineering strains are correctly determined by (4) and (5), regardless of surface slopes.

However, for exceptional studies in which the surface slope becomes large enough, the investigator might be interested in lengths measured on the specimen surface instead of its original plane abcd. This is where the publications deviate. In such cases, a mathematical transformation is required which involves additional variables, specifically the components of the surface slopes. These slopes must be measured (or otherwise determined) to augment the moiré data at each point where the strain is to be calculated. Thus, if the investigator feels the out-of-plane slopes are significant enough to degrade the accuracy that is sought for a specific analysis, then a Lagrangian strain or other three-dimensional strain should be calculated. Otherwise, the investigator accepts that the influence of the slope is negligible for calculating the strain.

Note that special apparatus and procedures may be required to record an image of the moiré fringes when surface slopes become large. Otherwise, the moiré pattern may display *black holes* in regions of large slopes [1]. In general, it is the engineering strain that is sought, not a transformed strain, and (4) and (5) are fully effective.

3 Advanced Techniques in Moiré Interferometry

3.1 Microscopic Moiré Interferometry

Special considerations arise for deformation measurements of tiny specimens or tiny regions of larger specimens. The relative displacements within a small field of view will be small (even if the strains are not small), and so the number of moiré fringes might not be enough for an accurate analysis. Perhaps the most important consideration, therefore, is the need for increased displacement sensitivity – enhanced sensitivity beyond the high sensitivity discussed above. In a method called *microscopic moiré interferometry*, the sensitivity is increased progressively by two techniques. The first is an *immersion interferometer*, whereby virtual reference gratings of 4800 lines/mm (122000 lines/in.) are produced in practice, thus doubling the usual basic sensitivity [11]. The second technique is optical/digital fringe multiplication (O/DFM), whereby fringe shifting and an efficient algorithm is used to generate an enhanced contour map of the displacement field; the map displays β times as many fringe contours as the original moiré pattern [12]. In practice, $\beta = 12$ has been achieved for microscopic moiré interferometry, which with the doubled sensitivity, represents a multiplication of 24.

A specific system used by the authors will be described briefly. The technique is described in much more detail in [1,13]. It is based on the premise that the moiré pattern encompassing the small field of view will contain only a few fringes. Accordingly, it is practical to record the pattern by a CCD camera. Good fringe resolution is preserved because the pattern is recorded with numerous pixels per fringe.

The apparatus is illustrated in Fig. 8. The specimen is mounted in a loading frame, which is fixed to a rotary table for angular adjustments. The imaging system comprises a long-working distance microscope lens and a CCD video camera (without lens). The interferometer is a compact four-beam unit that provides virtual reference gratings of 4800 lines/mm (122000 lines/in.). Its illumination is provided by optical fibers to two collimators. The microscope, interferometer and collimators are mounted on an x, y traverse that can move the system to observe any portion of the specimen. Additional equipment includes a personal computer with a frame-grabber board and TV monitors. A piezoelectric translation device is provided for phase stepping.

3.1.1 Immersion Interferometer

The immersion interferometer is illustrated in Fig. 9 in a cross-sectional view and a three-dimensional view. Portions of the collimated input beam intersect

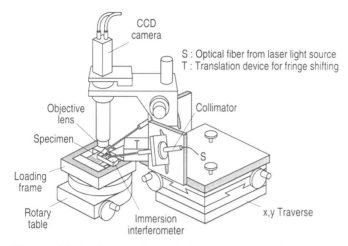

Fig. 8. Mechanical and optical arrangement for microscopic moiré interferometry [1]

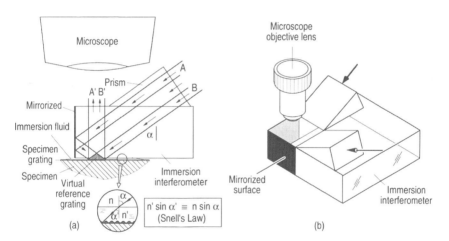

Fig. 9. (a) Optical paths in the immersion interferometer; (b) Immersion interferometer for U and V fields [1]

inside the glass medium to produce a virtual reference grating of frequency

$$f_\mathrm{m} = \frac{2n \sin \alpha}{\lambda}, \tag{6}$$

which exceeds the frequency produced in air by a factor of n, the refractive index of the glass. Thus, with $\alpha = 54.3°$, $n = 1.52$, and $\lambda = 514$ nm (in air), the virtual reference grating is 4800 lines/mm. The frequency is exactly the same in the immersion fluid – even when its index n' does not match n –

since the angle of refraction in the fluid compensates by Snell's law for any mismatch of the refractive index.

3.1.2 O/DFM

When the interferometer of Fig. 9a is translated horizontally by a fraction of the virtual reference grating pitch, g_m, the relative phase of rays A′ and B′ (from each point in the field) changes by the same fraction of 2π. Accordingly, when the interferometer is translated by g_m/β, the relative phase changes by $2\pi/\beta$ everywhere in the field of view. Thus, the fringe positions are shifted by any translation of the interferometer relative to the specimen, and a new intensity is present at every x, y point on the specimen.

In practice, the piezoelectric translation device is computer-controlled to move the interferometer by β equal steps of g_m/β each, where β is an even integer. For each step, the intensity at each pixel in the field is recorded by the CCD camera and compiled in a desktop computer. Thus, there are β intensities in the data bank for each pixel, all representing the same displacement field.

For each of the β patterns, there is a complementary pattern, i.e. another pattern with its phase shifted by π at every point in the field. The O/DFM algorithm is depicted by the graphs in Fig. 10 for one pair of complementary patterns. The result is two highly sharpened fringe contours for each fringe of the basic moiré pattern. The algorithm is extremely robust, and it cancels the influence of nonuniform illumination and noise in the moiré patterns. Each pair of complementary patterns in the series of β patterns is processed, and their sharpened fringe contours are displayed together in the final contour map. The output is a map with β correctly positioned fringe contours for each fringe in the basic moiré pattern. The mathematical analysis and a discussion of its implications are given in [1].

An example of microscopic moiré interferometry and O/DFM is given in Sect. 4, where the technique is applied to study of grain deformations of titanium in elastic tension. The O/DFM algorithm is also illustrated in Sect. 5.

3.2 Curved Surfaces

Moiré interferometry has been developed for routine analysis of flat surfaces. However, certain accommodations can be made for curved surfaces when the problem is important enough to invest extra effort. Such a case is the ply-by-ply deformation of multi-ply composite plates, in tension or compression, at the cylindrical surface of a central hole in the plate. Two studies were extremely effective, the first by *Boeman* [14] and the second by *Mollenhauer* [15].

The specimens were all thick laminated composite plates of graphite/epoxy or glass/epoxy, each with a 25.4 mm (1 in.) diameter central hole. To apply the specimen grating on the cylindrical surface of the hole,

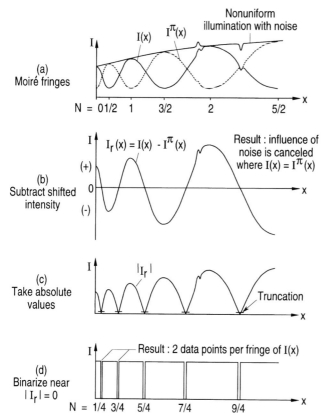

Fig. 10a–d. Graphical representation of steps in the O/DFM algorithm for $\beta = 2$. Additional pairs of complementary moiré patterns provide data for higher multiplication factors [1]

the mold used by Boeman was a cross-line grating on a thin, flexible steel substrate. He bent the steel to match the curvature of the hole and replicated the grating onto the specimen with a room-temperature curing epoxy. Subsequently, with the specimen subjected to a large compressive load, he replicated the deformed specimen grating onto another curved strip of steel; he straightened the steel strip, and observed the deformed grating in a moiré interferometer in the normal way.

In addition to investigating the ply-by-ply deformation of the hole, Boeman measured the deformation on the straight surface along the edge of the specimen. Examples of fringe patterns are shown in Fig. 11 for the straight edge of one specimen and for the hole boundary near $\theta = 90°$ for another specimen. The results showed definitive phenomena for the entire series of different laminate stacking sequences. Of special significance was the observation of markedly different strain distributions at the $\theta = 90°$ location from

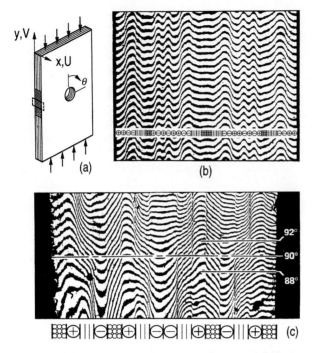

Fig. 11. (a) Composite laminate with central hole in compression; (b) N_y field along the straight free edge; and (c) N_θ field near $\theta = 90°$. The symbols represent fiber directions in each ply, where plus and minus indicate $+45°$ and $-45°$ fibers. (courtesy of R. G. Boeman)

that at the straight free edge of the same specimen. The strains at the hole were not merely equal to the strains at the free edge multiplied by a constant strain concentration factor. Instead, the strain multiplication factor varied dramatically on a ply-by-ply basis across the laminate. Also, interlaminar shear strains changed very rapidly with angular location θ along the hole. As an example, Fig. 11c shows interface regions (between plies with 0° and ±45° fiber directions) where the cross-derivative of displacement varies from negligibly low values at $\theta = 90°$ to very high values as close as 2° away. Interlaminar shear strains that were an order of magnitude higher than the applied compressive strain were measured, as well as the rapid gradients with angular position.

To apply the specimen grating, the mold used by *Mollenhauer* [15] was a cross-line grating on a solid cylindrical substrate whose radius matched the radius of the hole in the specimen. He replicated the deformed specimen grating onto another solid cylindrical substrate while the specimen was under (tensile) load. These curved grating replicas were not flattened for observation. Instead, the fringe patterns were photographed in a series of narrow strips, each perpendicular to the θ coordinate. With this procedure, Mollenhauer circumvented certain difficulties caused by small variations of the thickness of replicas made on flexible substrates.

Results are shown in Fig. 12 for a cross-ply tensile specimen with a 25.4 mm central hole. Fringe patterns at discrete locations, each $\theta = 3°$ apart, are shown as mosaics for a 90° portion of the hole. The ply-by-ply variations of displacement are clearly defined, allowing extraction of strains with a high degree of confidence. Data for strain analysis were taken along the centerline of each pattern. An example is shown in Fig. 13 for the $\theta = 75°$ location. This pattern is sufficient to determine the shear strains, since $\partial U_z / \partial \theta$ in Fig. 12c is negligibly small. Here, the interlaminar shear strains are very high – about six times larger than the average tensile strain at the 90° location.

In this work, the main objective was to evaluate a new three-dimensional numerical analysis technique by comparison with experimental results. The experiments provided a reliable basis for the evaluation.

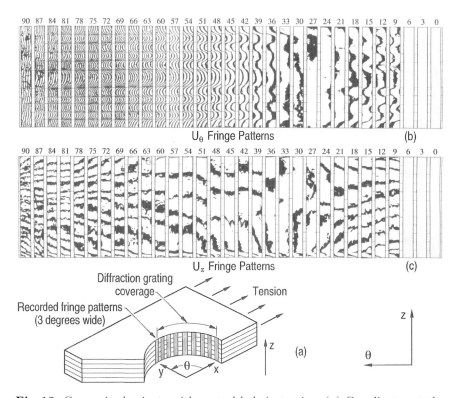

Fig. 12. Composite laminate with central hole in tension: (**a**) Coordinates at the hole; (**b**) mosaic of the U_θ fringe patterns along the cylindrical surface of the hole; and (**c**) mosaic of the U_z fringe patterns. The specimen material is a cross-ply laminate with $[0_4/90_4]_{3s}$ ply sequence. (courtesy D. H. Mollenhauer)

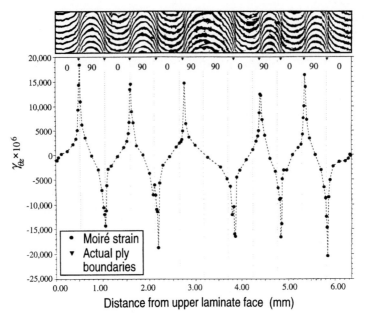

Fig. 13. Enlargement of the U_θ fringe pattern at the 75° location in Fig. 12b, and shear strains at that location. (courtesy D. H. Mollenhauer)

3.3 Replication of Deformed Gratings

Whereas real-time observation is desirable for many measurements (i.e. observation of fringe patterns at the same time as the loads are applied), there are important situations where replication of deformed gratings is preferred. Examples are measurements in a very noisy environment, measurements at a remote location, or measurements in a hostile environment. Replicas of deformed gratings can be made in many of these situations and then taken to the laboratory for analysis [1]. It is curious that this approach was practised long ago, when measurements on remote oil drilling platforms were desired [16].

The experience described above with curved surfaces further confirms the viability of this technique for both flat and cylindrical surfaces. Advances that are needed to make it a routine procedure lie in optimization of the replication process, particularly the choice of replication materials for relatively quick duplication of the deformed specimen grating. Investigations in this field are strongly recommended, whereby the technology and routine application of moiré interferometry can be significantly advanced.

4 Diverse Applications of Moiré Interferometry

4.1 Thermal Deformations of Microelectronics Devices

Microelectronics devices contain many electronic components within an active silicon chip, such as transistors, capacitors, resistors, etc. To form a usable device, a silicon chip requires protection from the environment as well as both electrical and mechanical connections to the surrounding components. The various conducting and insulating materials involved in the devices have different coefficients of thermal expansion (CTE). When the chip is powered up and thus the device is subjected to a temperature excursion, each material expands at a different rate. This non-uniform CTE distribution produces thermally induced mechanical stresses within the device assembly. As the components and structures involved in high-end microelectronics devices are made smaller, the thermal gradient increases and the strain concentrations become more serious. Hence, there is a continuously increasing activity in experimental analysis, both for specific studies and for guidance of numerical analyses [7,17,18,19,20,21]. Moiré interferometry is playing a leading role for experimental analysis. An example is presented here.

A special technique is required for replicating a specimen grating on the cross-sections of microelectronics devices, because they usually have such a complex geometry that the excess epoxy produced by the usual grating replication procedure cannot be removed [1]. The excess epoxy is critical since it could reinforce the specimen and change the local strain distribution. An effective replication technique was developed to circumvent the problem [7]. First, a tiny amount of liquid epoxy is dropped onto the grating mold; the viscosity of the epoxy should be extremely low at the replication temperature. Then, a lintless optical tissue (a lens tissue) is dragged over the surface of the mold, as illustrated in Fig. 14a. The tissue spreads the epoxy to produce a very thin layer of epoxy on the mold. The specimen is pressed gently into the epoxy, and it is prized off after the epoxy has polymerized. Before polymerization, the surface tension of the epoxy pulls the excess epoxy away from the edges of the specimen. The result is a specimen grating with a very clean edge.

An example is illustrated in Fig. 14b [21]. The specimen is a flip-chip plastic ball grid array (FC-PBGA) package assembly. In the assembly, a silicon chip (6.8 mm × 6.1 mm × 1.2 mm) was first attached to an organic substrate through tiny solder bumps called C4 interconnections. The gap between the chip and the substrate was filled with an epoxy underfill to reduce the strains of solder bumps. This subassembly was then surface-mounted to a typical FR-4 printed circuit board (PCB) through larger solder ball arrays to form a final assembly. The assembly was cut and its cross-section was ground to produce a flat, smooth, cross-sectional surface. The specimen grating was replicated at 82° C and the fringes were recorded at room temperature ($\Delta T = -60°$ C). Very clean edges of the specimen grating are evident.

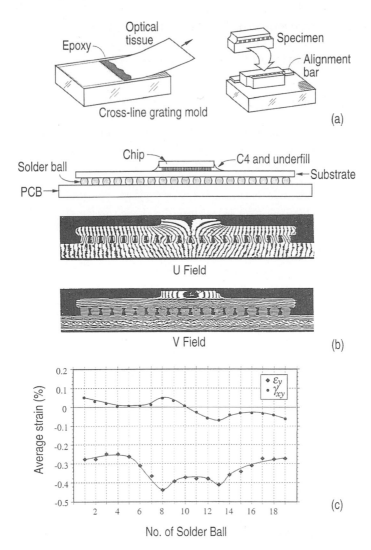

Fig. 14. (a) Procedure to replicate a specimen grating on a specimen with a complex geometry; (b) U and V displacement fields of FC-PBGA package assembly, induced by thermal loading of $\Delta T = -60°$ C; and (c) distribution of normal and shear strains (averaged along the vertical centerline) at each solder ball [1]

The V field fringe pattern reveals the detailed bending deformation of the substrate. Two distinct curvatures are observed, one in the area connected to the chip and the other in the rest of the substrate. The CTE of the substrate was higher than that of the PCB. The substrate contracted more than the PCB during cooling, while the deformation of the substrate covered by the chip was constrained by the low CTE of the chip. This complicated loading condition produced an uneven curvature of the substrate, which resulted in an inflection point below the edge of the chip. The substrate was connected to the PCB through the solder balls and the difference of curvature between the substrate and the PCB was accommodated by the deformation of the solder balls. The normal and shear strains (averaged along the vertical centerline) at each solder ball was calculated from the fringe patterns and they are plotted in Fig. 14c. The largest of these normal strains occurred in the solder ball located at the edge of the chip and its magnitude was nearly four times greater than the largest shear strain. This result is entirely different from the results of a previous study on a ceramic ball grid array package assembly [7]. The subsequent numerical parametric study quantified the effect of the substrate CTE on the solder ball strains. The study indicated that the substrate CTE was one of the most critical design parameters for optimum solder joint reliability [21]. The experimental evidence provided by moiré interferometry was essential for revealing this important design parameter.

4.2 Textile Composites

Polymer matrix composites made from carbon fiber textile preforms (fabric) were studied using moiré interferometry. Moiré was used to determine the strain fields on the surface of specimens loaded in tension, compression and shear. The results from these experiments were used to guide instrumentation practices with strain gages; to determine failure mechanisms; and to guide modeling efforts. Although numerous textile architectures were investigated during the course of the experimental program [22], only one example is illustrated here. A specimen made from a two-dimensional triaxial braid consisting of AS4 graphite yarns (each yarn is made up of many thousands of fibers), and 1895 epoxy resin was tested in tension. The textile preform, illustrated in Fig. 15, was braided in a 2/2 pattern where each braider yarn continuously passes over two opposing braider yarns and then under two. The axial yarns are sandwiched between the braider yarns during the braiding process. The preform was braided over a cylindrical mandrel and layering was achieved by overbraiding. Neighboring layers were not interconnected and nesting of the layers was a random process. After braiding, the preforms were cut lengthwise from the mandrel and flattened to form a plate with a nominal thickness of 3.2 mm. The epoxy resin was introduced by resin transfer molding. Two tension specimens were cut from the panel, one oriented along the axial direction (designated by the direction of the axial yarns) and one oriented perpendicular to the axial direction (transverse loading). The specimens were tabbed and a 1200 line/mm diffraction grating was replicated on the surface in the central region.

To illustrate the distinctive strain distribution on the surface of the textile composites, a moiré interferometry test was performed on a laminated cross-ply graphite/epoxy specimen for comparison. Figure 16 shows the vertical (V) and horizontal (U) displacement fields for this material. Notice that

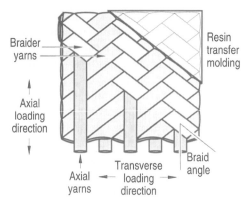

Fig. 15. Architecture of two-dimensional triaxial braid layer in composite specimens. Each yarn has thousands of fibers [1]

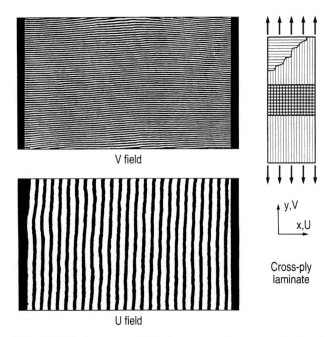

Fig. 16. Displacement fields for a cross-ply composite laminate in tension

the displacement contours are generally parallel and uniformly spaced in both the U and V fields, demonstrating the uniform strain throughout each field.

Figure 17 shows the moiré interferometry fringe patterns for the axially loaded textile specimen. In both fields the fringes are neither uniformly spaced nor parallel. The normal strain in the direction of loading varies as a function of the braid architecture, which is evident in the fringe pattern repetition. Peak normal strains are of the order of twice the average normal strain. Additionally, strong shear strains exist in resin-rich regions between the braider yarns. The variations follow a repetitive pattern which has a size scale equivalent to the smallest repeating unit in the textile architecture. Figure 18 shows the moiré fringe patterns for the textile specimen loaded in transverse tension. Again, the peak normal strains are two to three times the average normal strain. In the V field, the normal strain varies cyclically from high strain to low strain. Additionally, high shear strains exist between the braider yarns, as can be seen by the jogs in the V field fringe patterns.

From the moiré interferometry patterns, one can determine appropriate strain gage sizes to perform routine material property characterization of this class of materials. Otherwise, gages that are too small or too large give random estimates of the representative deformation, depending upon their location relative to the material architecture. Additionally, it was determined that the plane strain assumption was not valid for modeling the smallest repeating unit of the textile architecture. The failure mechanism was studied

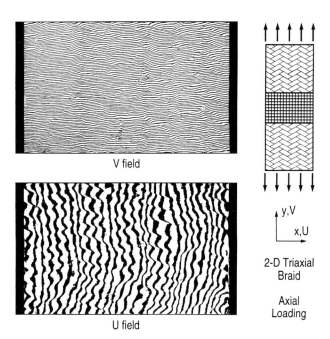

Fig. 17. Displacement fields on the surface of a braided composite for axial tension

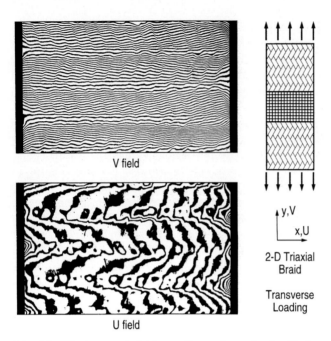

Fig. 18. Displacement fields on the surface of a braided composite for transverse tension [1]

by further loading of the specimens, whereby cracking was documented by the fringe patterns and corroborated with other methods such as edge replication and X-ray radiography.

4.3 Fracture Mechanics

Inelastic deformation appears to accompany the fracture process, even for brittle materials. It is recognized as a permanent deformation in the neighborhood of the crack path, and for nonmetals it is known as the damage wake. Moiré interferometry was employed to investigate the phenomenon.

The specimen was a notched coupon of pure alumina. It was subjected to a tensile loading until a crack initiated and extended to the length shown in Fig. 19 [23]. The pattern represents the V displacement field that remained after the specimen was fully unloaded. The plastic deformation along the crack path prevents a return to the original specimen geometry and prevents the complete closure of the crack. Thus, the fringes at some distance from the crack results from rigid-body rotation of the upper portion of the specimen relative to the lower portion, and also from the elastic strains required to accommodate the new geometry.

Moiré interferometry is exceptionally useful for the determination of crack opening displacements. They are established by a discontinuity of the fringe

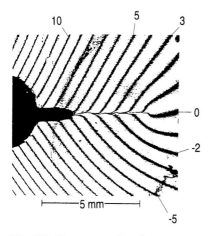

Fig. 19. Damage wake of permanent strain occurs near the crack surface in pure alumina (courtesy of J. S. Epstein) [1]

count across the crack. For example, at the notch tip, $N_y = 6$ above the crack and $N_y = -4.5$ below it. The crack opening displacement can be determined as $\Delta N_y/f = (10.5/2400)$ mm, or $4.38\,\mu$m.

4.4 Micromechanics: Grain Deformations

The heterogeneous deformation of titanium in elastic tension was investigated [24]. The specimen material was large-grain beta-alloy titanium. The specimen cross-section was 6 mm by 5 mm, its yield point stress (0.2% offset) was 1450 MPa (210000 psi), and its elastic modulus was 97 GPa (14×10^6 psi). Numerous grain sizes were present in the specimen, up to a maximum of 400 μm. The specimen was loaded as illustrated in Fig. 20, whereby a bending moment (plus a small compressive force) was applied, resulting in tension on the outside face of the specimen. The annealed copper gasket was designed to deform plastically to distribute contact forces. The forces, P, were applied by a displacement controlled loading fixture.

Microscopic moiré interferometry was employed. The specimen grating thickness was approximately 2 μm. The reference grating frequency was $f = 4800$ lines/mm. The V and U patterns (Fig. 21a,b) reveal point-to-point variations of tensile strains and the corresponding transverse strains. The average tensile strain ($\varepsilon_y^{\text{ave}}$) and the average transverse strain ($\varepsilon_x^{\text{ave}}$) was 0.85% and -0.30%, respectively. The corresponding average tensile stress was 825 MPa (119000 psi), which is 57% of the yield point stress. Thus, the deformation was nominally in the elastic range.

Carrier fringes were applied to subtract the average strain, with the results shown in Fig. 21c,d. Figures 21e,f are the same patterns processed by O/DFM with a fringe multiplication factor of $\beta = 6$. The contour interval is

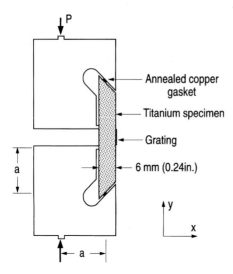

Fig. 20. Large-grain titanium specimen and loading geometry [1]

35 nm/contour. Contour numbers in the multiplied patterns define the relative displacements throughout the field. The V and U patterns of Fig. 21a,b were inspected carefully to assign correct fringe orders to the multiplied patterns.

The multiplied patterns reveal strong anomalies. Whereas elastic strains are uniform on a macroscopic scale, they are seen to vary on a microscopic scale. In order to evaluate the anomalous shear strains associated with individual grains, the shear strains in the tangential directions along grain boundaries were calculated from the multiplied patterns. Figure 22 shows an image of individual grains, where the dashed lines represent loci of the grain boundaries. The anomalous shear strain, $\gamma^a_{x',y'}$, is plotted for the irregular grain boundary path shown at the right. Since the carrier fringes cancel the average strain (acting along any line in the field), the multiplied patterns reveal the remaining anomalous strains.

The anomalous shear strain acting in the x',y' coordinate system (which is parallel and perpendicular to the grain boundary) was calculated from measurements of the x,y components of the normal and shear strains at each point. The strongest anomalies occurred near the boundary between grains A and E. The grain boundary was almost straight and it was inclined by $40°$ with respect to the positive x-axis. The anomalous part of the shear strain, $\gamma^a_{x',y'}$, at that location was 1.70%, which was twice as large as the average tensile strain. The uniform part of the tensile deformation generates shear on the $40°$ plane, too. Together with the anomalous part, the total shear strain on the $40°$ grain boundary was 2.84%.

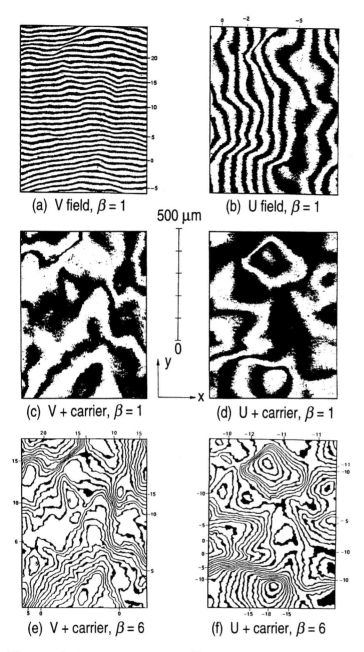

Fig. 21a–f. Fringe patterns at 57% of the yield point. The displacements are 208 nm/fringe for $\beta = 1$ and 35 nm/contour for $\beta = 6$

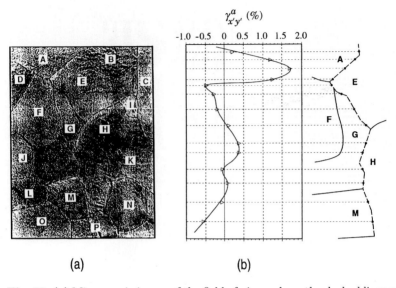

Fig. 22. (a) Microscopic image of the field of view, where the dashed lines represent the grain boundaries; (b) distribution of anomalous shear strain acting tangentially to the grain boundaries

5 Shadow Moiré

5.1 Basic Concepts

Shadow moiré has been used for engineering measurements of out-of-plane displacements W for several decades. It has been especially useful for applications where W exceeds 0.25 mm (0.01 in.) across the field of view. More recently, techniques utilizing phase stepping and computer manipulation of the data have extended its usefulness to applications where W is an order of magnitude smaller. For engineering problems requiring measurements that are two orders of magnitude smaller, e.g. where the change of W across the field is about 2.5 µm (100 µin.), the method of choice would usually be classical or holographic interferometry.

The most popular and practical shadow moiré arrangement is introduced here. Additional information is given in Chap. 3 and Appendix A of [1], which provides a more general treatment of shadow moiré. The competing method of projection moiré is also addressed in [1], although that technique is usually avoided because of its stringent requirements on optical resolution.

Figure 23 illustrates the principle of shadow moiré. The specimen surface is usually prepared by spraying it with a thin film of matt white paint. A linear reference grating of pitch g is fixed adjacent to the surface. This reference grating consists of black bars and clear spaces on a flat glass plate. A light source illuminates the grating and specimen, and the observer (or camera) receives the light that is scattered in its direction by the matt specimen.

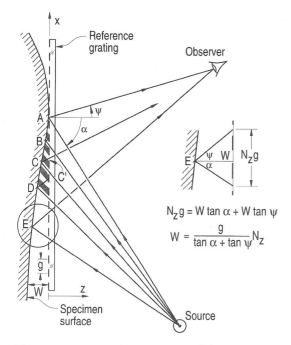

Fig. 23. Principle of shadow moiré [1]

When the gap W is small, a well-defined shadow of the reference grating is cast upon the specimen surface. This shadow, consisting of dark bars and bright bars, is itself a grating. The observer sees the shadow grating and the superimposed reference grating, which interact to form a moiré pattern.

In the cross-sectional view of Fig. 23, light that strikes points A, C and E on the specimen passes through clear spaces in the reference grating to reach the observer. Light that strikes points B and D are obstructed by the opaque bars of the reference grating. Thus, bright moiré fringes are formed in the regions near A, C, and E, while black fringes are formed near B and D. Of course, many more bars appear between A and B in an actual system, so the transition between bright and dark moiré fringes is gradual. With oblique illumination, the positions of the shadow on the specimen depend upon the gap W, and so the moiré pattern is a function of W.

Let the fringe order at A be $N_z = 0$, since the specimen and reference grating are in contact and $W = 0$. From the viewpoint of the observer, the number of shadow-grating lines between A and C is compared to the number of reference-grating lines between A and C'. The number of shadow grating lines is larger by one, and therefore the fringe order at C is $N_z = 1$. Similarly at E, the number of shadow grating lines is larger by two and the moiré fringe order at E is $N_z = 2$.

The relationship between the fringe order and the gap W is derived in the insert, with the result

$$W = \frac{g}{\tan\alpha + \tan\psi} N_z . \tag{7}$$

Note that angles α and ψ are variables that change with coordinate x. Therefore, the sensitivity of the measurement is not constant, i.e. the displacements are not directly proportional to the fringe orders of the moiré pattern. However, if the light source and observer are at the same distance L from the specimen and if D is the distance between the light source and observer, the variables reduce to

$$\tan\alpha + \tan\psi = \frac{D}{L} = K , \tag{8}$$

where K is a constant, and

$$W = \frac{g}{K} N_z . \tag{9}$$

Thus, W is directly proportional to the moiré fringe order. Consistent with this, the configuration of Fig. 24 is the usual arrangement of choice. In addition to constant sensitivity across the field, it has the advantage that the specimen is viewed at normal incidence, thus avoiding geometrical distortions. When $D = L$, the contour interval of the moiré fringe pattern (i.e. displacement W per fringe order) is equal to the pitch of the reference grating.

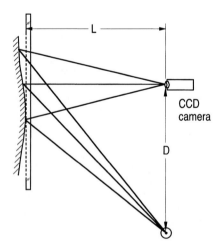

Fig. 24. Preferred arrangement for shadow moiré [1]

5.2 Additional Considerations

The equations relating W to N_z assume that W is negligibly small compared to L. Of course, W must be very small in order to form a well-defined shadow of the reference grating on the specimen. As a numerical guide, excellent conditions are achieved when W is within a small fraction of the Talbot self-imaging distance g^2/λ, e.g. within 5% of the Talbot distance. Thus, if a reference grating of 12 lines/mm (300 lines/in.) is used with white light, clear shadow moiré fringes are achieved when W is less than 0.65 mm (0.026 in.), which is a comfortable gap for most situations.

An incandescent white light source with a narrow filament is a good choice. The filament should be oriented parallel to the lines of the reference grating. A monochromatic source such as a mercury arc lamp is sometimes preferred, to take advantage of its narrower spectrum. Laser diodes have been used successfully. Lasers that produce highly coherent light are usually avoided since the concomitant speckle produces noise in the shadow moiré patterns.

A black-and-white CCD camera can be used very effectively for shadow moiré. In general, only a few moiré fringes appear across the field of view, and therefore the moiré pattern can be recorded with numerous pixels per fringe. Of course, no wet processing is required, and image processing by computer is readily accessible.

The intensity distribution in the moiré fringe pattern is complex. In the usual case of a reference grating consisting of dark bars and transparent spaces, the bright moiré fringes are crossed with the black bars of the reference grating. Everywhere between the center of a bright fringe and the center of the next dark fringe, the amount of light that passes through the transparent spaces of the reference grating varies linearly with the gap W. Consequently, if the light at each point in the pattern is averaged across a distance equal to the pitch of the reference grating, the intensity distribution is a triangular function of W.

These are idealized distributions. Realistic effects that distort the intensity distribution recorded by the CCD camera include nonlinearity of the CCD sensors and saturation at high intensity levels. Additionally, a small camera-lens aperture is used in order to reduce the resolution and average the intensity across the grating pitch. Frequently, an even smaller camera resolution is used, and this causes averaging that rounds-off the peaks and valleys of the triangular distribution. The objective (when the quasi-heterodyne method is used) is to approach a sinusoidal intensity distribution. It should be noted, however, that the degree of averaging required to transform closely spaced moiré fringes is very different from that required for broadly spaced fringes. In general, a moiré pattern exhibits both closely spaced and broadly spaced fringes. With regard to Fourier-series analysis of the intensity distribution, the camera resolution that filters out all but the first sinusoidal term of the series for closely spaced moiré fringes, may allow several terms for broadly spaced fringes.

6 Increased Sensitivity, Shadow Moiré

6.1 Phase Stepping (or Quasi-heterodyne) Method

The shadow moiré applications shown in subsequent sections illustrate cases in a wide range of out-of-plane displacements. In many present-day applications the deformation is small and very few moiré fringes are present. Consequently, supplementary image analysis techniques are used to extract sufficient data for determination of the deformation.

In the technique that is becoming better known as *phase stepping*, but previously called the *quasi-heterodyne* method [25], three or more images of the shadow moiré pattern are recorded by a CCD camera. In the case of three images, the reference grating is moved in the z direction by small increments $g/3K$ between recordings. By (9), g/K is the displacement per fringe order. The center of every moiré fringe shifts by $1/3$ of a fringe order, and the phase ϕ shifts by $1/3$ of 2π at every point in the field. The CCD camera records the intensity at every pixel in the field for the three images, and sends the data to a personal computer to evaluate the fringe order at every pixel.

In practice, a small camera lens aperture is used to suppress the resolution of the fringe pattern so that the intensity distribution approaches a sinusoidal function. The intensities at any single point in the three patterns are therefore approximated by

$$I_1 = I_0 + A\cos\phi,$$
$$I_2 = I_0 + A\cos\left(\phi - \frac{2\pi}{3}\right),$$
$$I_3 = I_0 + A\cos\left(\phi - \frac{4\pi}{3}\right), \tag{10}$$

where I_0 is the constant DC component of the intensity at the point and A is the intensity modulation, i.e., half the amplitude of the maximum change when ϕ is varied.

Therefore, the unknown phase at the point is determined from the three intensities measured by the CCD camera by

$$\phi = \arctan\frac{\sqrt{3}(I_2 - I_3)}{2I_1 - I_2 - I_3}. \tag{11}$$

Note that any effects of nonuniform background intensity and also any optical noise (if it is constant, i.e. does not change with phase stepping) is incorporated in I_0 and does not degrade the result. Since (11) yields the phase in the restricted range 0 to 2π, an unwrapping algorithm is used to establish the integral part M of the fringe order N at each pixel by

$$N = \frac{\phi}{2\pi} + M , \qquad M = \pm 0, 1, 2, 3, \ldots , \tag{12}$$

where M is determined from the point-to-point changes of phase throughout the field.

Accordingly the fringe order is determined at each pixel to within a small fraction of a fringe order, and by (9), the out-of-plane displacement field is determined with enhanced sensitivity. The displacement field can then be displayed as a contour map, pseudo-color map, or three-dimensional graph. Whole-field patterns corresponding to these steps are displayed below in Sect. 7 (Fig. 32). The three phase-stepped intensity distributions are shown, together with the wrapped-phase map and contour maps of the W displacement field.

6.2 Optical/Digital Fringe Multiplication (O/DFM)

The O/DFM method was introduced here in connection with microscopic moiré interferometry. It utilizes phase stepping of β equal steps to achieve fringe multiplication by a factor of β. The algorithm depicted in Fig. 10 illustrates its insensitivity to nonuniform illumination and noise. However, O/DFM is not based upon fringes of a sinusoidal intensity distribution. Instead, a broad class of periodic functions is applicable [1], including the case of shadow moiré represented in Fig. 25. Note that dark fringe contours in Fig. 25d are produced at points where the complementary fringe patterns have equal intensities, independent of the slopes at crossing points in Fig. 25a. Two fringe contours are produced for every fringe in the original moiré pattern. Whole-field patterns corresponding to Fig. 25a,c,d are shown later in Sect. 7 (Fig. 30).

If additional shadow moiré patterns are recorded with phase steps of $\pi/6$, $2\pi/6$, $4\pi/6$ and $5\pi/6$, they can be processed in complementary pairs to provide intermediate fringe contours. The result is portrayed in Fig. 25e, where the fringe contours are combined in a single computer output map. Six fringe contours are produced for every moiré fringe in the original pattern, providing fringe multiplication by a factor of 6. A multiplication factor of β can be produced by processing β phase-stepped shadow moiré patterns (where β is an even integer).

7 Applications of Shadow Moiré

7.1 Post-buckling Behavior of a Composite Column

The post-buckling behavior of an I-shaped laminated graphite/epoxy column was investigated using shadow moiré. The purpose of the experiment was to

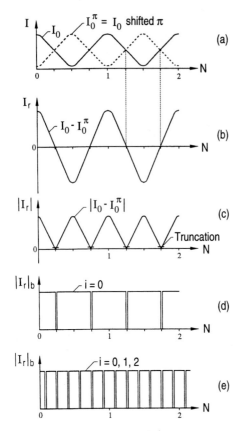

Fig. 25a–e. Steps in the O/DFM algorithm, applied with the quasi-triangular intensity distribution of shadow moiré [1]

verify a theoretical model and to obtain experimental insight into the post-buckling behavior of such structures. The out-of-plane deflections of both the web and the flange were measured, as a function of load.

Since the maximum out-of-plane displacement was expected to be quite high (approximately 6 mm), relatively low frequency linear gratings, called *grilles*, were used. The reference grilles were made by printing an image created with drawing software onto acetate transparency material, using a 600 dpi laser printer. The acetate material was bonded to a rigid transparent acrylic plate with acetone. Since it was desired to simultaneously measure the out-of-plane deflections on both the web and the flange of the column, two sets of reference grilles were produced. Deflection was monitored throughout the entire test, up to failure. Since the range of deflections was high and accuracy at each load was required, two grille frequencies were used. A frequency of 3.94 lines/mm (100 lines/in.) was used in the early stages of loading, and 1.97 lines/mm (50 lines/in.) at later stages. The experimental set-up is illus-

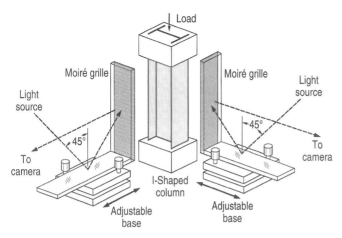

Fig. 26. Experimental arrangement for investigating post-buckling deformation of a column

trated in Fig. 26. The grilles were mounted on adjustable bases so that their positions relative to the specimen could be adjusted easily. Mirrors were used to direct the light to the reference grilles. This was necessary because the loading machine and fixtures would have blocked a large portion of the specimen if direct illumination had been used. Figure 26 shows the mirror/grille assemblies separated from the specimen for illustration purposes. During testing, both grilles were moved to near-contact with the specimen. The size of grille used for the web was chosen to fit between the flanges of the specimen. Two clear-glass incandescent light bulbs, mounted on tripods, were used for the light sources. The angle of illumination was 45°, and the viewing angle was normal to the specimen surface. Both the light source and camera position were located at the same perpendicular distance from the specimen surface, making the optical set-up equivalent to that of Fig. 24. The specimen was sprayed with flat white paint to provide a matt surface. Images were recorded by two 35 mm cameras using technical pan film.

In addition to shadow moiré other experimental techniques were incorporated. LVDTs were used to monitor the deflection of the web at two arbitrary locations along the center line on the back side of the web. This information was used to determine the absolute deflection of the specimen in order to obtain a baseline for the shadow moiré information. Otherwise, a live count of the fringes passing through a reference point during load application would be required for absolute displacement data. Strain gages were also mounted in the center of both sides of the web to monitor the load at which buckling occurred. Thin lead wires were used in order to minimize their appearance on the shadow moiré fringe patterns.

Figure 27 illustrates examples of the shadow moiré fringe patterns on the web and flange of the specimen. Nearly the entire web and flange were

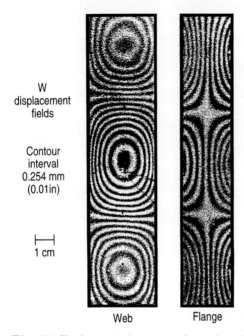

Fig. 27. Shadow moiré patterns for web and flange

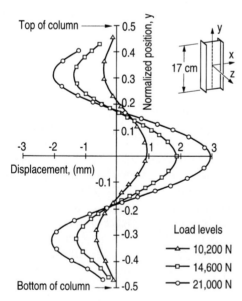

Fig. 28. Out-of-plane deflections, W, along the center line of the web for three load levels

observed. From these patterns, the out-of-plane deflections were determined as a function of load. Graphs of the out-of-plane deflection versus position along the vertical center line of the web, for various load levels, is shown in Fig. 28. These results were then compared to those obtained from numerical modeling to guide and validate the model.

7.2 Pre-buckling Behavior of an Aluminum Channel

The experiment depicted in Fig. 29 was performed to study the pre-buckling out-of-plane bending of a structural member [1]. The specimen was an aluminum channel with two 19 mm (0.75 in.) diameter holes, loaded in compression. An 11.8 lines/mm (300 lines/in.) reference grating was positioned in front of the specimen. The optical arrangement illustrated in Fig. 24 was used with $D = L$, which provided a basic sensitivity of 84.7 μm/fringe (3.3×10^{-3} in./fringe). Fringe shifting was accomplished by moving the reference grating in the z-direction, and enhanced sensitivity was achieved with the O/DFM algorithm.

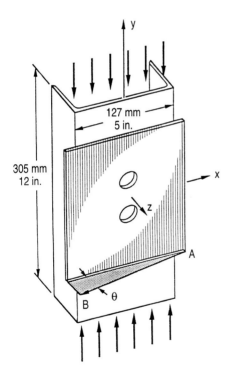

Fig. 29. Specimen and reference grating for pre-buckling deformation. For fringe shifting, the reference grating was translated in the direction AB in order to control the z component of translation more accurately [1]

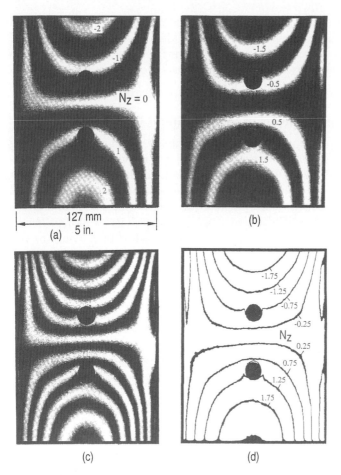

Fig. 30. (a) The initial shadow moiré pattern; (b) the complementary pattern; (c) and (d) Patterns corresponding to steps (c) and (d) in the O/DFM algorithm in Fig. 25 [1]

Figure 30 shows two complementary shadow moiré patterns from the fringe shifted recordings, followed by steps of the O/DFM algorithm illustrated in Fig. 25c,d. The fringe contours for multiplication by a factor of 6 are shown in Fig. 31. The left and right sides of the specimen were stiffened by the flanges of the channel and the out-of-plane distortions at the sides were very small. The web, or face, of the channel experienced a second bending mode wherein the upper portion deflected inwards and the lower portion deflected outwards. In the immediate vicinity of the holes, the multiplied pattern shows a ridge on the left and right side of each hole; the ridge was outwards, in the positive z-direction for both holes. At the top and bottom of each hole the surface curvature was inwards as indicated by the graph of

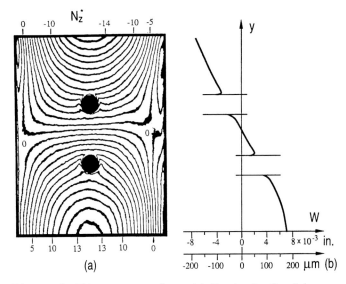

Fig. 31a,b. Fringe contours for multiplication by $\beta = 6$ (corresponding to Fig. 25e) and deflection W along the vertical center line [1]

displacements along the vertical center line. These features are not evident in the lower sensitivity pattern of Fig. 30a, but they are revealed effectively by fringe multiplication. The contour interval for the multiplied pattern is g/β, i.e. 14.1 µm/fringe (5.56×10^{-4} in./fringe).

7.3 Warpage of Electronic Devices

A plastic ball grid array (PBGA) package is composed of an active silicon chip on an organic carrier board called a *substrate*, which is usually a glass/epoxy composite. The warpage of the substrate is attributed to a large mismatch of CTE between the chip and substrate. In the subsequent assembly process, electrical and mechanical connections are made by solder balls between the substrate and a PCB. If the bottom side of the substrate warps significantly at the solder reflow temperature, it yields an uneven height of solder interconnections, which could cause premature failure of the assembly. Detailed knowledge of this out-of-plane deformation is essential to optimize design and process parameters for reliable assemblies.

Substrate warpage was documented by shadow moiré in two different experiments. The specimens were similar, but not identical. The phase stepping algorithm was used to enhance sensitivity in the first example, and the O/DFM algorithm in the second.

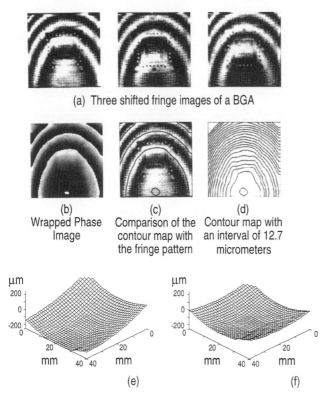

Fig. 32a–f. Warpage of an electronic package by the phase-stepping (quasi-heterodyne) algorithm (courtesy of Y. Wang)

7.3.1 Warpage by Phase Stepping

Figure 32a shows the three shadow moiré patterns, each one phase stepped by 1/3 of 2π with respect to its neighbor [27]. The reference grating was adjusted to a small angle with respect to the average plane of the specimen; so the fringes represent the wedge-shaped gap between the reference plane and the specimen. The array of dots in the images are copper pads on the bottom of the substrate (for subsequent solder connections to the PCB). The shadow moiré configuration of Fig. 24 was used, whith $D = L$; the reference grating frequency was 11.8 lines/mm (300 lines/in.). By (9), the contour interval of the shadow moiré pattern is equal to the pitch.

The patterns were processed by the phase-stepping (quasi-heterodyne) algorithm to produce Fig. 32b. Unwrapping yielded contours of constant fringe order, or constant W displacement. Figure 32c shows computed contours superimposed upon the original fringe pattern. Figure 32d shows the W field with a contour interval of 12.7 μm (0.0005 in.).

Since the resultant data are in the form of an orthogonal matrix of displacement at each pixel of the CCD camera, the displacement can be transformed readily to any other reference plane. A three-dimensional graph of the displacement data is shown in Fig. 32e, and the displacements relative to a horizontal plane are graphed in Fig. 32f.

7.3.2 Warpage by O/DFM

The shadow moiré configuration of Fig. 24 was used for this application too, but with different parameters. Here $D = 2L$; the reference grating frequency was 10 lines/mm (254 lines/in.) and its pitch was 100 μm (0.004 in.). By (9), the contour interval of the shadow moiré pattern was 50 μm (0.002 in.).

A moiré fringe multiplication factor of only $\beta = 4$ was required to achieve the corresponding sensitivity. The complementary patterns were recorded by the CCD camera and processed by the O/DFM algorithm to produce the contour map of Fig. 33a. The contour interval is 12.5 μm (0.0005 in.). The array of copper soldering pads was more distinct in this case because reduced camera resolution was not required. Remarkably, the height of the

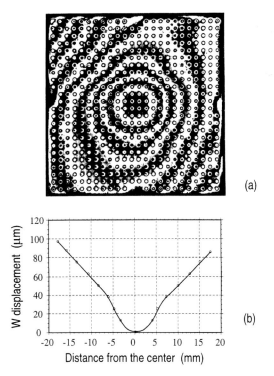

Fig. 33a,b. Warpage of an electronic package by the O/DFM algorithm, and graph of the deflection along the upper-left to lower-right diagonal

copper pads was resolved, although their size was distorted by the white paint applied to the specimen. A much thinner application of paint could be used, whereby fine surface details could be revealed with high fidelity.

The displacement along the diagonal (upper-left to lower-right) is plotted in Fig. 33b. Note that the displacement of the center relative to the corners is approximately the same in Fig. 32 and 33, which means they exhibit approximately the same number of contours when transformed to equal reference planes. Although both specimens are nominally symmetrical, the warpage deviates from pure symmetry in both cases. The deviations are typical, as corroborated by many examples.

These shadow moiré measurements are usually carried out in environmental chambers to monitor the progression of warpage with temperature. Examples of these real-time tests utilizing the phase stepping and O/DFM algorithms, are presented in [26] and [27], respectively.

Acknowledgments

The authors gratefully acknowledge the assistance of colleagues who provided figures, as cited in the figure captions: Raymond G. Boeman (Oak Ridge National Laboratory); Susan A. Chavez, Jonathan S. Epstein (Idaho National Engineering Laboratory); David H. Mollenhauer (Air Force Research Laboratory of Wright Patterson AFB); and Yinyan Wang (Electronic Packaging Services Ltd. Co.). We thank Springer-Verlag for permission to extract figures from [1]. We thank Xiaokai Niu for assistance with the post-buckling shadow moiré work (Fig. 26); and we thank Kaushal Verma and Derek Columbus for assistance with the O/DFM shadow moiré work (Fig. 33). The many contributions by co-workers and co-authors of references cited herein are gratefully acknowledged.

References

1. D. Post, B. Han, P. Ifju: *High Sensitivity Moiré: Experimental Analysis for Mechanics and Materials*, Springer, Berlin, Heidelberg (1994)
2. K. Patorski: *Handbook of the Moiré Fringe Techniques*, Elsevier, New York (1993)
3. O. Kafri and I. Glatt: *The Physics of Moiré Metrology*, Wiley, New York (1990)
4. G. Cloud: *Optical Methods of Engineering Analysis*, Cambridge Univ. Press., New York (1995)
5. A. Livnat, D. Post: The Governing Equations of Moiré Interferometry and Their Identity to Equations of Geometric Moiré, Exp. Mech. **25**, 360–366 (1985)
6. D. Post: Moiré Fringe Multiplication with a Non-symmetrical Doubly–blazed Reference Grating, Appl. Opt. **10**, 901–907 (1971)
7. Y. Guo, C. K. Lim, W. T. Chen, C. G. Woychik: Solder Ball Connect (SBC) Assemblies under Thermal Loading: I. Deformation Measurement via Moiré Interferometry, and Its Interpretation, IBM J. Res. Dev. **37**, 635–648 (1993)

8. D. Post, J. Wood: Determination of Thermal Strains by Moiré Interferometry, Exp. Mech. **29**, 318–322 (1989)
9. P. G. Ifju: Shear Testing of Textile Composite Materials, ASTM J. Compos. Technol. Res. **17**, 199–204 (1995)
10. Y. Guo, D. Post, B. Han: Thick Composites in Compression: An Experimental Study of Micromechanical Behavior and Smeared Engineering Properties, J. Compos. Mater. **26**, 1930–1944 (1992)
11. B. Han, D. Post: Immersion Interferometer for Microscopic Moiré Interferometry, Exp. Mech. **32**, 38–41 (1992)
12. B. Han: Interferometric Methods with Enhanced Sensitivity by Optical/Digital Fringe Multiplication, Appl. Opt. **32**, 4713–4718 (1993)
13. B. Han: Higher Sensitivity Moiré Interferometry for Micromechanics Studies, Opt. Eng. **31**, 1517–1526 (1992)
14. R. G. Boeman: Interlaminar Deformations on the Cylindrical Surface of a Hole in Laminated Composites: An Experimental Study, Center for Composite Materials and Structures Report 91-07, Virginia Polytechnic Institute and State University, Blacksburg, VA (1991)
15. D. H. Mollenhauer: Interlaminar Deformation at a Hole in Laminated Composites: A Detailed Experimental Investigation Using Moiré Interferometry, Ph.D. Thesis, Virginia Polytechnic Institute and State University, Blacksburg, Virginia (1997) [Available for download at http://scholar.lib.vt.edu/theses/theses.html]
16. J. McKelvie, C. A. Walker: A Practical Multiplied Moiré-Fringe Technique, Exp. Mech. **18**, 316–320 (1978)
17. B. Han, Y. Guo: Thermal Deformation Analysis of Various Electronic Packaging Products by Moiré and Microscopic Moiré Interferometry, Trans. ASME J. Electron. Packag. **117**, 185–191 (1995)
18. B. Han, Y. Guo, C. K. Lim, D. Caletka: Verification of Numerical Models Used in Microelectronics Packaging Design by Interferometric Displacement Measurement Methods, Trans. ASME J. Electron. Packag. **118**, 157–163 (1996)
19. B. Han, Y. Guo: Determination of Effective Coefficient of Thermal Expansion of Electronic Packaging Components: A Whole-field Approach, IEEE Trans. Compon. Packaging Manuf. Technol. A **19**, 240–247 (1996)
20. B. Han: Deformation Mechanism of Two-Phase Solder Column Interconnections under Highly Accelerated Thermal Cycling Condition: An Experimental Study, Trans. ASME J. Electron. Packag. **119**, 189–196 (1997)
21. B. Han, Y. Guo: Photomechanics Tools as Applied to Electronic Packaging Product Development, in B. Han, R. Mahajan, D. Barker (Eds.): *Experimental/Numerical Mechanics in Electronics Packaging* **1**, Society for Experimental Mechanics, Bethel, CT (1997) pp. 11–15
22. B. Han, M. Chopra, S. Park, L. Li, K. Verma: Effect of Substrate CTE on Solder Ball Reliability of Flip-Chip PBGA Package Assembly, J. Surf. Mount Technol. **9** 43–52 (1996)
23. P. G. Ifju, J. E. Masters, W. C. Jackson: Using Moiré Interferometry to Aid in Standard Test Method Development for Textile Composite Materials, Compos. Sci. Technol. **53**, 155–163 (1995)
24. J. S. Epstein, M. S. Dadkhah: Moiré Interferometry in Fracture Mechanics, in J. S. Epstein (Ed.): *Experimental Techniques in Fracture Mechanics, Vol. III*, VHC, New York (1993) pp. 427–508

25. B. Han: Micromechanical Deformation Analysis of Beta Alloy Titanium in Elastic and Elastic/Plastic Tension, Exp. Mech. **36**, 120–126 (1996)
26. P. Hariharan: Quasi-heterodyne Hologram Interferometry, Opt. Eng. **24**, 632–638 (1985)
27. Y. Wang, P. Hassell: On-Line Measurement of Thermally Induced Warpage of BGAs with High Sensitivity Shadow Moiré, Int. J. Microcircuits Electron. Packag. **21**, 191–196 (1998)
28. K. Verma, D. Columbus, B. Han, B. Chandran: Real-time Warpage Measurement of Electronic Components with Variable Sensitivity, *Proc. 48th Electronic Component & Technology Conf.*, Seattle, WA (1998) pp. 975–980

Digital Photoelasticity

Terry Y. Chen

Department of Mechanical Engineering, National Cheng Kung University
Tainan, Taiwan 701, Republic of China
ctyf@mail.ncku.edu.tw

Abstract. Photoelasticity is an experimental technique for stress and strain analysis. The method is based upon an optical property called double refraction, or birefringence, of some transparent materials. The birefringence in a stressed photoelastic model is controlled by the state of stress at each point in the model. It is very useful for problems in which stress or strain information is required for extended regions of the structure or member, and particularly for those having complicated geometry, complicated loading conditions, or both. While the traditional areas of application have largely been taken over by numerical techniques, advances in computer technology and digital image processing techniques have made photoelastic analysis more efficient and reliable for solving engineering problems. The main aim of this review is to provide the reader with a brief background of the computer-based digital image processing approaches for evaluation of photoelastic fringe patterns, and for the determination of isochromatic fringe orders and principal stress directions from photoelastic images.

1 Basic Principles of Photoelasticity

1.1 Light and Complex Notation

The photoelastic effect can be adequately described by the electromagnetic theory of light [1,2,3]. According to this theory, light can be regarded as a

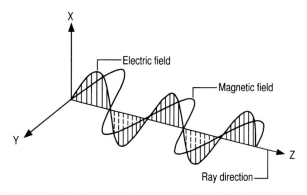

Fig. 1. The electric and magnetic fields of a linearly polarized wave

transverse electromagnetic wave. As shown in Fig. 1, the electric and magnetic fields are perpendicular to each other and to the direction of propagation. Accordingly, the magnitude of the light (electric) vector, E, can be expressed in terms of a one-dimensional wave equation

$$E = a\cos(2\pi/\lambda)ct = a\cos 2\pi ft = a\cos\omega t , \qquad (1)$$

where a is the amplitude, λ is the wavelength, c is the velocity of propagation (3×10^8 m/s in vacuum), t is the time, f is the frequency of the light, and $\omega = 2\pi f$ is the circular frequency of the light.

For convenience, both the amplitude and phase of a light wave are represented by the real part of the complex expression

$$E = ae^{i(2\pi/\lambda)ct} = ae^{i\omega t} , \qquad (2)$$

The imaginary part in (2) could also be used to represent the wave. Furthermore, a wave with an initial optical path δ can be expressed in exponential form as

$$E = ae^{i(2\pi/\lambda)(ct-\delta)} = ae^{i(\omega t - \Delta)} , \qquad (3)$$

where $\Delta = 2\pi\delta/\lambda$. Superposition of two or more waves having the same frequency but different amplitude and phase can be easily performed using the exponential representation.

1.2 Polarization of Light

Most light sources consist of a large number of randomly oriented atomic or molecular emitters. The light rays emitted from such a source will have no preferred orientation. These rays make up an ordinary (unpolarized) light beam. If a light beam is made up of rays with their electric fields restricted in a single plane, the light beam is said to be plane polarized or linearly polarized. This is shown diagramatically in Fig. 2a. The direction of the light vector is used to describe the polarization. Linearly polarized light can be resolved into its components along any pair of arbitrary orthogonal axes by the usual vector-sum rules. Two linearly polarized light waves having the same frequency but lying in mutually perpendicular planes are also superposed by the vector-sum rules. For two light waves having different amplitude and/or phase the result is an elliptically polarized light. If the amplitudes of the two waves are equal and the relative phase is $(2m + 1)\pi/2$, where $m = 0, 1, 2, \ldots$ a circularly polarized light is obtained. Figure 2b,c depict the two forms of polarization. Polarized light can be produced from a natural (randomly polarized) source by (1) reflection, (2) scattering, (3) fine grids, (4) use of Polaroid sheets, and (5) passing through double refraction or birefringence [4].

Digital Photoelasticity 199

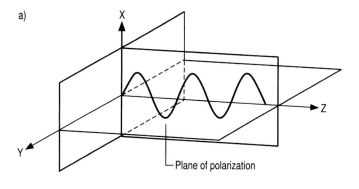

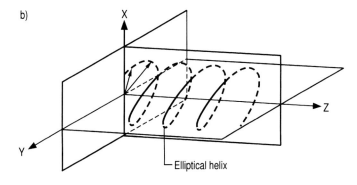

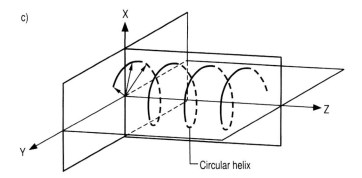

Fig. 2. Motion of the polarized light vectors: (**a**) plane or linear (**b**) elliptical (**c**) circular

1.3 Retardation

In free space, light propagates with a velocity c that is faster than the velocity in any other medium. The ratio of the velocity of light in free space to the velocity in a medium, n, is called the index of refraction. A wave propagating in a material will develop a linear optical path shift δ, called retardation, with respect to a similar wave propagating in free space. The absolute retardation is given by

$$\delta = h(n-1), \qquad (4)$$

where h is the thickness of the material along the path of light propagation. The retardation δ is a positive quantity since the index of refraction of a material is always greater than unity. The retardation of one wave with respect to another wave is called the relative retardation.

1.4 Optical Media

In photoelasticity, an optical instrument, called a polariscope, is used to study the stress effect on models made of birefringent materials. In experimental work, two types of polariscope are frequently employed - the plane polariscope and the circular polariscope. In the plane polariscope, a plane-polarized light is produced with an optical element known as a plane or linear polarizer. Production of circularly or elliptically polarized lights requires the use of a linear polarizer together with an optical element known as a wave plate.

1.4.1 Linear or Plane Polarizers

When a light wave strikes a plane polarizer, this optical element resolves the wave into two mutually perpendicular components, as shown in Fig. 3. The component parallel to the axis of polarization is transmitted while the component perpendicular to the axis of polarization is absorbed. The transmitted component of the light vector is

$$E_{tr} = a\cos\theta\, e^{i(\omega t)}, \qquad (5)$$

where θ is the angle between the axis of polarization and the incident light vector. Polarizers or Polaroid sheets are often used for producing polarized light in polariscopes.

1.4.2 Birefringent Materials

Birefringent or doubly refracting materials are optical media which have the ability to resolve a light vector into two orthogonal components and to transmit these components with different velocities. Therefore, the two orthogonal components emerge with different retardation. These materials include both

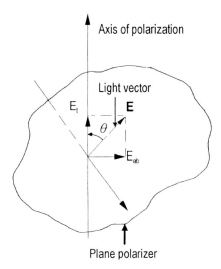

Fig. 3. Transmitting and absorbing characteristic of a plane polarizer

photoelastic models and wave plates. For models, the orientation of the two orthogonal components and relative retardation vary from point to point. For wave plates, the orientation of the two orthogonal components and relative retardation are constant throughout the plate.

As illustrated in Fig. 4, a doubly refracting plate has two principal axes labeled 1 and 2. If the velocity of light transmitted along axis 1 is greater than the velocity transmitted along axis 2, axis 1 is called the fast axis, and axis 2 the slow axis. When a field of plane-polarized light, E, enters the plate, the light vector is resolved into two components E_1 and E_2 along axes 1 and 2, given by

$$E_1 = E\cos\theta = a\cos\theta e^{i\omega t} , \tag{6}$$
$$E_2 = E\sin\theta = a\sin\theta e^{i\omega t} , \tag{7}$$

where θ is the angle between the light vector E and the fast axis 1.

Since the light components E_1 and E_2 travel through the plate with different velocities, the two components emerge from the plate with different retardation with respect to a wave in air. From (4), the retardation can be expressed as

$$\delta_1 = h(n_1 - 1) , \tag{8}$$
$$\delta_2 = h(n_2 - 1) , \tag{9}$$

where h is the thickness of the plate, and n_1 and n_2 are the indices of refraction for waves vibrating parallel to the principal axes 1 and 2, respectively. The relative retardation is then computed as

$$\delta = \delta_2 - \delta_1 = h(n_2 - n_1) . \tag{10}$$

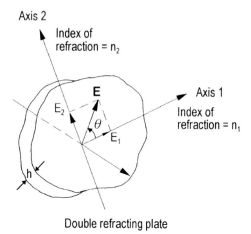

Fig. 4. A plane-polarized light wave entering a double refracting plate

The relative phase shift Δ between the two components as they emerge from the plate is given by

$$\Delta = \Delta_2 - \Delta_1 = (2\pi/\lambda)h(n_2 - n_1) \ . \tag{11}$$

Since only the relative phase shift Δ affects the interference of the two components of the light, we usually set $\Delta_1 = 0$ and $\Delta_2 = \Delta$. Thus the two components are described by

$$E_1' = a\cos\theta\, e^{i\omega t} \ , \tag{12}$$
$$E_2' = a\sin\theta\, e^{i(\omega t - \Delta)} \ , \tag{13}$$

1.4.3 Quarter-Wave Plate

When the doubly refracting plate is designed to give $\Delta = \pi/2$, it is called a quarter-wave plate. If a quarter-wave plate is used in Fig. 4 with an angle $\theta = \pi/4$, the magnitude of the light vector emerging from the plate has a constant magnitude and the tip of the light vector traces out a circle as the wave propagates in the z direction. Such a light wave is known as circularly polarized light. If the angle $\theta \neq m\pi/4$, for $m = 0, 1, 2, 3$, the magnitude of the light vector varies with angular position in such a way that the tip of the light vector traces out an ellipse as it propagates in the z direction. Such a light wave is known as elliptically polarized light. It is to be noted that a quarter-wave plate suitable for a particular wavelength of monochromatic light will not be suitable for a different wavelength of light.

1.5 The Stress-Optic Law

Many noncrystalline transparent materials are optically isotropic under normal conditions but become doubly refractive or birefringent like crystals when stressed. This effect persists while the loads are maintained but disappears almost instantaneously when the loads are removed. In a three-dimensional stress system, there are three principal stresses and three principal stress directions at any point in a stressed birefringent material. Relations between stresses and the indices of refraction for a temporary birefringent material were formulated by Maxwell in 1852 as

$$n_1 - n_0 = C_1\sigma_1 + C_2(\sigma_2 + \sigma_3) , \tag{14}$$
$$n_2 - n_0 = C_1\sigma_2 + C_2(\sigma_1 + \sigma_3) , \tag{15}$$
$$n_3 - n_0 = C_1\sigma_3 + C_2(\sigma_1 + \sigma_2) , \tag{16}$$

where n_0 is the index of refraction of material in the unstressed state, n_1, n_2 and n_3 are the principal indices of refraction which coincide with the principal stress directions, σ_1, σ_2, and σ_3 are the principal stresses at the point, and C_1 and C_2 are material constants known as "stress-optic coefficients". Equations (14,15,16) are the fundamental relationships between the stress and optical effects and are known as the stress-optic law. Eliminating n_0 from (14,15) gives the stress-optic law for a material under a general triaxial stress system as

$$n_2 - n_1 = (C_2 - C_1)(\sigma_1 - \sigma_2) = C(\sigma_1 - \sigma_2) , \tag{17}$$

where C is the relative stress-optic coefficient. Similarly,

$$n_1 - n_3 = C(\sigma_3 - \sigma_1); \quad and \quad n_3 - n_2 = C(\sigma_2 - \sigma_3) . \tag{18}$$

The relative stress-optic coefficient C is usually assumed to be a material constant. However, various studies [5,6,7] have shown that this coefficient depends on wavelength and should be used with care. The dependence of the relative stress-optic coefficient on the wavelength of the light being used is referred to as photoelastic dispersion or dispersion of birefringence.

For a two-dimensional plane-stressed birefringent body, the incident light will propagate through the body polarized in the principal planes, and with emerge with a stress-induced relative retardation. From (10,11,17) we have

$$\delta = hC(\sigma_1 - \sigma_2) , \tag{19}$$

or

$$\Delta = 2\pi hC(\sigma_1 - \sigma_2)/\lambda . \tag{20}$$

Equation (19) is more commonly written as

$$\sigma_1 - \sigma_2 = Nf_\sigma/h , \tag{21}$$

where $N(=\delta/\lambda = \Delta/2\pi)$ is the relative retardation in terms of cycles of retardation often called the "fringe order", and $f_\sigma(=\lambda/C)$ is the material fringe value with typical units of N/m-fringe or lb/in-fringe. Equation (21) is used frequently in photoelastic stress analysis. The principal stress difference can be determined if the material fringe value of the model is calibrated and the fringe order N can be measured at each point.

The stress-optic law in terms of secondary principal stresses works well for three-dimensional photoelasticity. The secondary principal stresses at the point of interest lie in the plane whose normal vector is coincident with the path of the light beam, but not the direction of the principal stress σ_1, or σ_2, or σ_3. Equations (20,21) can be expressed for the secondary principal stresses σ'_1 and σ'_2 as

$$\Delta = 2\pi hC(\sigma'_1 - \sigma'_2)/\lambda , \tag{22}$$
$$\sigma'_1 - \sigma'_2 = Nf_\sigma/h . \tag{23}$$

1.6 Plane Polariscope

A polariscope is an instrument that measures the relative retardations and the principal-stress directions that occur when polarized light passes through a stressed photoelastic model. The plane polariscope arrangement, as shown in Fig. 5, consists of a plane-stressed model placed between two linear polarizers, called the polarizer and analyzer, respectively. The principal-stress direction at the point under consideration in the model makes an angle θ with the polarization axis of the polarizer.

When a light wave passes through the plane polarizer, the plane-polarized light wave emerging from the polarizer is the same as in (5):

$$E = a\cos\theta e^{i\omega t} , \tag{24}$$

After leaving the polarizer, this plane-polarized light wave enters the stressed model, as shown in Fig. 6, and is resolved into two components, E_1 and E_2, with vibrations parallel to the principal stress directions, σ_1 and σ_2, at the point. Since the two components propagate through the model with different velocities ($v_1 > v_2$), the waves emerge from the model with a relative phase shift Δ between them in accordance with (12,13). Upon entering the analyzer

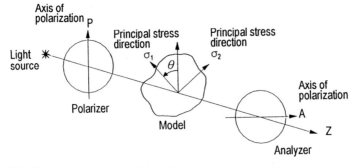

Fig. 5. Arrangement of the optical elements in a plane polariscope

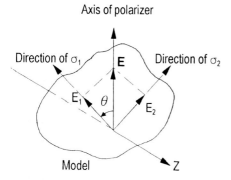

Fig. 6. Resolution of the light vector as it enters a stressed photoelastic model

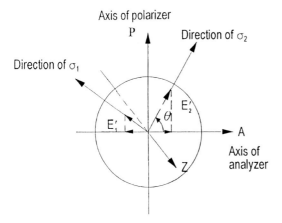

Fig. 7. Transmission of the components of the light vectors through the analyzer

(Fig. 7), the light components resolve into components that vibrate parallel and perpendicular to the axis of the analyzer. The parallel components transmitted by the analyzer combine to produce an emerging light vector, A, given by

$$A = E'_2 \cos\theta - E'_1 \sin\theta = ia \sin(2\theta) \sin(\Delta/2) e^{i(\omega t - \Delta/2)} . \tag{25}$$

Since the intensity of light is equal to the product of the light vector and its conjugate, the intensity of the light emerging from the analyzer is

$$I_{\text{PD}} = AA^* = a^2 \sin^2(2\theta) \sin^2(\Delta/2) = I_m \sin^2(2\theta) \sin^2(\Delta/2) , \tag{26}$$

where I_m is the maximum intensity of the light emerging from the analyzer. Because the unstressed background appears dark, this arrangement is called a dark-field plane polariscope. Rotating the analyzer through 90°, a light-field intensity, I_{PL}, can be derived as

$$I_{\text{PL}} = I_m [1 - \sin^2(2\theta) \sin^2(\Delta/2)] . \tag{27}$$

Equation (26) indicates that extinction ($I_{PD} = 0$) occurs for $\sin^2 \theta = 0$ or $\sin(\Delta/2) = 0$. These two conditions are discussed in turn as follows.

When $\sin^2 \theta = 0$, then $2\theta = m\pi$, where $m = 0, 1, 2, \ldots$. This result indicates that one of the principal stress directions coincides with the axis of the polarizer ($\theta = 0, \pi/2$, or any exact multiple of $\pi/2$). The fringe pattern produced by the $\sin^2 \theta$ term, called the isoclinic fringe pattern, is used to determine the principal stress directions at all points of a photoelastic model.

When $\sin(\Delta/2) = 0$, then $\Delta/2 = m\pi$, where $m = 0, 1, 2, 3, \ldots$. This result indicates that the principal-stress difference is either zero ($m = 0$) or is sufficient to produce an integral number of wavelengths of retardation ($m = 1, 2, 3, \ldots$). The fringe pattern produced by the $\sin(\Delta/2)$ term is called the isochromatic fringe pattern. The order of extinction, m, can be related to the stress-optic law as

$$m = \Delta/2\pi = N = h(\sigma_1 - \sigma_2)/f_\sigma \ . \tag{28}$$

Equation (28) shows that the order of extinction m is equivalent to the fringe order N. When a model is viewed in monochromatic light, the isochromatic fringe pattern appears as a series of dark bands. However, when a model is viewed in white light, the isochromatic fringe pattern appears as a series of colored bands. The black fringe appears only when the principal stress difference is zero and the extinction occurs for all wavelengths of light. Figure 8 shows a typical fringe pattern of the superimposed isoclinic fringes and the isochromatic fringes.

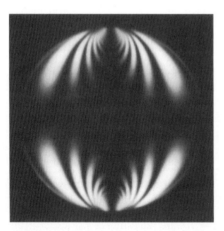

Fig. 8. Superimposed isochromatic and isoclinic fringe patterns from a disk loaded in diametral compression

1.7 Circular Polariscope

Since isoclinic and isochromatic fringe patterns are superimposed in a plane polariscope, an accurate examination of the isochromatic fringes is difficult. Placing a stressed photoelastic model in the field of a circular polariscope can eliminate the isoclinic fringes and still maintain the isochromatic fringes. To illustrate this effect, consider a stressed model in the circular polariscope shown in Fig. 9. The plane polarized light beam emerging from the polarizer is also given by (24). As the light enters the first quarter-wave plate, it is resolved into components with vibrations parallel to the fast (E_f) and slow (E_s) axes, respectively. Since a relative angular phase shift $\Delta = \pi/2$ is developed between the components, these emerge from the plate as

$$E_f = a \sin 45° e^{i\omega t} = (1/\sqrt{2})ae^{i\omega t} , \tag{29}$$
$$E_s = a \cos 45° e^{i\omega t} e^{(i-\pi/2)} = -i(1/\sqrt{2})ae^{i\omega t} . \tag{30}$$

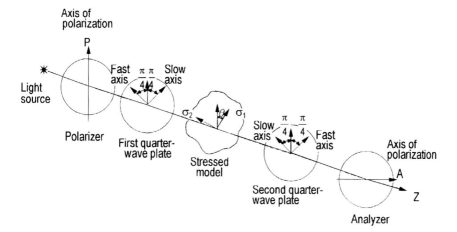

Fig. 9. A stressed photoelastic model in a circular-crossed polariscope

After leaving the quarter-wave plate, the components of the light vector enter the model and resolve into components E_1 and E_2, as illustrated in Fig. 10. Let ϕ be the angle between the slow axis and the principal stress direction σ_1. The components emerging from the model with a relative retardation Δ are

$$E_1 = E_f \sin\phi + E_s \cos\phi = -ia(1/\sqrt{2})ae^{i(\omega t + \phi)} , \tag{31}$$
$$E_2 = (E_f \cos\phi - E_s \sin\phi)e^{-i\Delta} = a(1/\sqrt{2})ae^{i(\omega t + \phi - \Delta)} . \tag{32}$$

The light emerging from the model propagates to the second quarter-wave plate (Fig. 11). It is noted here that the direction of the slow and fast

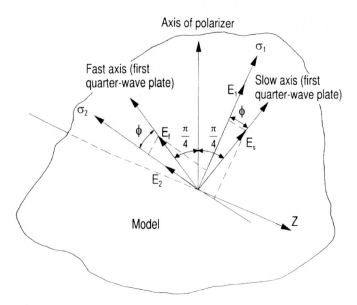

Fig. 10. Resolution of the light vectors as they enter a stressed photoelastic model

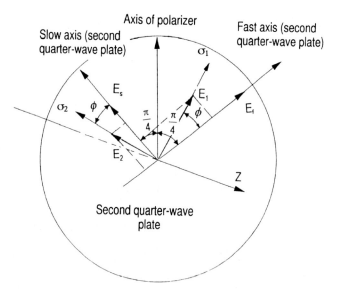

Fig. 11. Resolution of the light vectors as they enter the second quarter-wave plate

axes is switched as compared to the first quarter-wave plate. Thus the waves emerging from the plate can be expressed as

$$E'_f = E_1 \cos\phi - E_2 \sin\phi ,$$
$$= -a(1/\sqrt{2})e^{i(\omega t+\phi)}(-i\cos\phi - e^{-i\Delta}\sin\phi) , \qquad (33)$$
$$E'_s = (E_2 \cos\phi + E_1 \sin\phi)e^{-i(\pi/2)} ,$$
$$= -ia(1/\sqrt{2})e^{i(\omega t+\phi)}(e^{-i\Delta}\cos\phi - i\sin\phi) . \qquad (34)$$

When the light enters the analyzer (Fig. 12), the component parallel to the analyzer alone is transmitted. The amplitude of light emerging from the analyzer is

$$A = (E'_f - E'_s)\cos 45° ,$$
$$= ae^{i(\omega t+2\phi-\Delta/2)}\sin(\Delta/2) . \qquad (35)$$

The intensity of light emerging from the analyzer of a circular polariscope follows as

$$I_D = AA^* = a^2 \sin^2(\Delta/2) = I_m \sin^2(\Delta/2) . \qquad (36)$$

Since the angle θ does not appear in the above expression, the isoclinics have been eliminated from the fringe pattern. The intensity of the light beam emerging from the circular polariscope is a function of only the principal stress difference. The fringes are counted in the sequence $0, 1, 2, 3$. Since the unstressed background appears dark, this arrangement is called a dark-field circular polariscope.

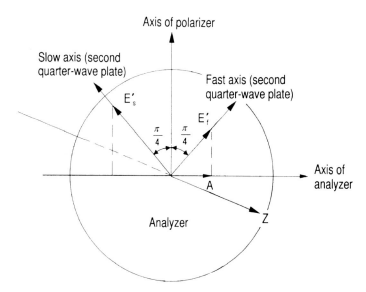

Fig. 12. Transmission of the components of the light vectors through the analyzer

The circular polariscope can be converted from dark field to light field simply by rotating the analyzer by 90°. It can be shown that the intensity I_L is given by

$$I_\mathrm{L} = I_\mathrm{m} \cos^2(\Delta/2) \,. \tag{37}$$

Equation (37) shows that extinction ($I_\mathrm{L} = 0$) will occur when

$$\Delta/2 = (2m+1)\pi/2 \quad for \quad m = 0, 1, 2, 3, \ldots ,$$

thus

$$N = \Delta/2\pi = m + 1/2 \,,$$

which implies that the order of the fringes observed in a light-field polariscope is $1/2, 3/2, 5/2$, etc. An example of dark- and light-field isochromatic fringe patterns is shown in Fig. 13.

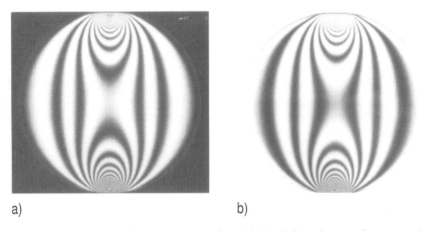

a)   b)

Fig. 13. Isochromatic fringe patterns of a disk loaded in diametral compression: (a) dark field (b) light field

2 Computer Evaluation of Photoelastic Fringe Patterns

In conventional photoelastic analysis, light- and dark-field isochromatic fringe patterns are obtained to allow for the determination of half- and whole-order fringes. The measurement of the principal stress directions is made by manually rotating the polarizer and analyzer of a plane-crossed polariscope at the same time. In order to obtain an accurate measurement of the stress, especially in the region where the fringes are sparse, the fractional fringe orders are measured either by the Tardy method, or by using a compensator, or by the fringe multiplication method [1,8,9]. The compensation methods are basically point-by-point methods; fringe multiplication by using partial mirrors produces a full-field measurement, but extra optical equipment is required and a considerable loss of light intensity occurs.

In whole-field analysis, conventional measurements can be very tedious and time-consuming, and require skill in the identification and measurement of isochromatic and isoclinic fringes. Therefore, various systems that incorporate image processing boards, electronic cameras (video or CCD) and computers, have been developed in the past twenty years to overcome this difficulty [7,10,11,12,13,14,15,16,17,18,19]. Earlier developments mainly involved the use of mini-computer or mainframe and various digital image-processing techniques to find the location of the fringe points and to interactively assign the fringe orders. Recent studies propose to process intensity data to obtain the whole field experimental parameters, both quantitatively and automatically, with a PC-based image processing system. In the following sections, digital image processing is introduced, and various methods developed for the whole-field analysis of photoelastic fringe patterns are described and discussed.

2.1 Digital Image Processing

Digital image processing [20,21,22] provides high-speed analysis, and helps to extract useful information that might otherwise be suppressed when conventional methods are used. A commonly used PC-based digital image processing system shown in Fig. 14, consists of video or CCD camera for image input, a frame grabber to digitize the images, a personal computer with enough memory to process and store the digitized data, and some peripheral devices for image display or output. Usually the image is presented to the CCD camera directly and is digitized at a rate of 30 frames per second (NTSC).

The frame grabber or digitizer samples the camera signal into an array of discrete picture elements, called pixels. Common digital image sizes are $256 \times 256, 512 \times 512$, and 1024×1024 pixels. The intensity at each pixel is quantized into 256 gray-levels, which correspond to 8 bits. The gray levels are denoted by integers, with 0 as the darkest level and 255 as the brightest

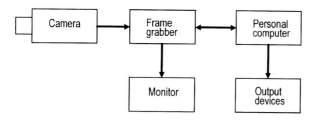

Fig. 14. Diagram of a PC-based digital image processing system

level. The output of the digitizer is stored and processed by digital image processing algorithms. The results may be stored or displayed on a monitor.

With advancements in digital technology, color image-processing systems with RGB (Red, Green, Blue) cameras and RGB frame grabbers are now available at affordable prices. The color image-processing system acquires images in RGB mode, and the intensity is quantized for each of the image planes of R, G, and B [23,24,25].

Usually the digitized gray-level value is not proportional to the brightness because of the nonlinearity of the image sensor. The dark current generated by thermal excitation also limits the readout in areas with low-light. Thus, a careful calibration of the system should be made in order to obtain accurate results. Figure 15 shows the perfectly linear response, and the original and calibrated digital gray-level responses of a system [26]. The minimum resolvable gray-level value is about 30.

Some typical algorithms or operations in digital image processing of photoelastic fringe patterns are summarized as follows [26,27,28,29]:
1. Histogram and gray-level transformation. A histogram is a plot of the frequency of gray-level versus gray levels. It is used to check whether a recorded image is appropriate. The aperture of the camera may be changed if the frequency of the minimum and/or maximum gray level is too large or too small. The method for gray-level transformation can be chosen based on the histogram. The original gray-level range can be stretched or compressed into

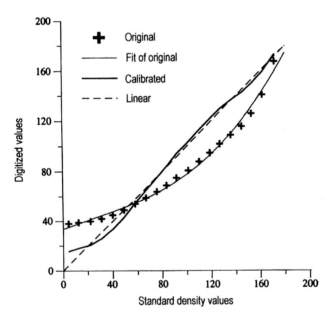

Fig. 15. Gray-level response of a standard density-step tablet

an ideal range by a linear transformation. A proper gray level can be selected from the histogram to binarize or slice the image for better examination of the image.

2. Algebraic operations produce an output image which is the pixel-by-pixel summation, difference, product, or quotient of two or more images. Typical examples include averaging over several consecutively recorded fringe patterns to suppress time-varying intensity fluctuations and normalization of an image by division.

3. Operation Masks. A mask is used to operate all pixels in the neighborhood of an input pixel to determine the gray value of the corresponding output pixel. Typical operations include smoothing of the fringe pattern by low-pass filtering or median filtering to minimize the noise and high-pass filtering to find the boundary of an object.

These fundamental digital image-processing techniques can be applied to analyze photoelastic fringe patterns. Other operations, such as the Fourier-transform [30,31], pattern segmentation, thinning, and clustering of the fringe patterns, are also useful for digital evaluations of the photoelastic fringes.

2.2 Extraction of Fringe Point

The first application of digital image processing in photoelasticity concerns the recognition of fringes in a single photoelastic fringe pattern. The fringe pattern is binarized [16,17] and then the geometrical centers of the binarized fringes are determined by using iterative circle fitting [16] or thinning processes [17]. The associated fringe orders are assigned to the determined fringe points interactively. These methods enable the points having integral or half-fringe orders to be measured. However, larger errors are reported in zones with fringe gradients that are too high or too low or in the case of broad fringes with large curvature [17]. The minima of the intensity of the original fringe pattern can also be found by using predefined criteria [18,19]. These methods eliminate the need for iterative procedures and determine the fringe locations quite accurately [13]. Since only one image needs to be processed, these methods are applicable to both static and dynamic problems.

2.3 Fringe Multiplication

Digital fringe multiplication of photoelastic fringe patterns was proposed first by *Toh* et al. [32]. They multiplied the fringe by subtracting the pixel intensities of the dark-field image from the light-field image. In their method, the circular polariscope is used to obtain the isochromatic field. Considering the variation of light recorded in dark- and light-field, the intensity of dark- and light-field isochromatic fringe patterns given by (36) and (37) is modified as

$$I_\mathrm{D} = I_\mathrm{Dm} \sin^2(\Delta/2) \ , \tag{38}$$
$$I_\mathrm{L} = I_\mathrm{Lm} \cos^2(\Delta/2) \ , \tag{39}$$

where I_{Dm} and I_{Lm} are the maximum dark- and light-field intensities of light emerging from the analyzer, respectively. Firstly, the dark-field intensity is adjusted as

$$I'_D = I_D(I_{Lm}/I_{Dm}) = I_{Lm}\sin^2(\Delta/2) . \tag{40}$$

Then subtracting the modified intensity,

$$I_r = |I_L - I'_D| = |I_{Lm}\cos\Delta| . \tag{41}$$

Since the gray levels are represented by positive values in digital image-processing systems, only the absolute values of the differences in (41) can be used. Extinction occurs when $\Delta = (2m+1)\pi/2$ for $m = 0, 1, 2, \ldots$; i.e. the fringe order observed on the resultant image is $1/4$, $3/4$, etc., and the fringes are doubled. Since the resultant fringe pattern does not have the original cosine feature of a gray-level distribution, a square operation as follows can be performed on the resultant image to obtain the original characteristic [33]

$$I'_r = (I_L - I'_D)^2/I_{Lm} = I_{Lm}\cos^2\Delta . \tag{42}$$

The squared resultant image, I'_r, can be reversed by using standard image-processing techniques. Further subtraction of the reversed squared resultant image from the squared resultant image increases the number of fringes by a factor of two. The procedures can be repeated until the resolution limit of the image system is reached. Although the intensity I_{Dm} is adjusted to be equal to I_{Lm}, errors might occur due to local variations.

A simple approach based on the image-division (or normalization) technique and the multiple-angle relation of the cosine function for fringe multiplication is described in [34]. This method uses only light-field isochromatic images. Let I_U be the light-field intensity of the unloaded photoelastic model. Using (37) and dividing I_L by I_U gives the resultant intensity

$$I_r = I_L/I_U = \cos^2(\Delta/2) . \tag{43}$$

For the purpose of fringe multiplication, (43) can be substituted directly into the squared trigonometric identities of multiple-angle relations of cosine function. For example, the following relation (44) is obtained by making use of the trigonometric identity of the double-angle relation

$$I_s = \cos^2\Delta = 4I_r^2 - 4I_r + 1 . \tag{44}$$

The extinction of the intensity I_s occurs when $\Delta = (2m+1)\pi/2$ for $m = 0, 1, 2, \ldots$. The order of the fringe observed on the image is $N = (2m+1)/4$, i.e. the fringe order is $1/4$, $3/4$, etc. Therefore the number of fringes is doubled.

The multiplication process can be performed repeatedly until the resolution limit is reached. Generally, the fringes can be increased by a factor of k^n utilizing the multiple-angle relation of $k\Delta$ with n iterations. A mixed use of multiple-angle relations of $k\Delta$ and $m\Delta$ increases the number of the fringe $m \times k$ times. Figure 16 shows the fringe patterns of a circular disk under a diametral compression load [34]. This method, with slight modification, has been applied to photoelastic slices obtained through the frozen stress method [26], and to photoelastic coatings [35].

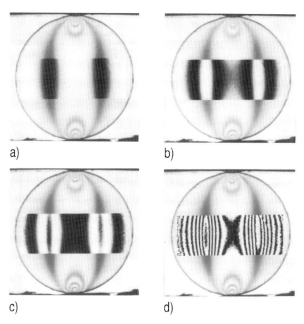

Fig. 16. Light-field images of a circular disk with the window area being (**a**) normalized and adjusted (**b**) multiplied by 2 (**c**) multiplied by 3 (**d**) multiplied by 16

2.4 Determination of the Isochromatic Fringe Order

Approaches for whole-field digital determination of photoelastic birefringence, or fringe order, can be grouped into two categories: (1) point-by-point analysis and (2) full-field analysis. The point-by-point analysis, developed in [36,37,38] determines the isochromatic fringe orders automatically by using spectral contents. In full-field analysis, the methods developed in [27,34,39] are limited to half-fringe orders. The use of two wavelengths of light to automatically determine the isochromatic fringe order are reported in [7,18]. Phase-stepping algorithms to determine the phase in modulo π are reported in [29,40,41]. Recent studies that use color image processing systems are reported in [23,24,25]. The red-green-blue (RGB) or hue-saturation-intensity (HSI) values recorded by a color image-processing system are used to determine the fringe orders.

2.4.1 Half-Fringe Photoelasticity

The method, called "half-fringe photoelasticity" (HFP), operates effectively with less than 0.5 wavelength of relative retardation because it cannot distinguish the gray levels in any interval from 0 to $\pi/2$, $\pi/2$ to π, and so on [39]. In this method, the response Z of the camera of the image analysis system is related to the light intensity by

$$Z = kI^\gamma ,$$

where Z is the digitized pixel value, γ is the log-linear slope of the camera sensitivity, and k is a proportionality constant. If the response to the brightest point I_{Dm} (with fringe order $N = 1/2$) in the dark-field circular polariscope is Z_m, then the fringe order at any point in the image can be calibrated with respect to that point by

$$N = \sin^{-1}(Z/Z_m)^{1/2\gamma}/\pi . \tag{45}$$

The values of N and Z on an unloaded specimen can be obtained by employing the Tardy compensation method. The least-squares method is then used to establish the relationship between N and Z

$$N = \sin^{-1}(Z/bZ_m)^{1/2\gamma}/\pi , \tag{46}$$

where b is the calibration constant, and γ is the slope of the best fitted line. *Wang* and *Chen* [42] extended the range of half-fringe photoelasticity to orders greater than 1/2 by assigning the total fringe order manually.

2.4.2 Phase-Stepping Method

In most classical interferometers, phase differences are added by altering the phase of the reference light beam in known steps and measuring the local light intensity after each step [43]. In photoelasticity, the change in phase is achieved by rotation of the optical elements of the polariscope. Patterson and Wang [29] presented an algorithm that determines the fractional fringe order and principal stress direction from six equations. The circular polariscope setup used is illustrated in Fig. 17; β, ϕ and θ are angles between the horizontal and the slow axis of the analyzer, the output quarter-wave plate, and the photoelastic model, respectively. The six steps used in pairs of ($\phi = 0, \beta = \pi/4$), ($\phi = 0, \beta = 3\pi/4$), ($\phi = 0, \beta = 0$), ($\phi = \pi/2, \beta = \pi/2$), ($\phi = \pi/4, \beta = \pi/4$), and ($\phi = 3\pi/4, \beta = 3\pi/4$) give rise to six sets of intensity measurements

$$I_1 = I_b + I_m \cos \Delta , \quad \text{(light-field circular polariscope)} \tag{47}$$
$$I_2 = I_b - I_m \cos \Delta , \quad \text{(dark-field circular polariscope)} \tag{48}$$
$$I_3 = I_b - I_m \sin 2\theta \sin \Delta , \tag{49}$$
$$I_4 = I_b + I_m \cos 2\theta \sin \Delta , \tag{50}$$
$$I_5 = I_b + I_m \sin 2\theta \sin \Delta , \tag{51}$$
$$I_6 = I_b - I_m \cos 2\theta \sin \Delta , \tag{52}$$

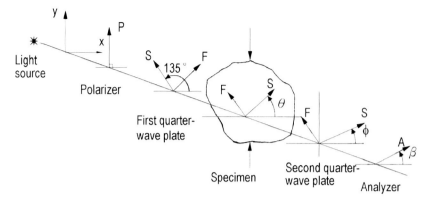

Fig. 17. Diagramatic representation of the optical arrangement for phase stepping

where I_b represents the background/stray light intensity. Using the above equations, the fractional retardation and principal stress direction are determined as

$$\theta = \{\tan^{-1}[(I_5 - I_3)/(I_4 - I_6)]\}/2 ,\tag{53}$$
$$\Delta = \tan^{-1}\{(I_5 - I_3)/[(I_1 - I_2)\sin 2\theta]\} ,$$
$$= \tan^{-1}\{(I_4 - I_6)/[(I_1 - I_2)\cos 2\theta]\} .\tag{54}$$

From (54), the retardation Δ can be expressed either in the range $0 - \pi$ or $0 - 2\pi$ if the sign of the numerator and denominator are known. Figure 18 shows a sequence of six images of a circular disk under diametral compression and the determined fractional fringe order as a phase map. It is to be noted that when $I_3 = I_4 = I_5 = I_6 = 0$, the isoclinic parameter θ is undefined. However, it can be shown that the fractional retardation is zero at these undefined points. In order to find the total fringe order, phase unwrapping has to be performed. The unwrapping process requires the determination of the total fringe order for at least one point along the scanned line by auxiliary means.

A method has been proposed that makes use of the whole-field intensity data for three analyzer positions in a plane polariscope [40]. The optical arrangement is shown in Fig. 19. The three analyzer positions used ($\beta = 0, \pi/2, \pi/4$) gives rise to three sets of intensity measurements

$$I_1 = I_m[\cos^2(\Delta/2) + \cos^2 2\theta \sin^2(\Delta/2)] \tag{55}$$
$$I_2 = I_m \sin^2 2\theta \sin^2(\Delta/2) \quad \text{(dark-field plane polariscope)} \tag{56}$$
$$I_3 = I_m[1 - \sin 4\theta \sin^2(\Delta/2)]/2 . \tag{57}$$

The principal stress direction and the fractional retardation are obtained as

$$\theta = \{\tan^{-1}[2I_2/(I_1 + I_2 - 2I_3)]\}/2 , \tag{58}$$
$$\Delta = \cos^{-1}[(A - A^2 - B^2)/(1 - A)] , \tag{59}$$

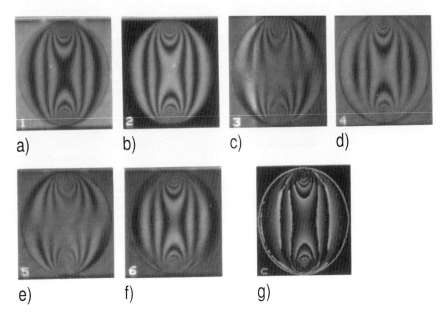

Fig. 18. The sequence of six images (**a-f**), and the phase map (**g**) of the determined fractional fringe order of a circular disk under diametral compression [47]

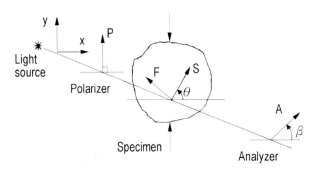

Fig. 19. Diagramatic representation of a plane polariscope for phase stepping

where

$$A = (I_1 - I_2)/(I_1 + I_2) ,$$
$$B = (I_1 + I_2 - 2I_3)/(I_1 + I_2) .$$

Owing to the background/stray light effect, the intensity values of the unstressed model captured in the dark-field plane polariscope should be subtracted from all other images before using the above equations. Since Δ is evaluated from the inverse cosine function, ambiguity exists regarding the sign of Δ. An auxiliary method is required to obtain the total fringe order.

2.4.3 Two-Wavelength Method

The use of two wavelengths of light to determine the isochromatic fringe order is reported in [18,44]. The zero-order fringe can be identified through an image processing method in which two patterns of different wavelengths are superimposed on each other [18]. A full-field scheme has been proposed that uses the intensity data of two light-field isochromatic fringe patterns with different wavelengths to determine the total fringe order automatically [7]. In this method the total fringe order is determined regardless of whether zero-order fringes are present or not, and phase unwrapping is not needed. The experimental set-up is shown schematically in Fig. 20. The two wavelengths are produced by using two narrow-band filters put in front of the camera. The intensity of a light-field isochromatic fringe pattern, given by (37), is expressed in terms of the fringe order as

$$I_\mathrm{L} = I_\mathrm{m} \cos^2(N\pi) \,. \tag{60}$$

The normalized intensity can therefore be obtained from

$$I_\mathrm{r} = I_\mathrm{L}/I_\mathrm{U} = \cos^2(N\pi) \,. \tag{61}$$

By inverting (61), the fringe order obtained is multiple-valued as

$$\begin{aligned} N &= m + \cos^{-1}(I_\mathrm{L}/I_\mathrm{U})^{1/2}/\pi \,, & for \quad m = 0, 1, 2, \dots \quad \text{and} \\ N &= m - \cos^{-1}(I_\mathrm{L}/I_\mathrm{U})^{1/2}/\pi \,, & for \quad m = 1, 2, 3, \dots \end{aligned} \tag{62}$$

If the maximum fringe order in the field of interest does not exceed $1/2$, the fringe order can be determined without ambiguity. This is termed as half-fringe photoelasticity.

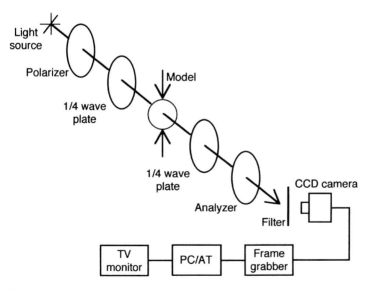

Fig. 20. Schematic of the experimental set-up for the two-wavelength method

For a maximum fringe order, in an isochromatic fringe pattern greater than 1/2, two sets of light-field intensities can be used to determine the total fringe orders. For the two wavelengths λ_1 and λ_2, two sets of multiple-valued fringe orders are obtained from (62) as

$$N_i = m + \cos^{-1}(I_{Li}/I_{Ui})^{1/2}/\pi, \qquad \text{for} \quad m = 0, 1, 2, \ldots \quad \text{and}$$
$$N_i = m - \cos^{-1}(I_{Li}/I_{Ui})^{1/2}/\pi, \qquad \text{for} \quad m = 1, 2, 3, \ldots \qquad (63)$$

where the subscript $i (= 1, 2)$ corresponds to the wavelength λ_i. Since the state of stress at the same point in a model is the same for different wavelengths, the following relationship is obtained from the stress-optic law

$$N_1 f_{\sigma_1} = N_2 f_{\sigma_2}, \qquad (64)$$

where f_{σ_1} and f_{σ_2} are the material fringe values with respect to wavelengths λ_1 and λ_2. Therefore, the actual values of N_1 and N_2 at any point must satisfy (64). The search for the exact fringe orders from the multiple-valued N_1 and N_2 can be done by using a computer to calculate the difference function D

$$D(N_1, N_2) = |N_1 f_{\sigma_1} - N_2 f_{\sigma_2}|. \qquad (65)$$

The exact fringe orders, (N_1, N_2), are determined by finding a pair that minimizes the difference D. This simple procedure works well for cases where the maximum fringe order is not greater than 1. However, in cases where the maximum fringe order is greater than 1, errors in the digitized value mean that exact fringe orders may not be those with a minimum difference. In order to ensure the correctness of the fringe orders, a digital procedure that uses 5×5 cross-shaped points, P1, P2,..., P9, as shown in Fig. 21, can be used. The nine points are equally spaced either horizontally or vertically across one fringe width approximately. The procedures for finding the exact fringe order of the central (seed) point P5, N5, for one load are summarized as follows:
1. Calculate the difference function $D(N_1, N_2)$ for points P1, P2,..., P9.

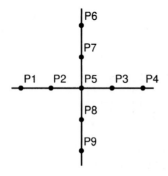

Fig. 21. 5×5 cross-shaped points for exact fringe order searching

2. Search over the function $D(N_1, N_2)$ and store three pairs of fringe orders having the smallest values for each point.
3. Fit the fringe orders of horizontal and vertical points respectively, and calculate the error between the original and the fitted new fringe orders (N_h, N_v) of P5, $(|N_h - N5| + |N_v - N5|)$.
4. Search for the set $(|N_h - N5| + |N_v - N5|)$ with the lowest value and designate $N5$ as the exact fringe order of point P5.

After the exact fringe order of point P5 is determined, this point is used as a seed point to search for the exact fringe orders of the rest of the points through a tree search [17].

Since the light intensity function $\cos^2(N\pi)$ has a period of π, the difference of the two fringe orders between the two images is limited to $1/2$. Hence the range of measurement can be derived as

$$N_1 \leq f_{\sigma_2}/2(f_{\sigma_1} - f_{\sigma_2}) . \tag{66}$$

For simplicity, the range of measurement can be estimated without calibration of the model:

$$N_1 \leq \lambda_2/2(\lambda_1 - \lambda_2) . \tag{67}$$

Moreover, the range of measurement may be extended by setting the difference of fringe orders as less than $1/2$ between two neighboring points. Figure 22a,b shows the fringe patterns of a disk under diametral compression with wavelengths $\lambda_1 = 632.8$ nm and $\lambda_2 = 551.4$ nm. A gray-level representation of the determined fringe orders of the disk is shown in Fig. 22c for $\lambda_1 = 632.8$ nm.

2.4.4 RGB Color Method

Recently, color image-processing systems have been applied to determine isochromatic fringe orders. These are most often based on the direct comparison of the RGB values at each pixel of the specimen to the corresponding color components of a calibration specimen [23,24]. A carefully calibrated data table that relates the fringe order N_i and its values (R_i, G_i, B_i) should be first established. Then an error function is used to search the fringe order of the levels (R_j, G_j, B_j) measured at point j in the specimen:

$$E_i = (R_i - R_j)^2 + (G_i - G_j)^2 + (B_i - B_j)^2 . \tag{68}$$

After the values (R_i, G_i, B_i) of the index i that minimize the error function E_i are found, the corresponding fringe value N_i is determined as the fringe order of point j. A typical RGB component along a line in a pure bending beam under white-light illumination is shown in Fig. 23. The feasibility of using the HIS values for fringe-order determination has also been reported [25].

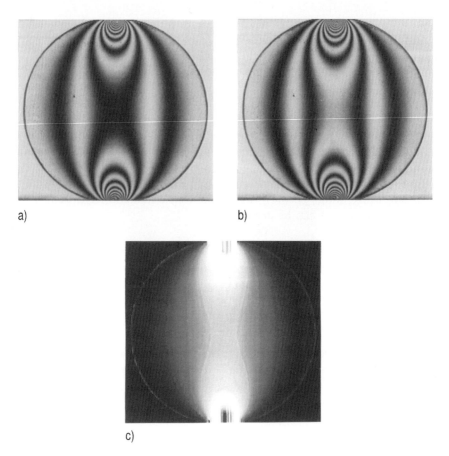

Fig. 22. Fringe patterns of a disk under diametral compression for wavelengths (**a**) $\lambda_1 = 632.8$ nm and (**b**) $\lambda_2 = 551.4$ nm (**c**) gray-level representation of the determined fringe order for wavelength $\lambda_1 = 632.8$ nm

2.5 Determination of Principal Stress Direction

Recent developments in the automated measurement of isoclinic parameters include using CCD (or TV) cameras, intensity measurement, polarization stepping, phase stepping, and digital image processing. *Yao* [45] used image-division and image-differentiation methods to extract the isoclinics. Approaches that use the polarization- or phase-stepping images are described below.

2.5.1 Polarization-Stepping Method

Brown and *Sullivan* [27] used four equally stepped isoclinic images from a plane-crossed polariscope to measure the whole-field principal stress direc-

 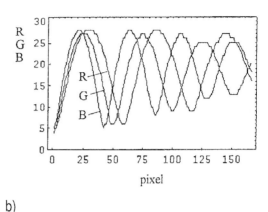

Fig. 23. (a) Isochromatic fringe pattern and (b) RGB color components along the transverse section of the calibration beam [23]

tions. The light intensity is given by

$$I_i = I_m \sin^2(\Delta/2)[1 - \cos 4(\theta - \eta_i)]/2 , \qquad (69)$$

where η_i is the angle between the polarizer and the horizontal. By taking four polarization-stepped images at $\eta_i = \pi(i-1)/8$, the isoclinic angle can be determined from

$$\theta = \{\tan^{-1}[(I_4 - I_2)/(I_3 - I_1)]\}/4 . \qquad (70)$$

The angles determined from (70) are in the range $(\pi/8, -\pi/8)$. Unwrapping of the results is thus required. Since the amplitude of light varies with the polarizer angle, the isoclinic images are normalized by using a separate set of four light-field plane polariscope images of the unloaded model.

Mawatari et al. [28] derived a single-valued representative function to determine the principal stress direction. They demonstrated the method by using four polarization-stepped isoclinic images $(0, \pi/12, \pi/3, \pi/4)$. The distribution of the intensity of the images was assumed to be independent of the polarizer angle in their study.

In consideration of the variation of light that occurs at different polarizer angles and the dispersion in the quarter-wave plates, *Chen* and *Lin* [46] have proposed a method that uses three light-field isoclinic images from a plane polariscope to directly determine the principal stress directions in the range $0-\pi/2$. Phase unwrapping is eliminated. In this method, the light-field intensity given by (27) is modified as

$$I_{\mathrm{PL}}(\phi) = I_{\mathrm{m}}(\phi)[1 - \sin^2 2(\theta - \phi)\sin^2(N\pi)] , \tag{71}$$

where ϕ is the angle between the polarizer and an arbitrarily selected reference axis, and θ is the angle between a principal stress direction and the reference axis. By using the unloaded model image, $I_{\mathrm{U}}(\phi)$, the dark-field normalized intensity of the loaded model I_1 can be calculated as

$$I_1(\phi) = 1 - I_{\mathrm{L}}(\phi)/I_{\mathrm{U}}(\phi) = \sin^2 2(\theta - \phi)\sin^2(N\pi) . \tag{72}$$

Similarly, for two light-field isoclinic images acquired at angles $\phi + \pi/8$ and $\phi + \pi/4$, the dark-field normalized intensity of the two images are

$$I_2(\phi + \pi/8) = \sin^2 2(\theta - \phi - \pi/8)\sin^2(N\pi) , \tag{73}$$
$$I_3(\phi + \pi/4) = \sin^2 2(\theta - \phi - \pi/4)\sin^2(N\pi) ,$$
$$= \cos^2 2(\theta - \phi)\sin^2(N\pi) . \tag{74}$$

From (72 – 74):

$$\cos 4(\theta - \phi) = (I_3 - I_1)/(I_1 + I_3) , \tag{75}$$
$$\sin 4(\theta - \phi) = (I_1 + I_3 - 2I_2)/(I_1 + I_3) . \tag{76}$$

If the denominator in (75) and (76) is equal to zero, no usable isoclinic information exists. It is assumed that the denominator is not equal to zero, and $V_{\mathrm{s}} = \sin 4(\theta - \phi)$, and $V_{\mathrm{c}} = \cos 4(\theta - \phi)$.

Referring to the signs in (75) and (76), the principal stress direction can be determined distinctly in the range $0 - \pi/2$ by the following six conditions:

$$\begin{aligned}
\theta &= \phi + 0.25 \tan^{-1}(V_{\mathrm{s}}/V_{\mathrm{c}}) & if \quad V_{\mathrm{s}} \geq 0; \quad V_{\mathrm{c}} > 0 \\
\theta &= \phi + 0.25 \tan^{-1}(V_{\mathrm{s}}/V_{\mathrm{c}}) + \pi/4 & if \quad V_{\mathrm{s}} \geq 0; \quad V_{\mathrm{c}} < 0 \\
\theta &= \phi + 0.25 \tan^{-1}(V_{\mathrm{s}}/V_{\mathrm{c}}) + \pi/4 & if \quad V_{\mathrm{s}} \leq 0; \quad V_{\mathrm{c}} < 0 \\
\theta &= \phi + 0.25 \tan^{-1}(V_{\mathrm{s}}/V_{\mathrm{c}}) + \pi/2 & if \quad V_{\mathrm{s}} \leq 0; \quad V_{\mathrm{c}} > 0 \\
\theta &= \phi + \pi/8 & if \quad V_{\mathrm{s}} > 0; \quad V_{\mathrm{c}} = 0 \\
\theta &= \phi + 3\pi/8 & if \quad V_{\mathrm{s}} < 0; \quad V_{\mathrm{c}} = 0 .
\end{aligned} \tag{77}$$

If the determined θ is greater than $\pi/2$, θ should be reduced by $\pi/2$. With $\phi = 0$, the three isoclinic images and the three associated unloaded images are acquired for the angle of 0, $\pi/8$, and $\pi/4$. Figure 24a,b shows the light-field and the normalized and adjusted dark-field isoclinic images of a disk. The determined principal stress direction, represented by gray levels, is given in Fig. 24c. It can be observed that the isoclinic contours appear to zigzag at and near the isochromatic fringes where no usable isoclinic information exist.

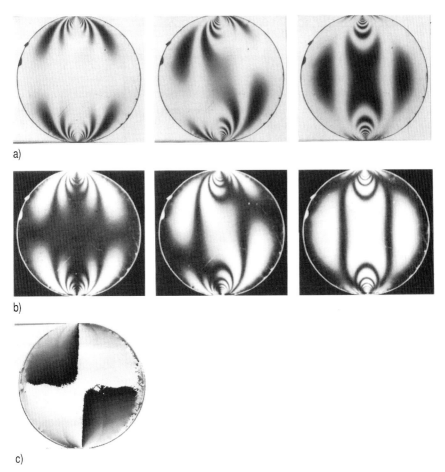

Fig. 24. (a) Light-field (b) normalized and adjusted dark-field isoclinic images ($\theta = 0, \pi/8, \pi/4$, from left to right) of a disk under diametral compression (c) digitally determined principal stress directions of the disk represented by gray levels

2.5.2 Phase-Stepping Method

The use of phase-stepped images to determine the isoclinic parameters has been reported using a circular polariscope [29], and using a plane polariscope [40]. The intensity forms and the images used have been described in the previous section. The resulting equations are given by (53) and (58). Both methods give satisfactory results [47]. Since the angles determined are in the range of $(\pi/4, -\pi/4)$, unwrapping of the results is required.

3 Applications of Evaluated Data

3.1 Stress Analysis

In two-dimensional photoelasticity, a properly scaled model is machined from a flat, optically isotropic plate of a suitable transparent material. The model is placed in a polariscope and is loaded in directions along the plane of the model. By using digital methods, the fringe orders and the principal stress directions in the field of the model are determined. The difference between the in-plane principal stresses at every point is then found from the stress-optic law using (21).

If the in-plane principal stresses have opposite signs ($\sigma_1 > 0 > \sigma_2$), the isochromatics yield directly the maximum shear stress at every point in the model

$$\tau_{mas} = (\sigma_1 - \sigma_2)/2 = Nf_\sigma/2h \ . \tag{78}$$

If the two in-plane principal stresses have the same sign, (78) gives only the in-plane maximum shear stress at each point.

At the free boundaries of the model, the in-plane principal stress normal to the boundary is zero. By knowing the isochromatic fringe order at the boundary, the principal stress tangential to the boundary can be determined from one of the following equations:

if $\sigma_2 = 0$, then $\sigma_1 = Nf_\sigma/h$,
if $\sigma_1 = 0$, then $\sigma_2 = -Nf_\sigma/h$.

The sign of the principal stress can be found by using a compensator. Since the principal-stress directions represented by isoclinics are not well understood, it can be better interpreted by presenting the principal-stress directions in the form of a stress trajectory or isostatic diagram where the principal stresses are tangent or normal to the stress trajectory lines at each point. The isoclinic parameters are also employed to determine the shear stresses on an arbitrary plane from

$$\tau_{xy} = -(\sigma_1 - \sigma_2)\sin 2\theta_1/2 = Nf_\sigma \sin 2\theta_1/2h \ , \quad \text{or} \tag{79}$$
$$\tau_{xy} = -(\sigma_1 - \sigma_2)\sin 2\theta_2/2 = Nf_\sigma \sin 2\theta_2/2h \ , \tag{80}$$

where θ_1 or θ_2 is the angle between the direction of σ_1 or σ_2 and the x axis in the coordinate system. The shear stress τ_{xy} is used in the application of the shear-difference method for individually determining the values of σ_1 and σ_2.

In three-dimensional analysis, a stress-frozen model is sliced for photoelastic examination. The digital measurement of photoelastic parameters is almost the same as that for the two-dimensional analysis. The complete determination of the state of stress at an arbitrary interior point usually requires the use of the shear-difference method. The principal stresses can be directly determined only for some particular situations, for example on the free surface of a three-dimensional model.

For plastic yielding or fatigue failure analysis using Von-Mises criterion, the effective stress, σ_e, can be evaluated from a cube cut from a photoelastic model as [1]

$$\sigma_e^2 = \frac{1}{4}\left[\left(\frac{N_x f_\sigma}{h_x}\right)(2+\sin^2 2\theta_x) + \left(\frac{N_y f_\sigma}{h_y}\right)(2+\sin^2 2\theta_y)\right.$$
$$\left. + \left(\frac{N_z f_\sigma}{h_z}\right)(2+\sin^2 2\theta_z)\right], \tag{81}$$

where N_x, N_y, and N_z are the fringe orders, θ_x, θ_y, and θ_z are the isoclinic parameters recorded by observing the three mutually orthogonal faces of the cube at normal incidence, and h_x, h_y, and h_z represent the thickness of the cube.

3.2 Examples

The use of digital fringe multiplication to the analysis of two stress-frozen slices of a full-scale variable-pitch screw has been reported [48]. Figure 25 shows the original and the multiplied isochromatic fringe patterns. The multiplied fringe patterns show details of fringe distribution that cannot be discerned in the original fringe pattern. Half-fringe photoelasticity has been applied to a wide range of studies, including fracture-mechanics problems under quasi-static, or transient thermal stresses of various materials [49,50,51], and stress analysis in composite materials [52]. The determined fringe orders are used to calculate the stress intensity factors in fracture mechanics.

The phase-stepping procedure combined with max-min scanning have been used in the analysis of a central slice taken from a full-scale frozen-stressed model of an engine connecting rod [14]. A complete analysis of a two-dimensional turbine blade and slot models using phase-shifting and spectral-content analysis is useful for designing turbine-blade structures [53]. The application of the two-wavelength method and polarization-stepping method to analyze a three-dimensional model has been reported [46,54]. The fringe patterns and the determined fringe order and principal stress direction maps are shown in Fig. 26. The determined photoelastic data were used to calculate the effective stresses to verify design optimization computer programs.

4 System and Error Assessment

The accuracy of the system may be affected by the following factors: (1) linearity, (2) quantization errors, and (3) electrical and optical noise. Linearity of the system is the most important factor. A difference of up to 2 gray levels can easily be found between the perfectly linear response and the calibrated response for over most of the operating region of the system [26,34]. For a

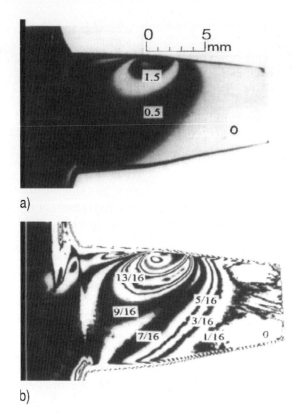

Fig. 25. Light-field isochromatic fringe patterns of a photoelastic slice: (a) original (b) multiplied by 8

maximum gray-level of 200, the error caused is 1%. In the low-level region where the gray level is lower than 50, the difference could be up to 5 gray levels. That is a 2.5% error.

The quantization error caused on the gray-level value is estimated to be less than 0.5. It is relatively small in comparison to the system's linearity. Electrical and optical noise can be reduced by frame averaging and through the normalization process, respectively. Further reduction of the noise is obtained by digital filtering.

The discrepancy between digital and manual analysis might be due to the incorrect positioning, orientation, or alignment of the optical components or the model. The repeatability of a system has been examined in [55]. The paper shows that for 20 randomly chosen points on a model, the worst error is ±0.007 fringe for isochromatics and ±0.75° for isoclinics.

The spatial resolution, defined as the distance between the detection points, affects the precision of a system. Generally, the higher the resolution, the more the number of points of a fringe that can be digitized. A higher

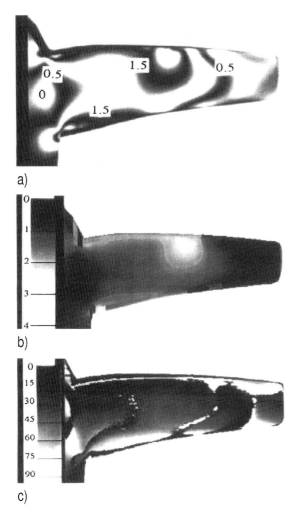

Fig. 26. (a) Fringe pattern of a photoelastic slice and gray-level representation of (**b**) the determined fringe order (**c**) the determined principal stress direction of the slice

pixel resolution can be obtained by choosing a proper lens or moving the camera closer to the model. This modification is particularly useful in the processing of regions with high stress gradients. To increase the resolvable intensity range of a system, the use of a proper light source and/or a proper lens aperture is helpful.

5 Conclusions

This contribution has reviewed some of the leading digital techniques for whole-field analysis of photoelastic fringe patterns. Relevant photoelastic theories and procedures have been described. The success of using digital approaches to solve two-dimensional or three-dimensional problems has been shown in various areas of experimental mechanics. It is expected that future technological improvements in the narrow-bandwidth CCD color camera with improved software algorithms are bound to make digital photoelasticity a fully automatic tool in the not-too-distant future.

References

1. J. W. Dally, W. F. Riley: *Experimental Stress Analysis*, McGraw-Hill, New York (1991)
2. H. T. Jessop, F. C. Harris: *Photoelasticity: Principals and Methods*, Dover, New York (1949)
3. F. A. Jenkins, H. E. White: *Fundamentals of Optics*, McGraw-Hill, New York (1976)
4. A. S. Kobayashi (Ed.): *Handbook on Experimental Mechanics*, VCH, New York (1993)
5. J. T. Pindera, G. Cloud: On Dispersion of Birefringence of Photoelastic Materials, Exp. Mech. **6**, 470–480 (1966)
6. L. S. Srinath, A. Sarma: Determination of Integral Fringe Order in Photoelasticity, Exp. Mech. **13**, 138–141 (1973)
7. T. Y. Chen: Digital Determination of Photoelastic Birefringence Using Two Wavelengths, Exp. Mech. **37**, 232–236 (1997)
8. D. Post: Isochromatic Fringe Sharpening and Fringe Multiplication in Photoelasticity, Proc. SESA, **12**, 143–157 (1955)
9. D. J. Bynum: On the Accuracy of Fringe Multiplication with Mirrored Birefringent Coatings, Exp. Mech. **6**, 381–382 (1966)
10. E. A. Patterson: Automated Photoelastic Analysis, Strain **24**, 15–20 (1988)
11. T. Kihara: Automatic Whole-Field Measurement of Principal Stress Directions Using Three Wavelengths, in S. Gomes et al. (Eds.): *Recent advances in experimental mechanics*, Balkema, Rotterdam (1994) pp. 95–99
12. J. Cazaro-Alvarez, S. J. Haake, E. A. Patterson: Completely Automated Photoelastic Fringe Analysis, Opt. Lasers Eng. **21**, 133–149 (1994)
13. K. Ramesh, K. S. Rajeev: Comparative Performance Evaluation of Various Fringe Thinning Algorithms in Photoelastic Mechanics, Electron. Imaging **14**, 71–83 (1995)
14. C. Buckberry, D. Towers: New Approaches to the Full-Field Analysis of Photoelastic Stress Patterns, Opt. Lasers Eng. **24**, 415–428 (1996)
15. J. S. Sirkis, Y. M. Chen, S. Harmeet, A. Y. Cheng: Computerized Optical Fringe Pattern Analysis in Photomechanics: A Review, Opt. Eng. **31**, 304–314 (1992)
16. R. K. Müller, L. R. Saackle: Complete Automatic Analysis of Photoelastic Fringes, Exp. Mech. **19**, 245–251 (1979)

17. T. Y. Chen, C. E. Taylor: Computerized Fringe Analysis in Photomechanics, Exp. Mech. **29**, 323–329. (1989)
18. B. Umezaki, T. Tamaki, S. Takahashi: Automatic Stress Analysis from Photoelastic Fringes due to Image Processing Using a Personal Computer, SPIE Proc. **504**, 127–134 (1984)
19. K. Ramesh, V. R. Ganesan, S. K. Mullick: Digital Image Processing of Photoelastic Fringes - A New Approach, Exp. Tech. **15**, 41–46 (1991)
20. W. K. Pratt: *Digital Image Processing*, Wiley-Interscience, New York (1978)
21. A. Rosenfeld, A. C. KaK: *Digital Picture Processing*, Academic, New York (1982)
22. R. C. Gonzalez, P. Wintz: *Digital Image Processing*, Addison-Wesley, Reading, MA (1987)
23. A. Ajovalasit, S. Barone, G. Petrucci: Toward RGB Photoelasticity: Full-Field Automated Photoelasticity in White Light, Exp. Mech. **35**, 193–200 (1995)
24. K. Ramesh, S. Deshmukh: Three Fringe Photoelasticity - Use of Color Image Processing Hardware to Automate Ordering of Isochromatics, Strain **32**, 79–86 (1996)
25. D. E. P. Hoy, F. Yu: Fuzzy Logic Approach for Analysis of White-Light Isochromatic Fringes, Post Conf. Proc. SEM VIII, Int. Cong. on Exp Mech., Nashville, TN (1996) pp. 279–284
26. T. Y. Chen: Digital Fringe Multiplication in Three-Dimensional Photoelasticity, J. Strain Anal. **30**, 1–7 (1995)
27. G. M. Brown, J. L. Sullivan: The Computer-Aided Holophotoelastic Method, Exp. Mech. **30**, 135–144 (1990)
28. S. Mawatari, M. Takashi, Y. Toyoda, T. Kunio: A Single-Valued Representative Function for Determination of Principal Stress Direction in Photoelastic Analysis, Proc. 9th Intl. Conf Exp Mech., Copenhagen, **5**, (1990) pp. 2069–2078
29. E. A. Patterson, Z. F. Wang: Towards Full-Field Automated Photoelastic Analysis of Complex Components, Strain **27**, 49–56 (1991)
30. C. Quan, P. J. Bryanston-cross, T. R. Judge: Photoelasticity Stress Analysis Using Carrier Fringe and FFT Techniques, Opt. Lasers Eng. **18**, 79–108 (1993)
31. Y. Morimoto, Y. Morimoto Jr., T. Hayashi: Separation of Isochromatics and Isoclinics Using Fourier Transform, Exp. Tech. **18**, 13–17 (1994)
32. S. L. Toh, S. H. Tang, J. D. Hovanesian: Computerized Photoelastic Fringe Multiplication, Exp. Tech. **14**, 21–23 (1990)
33. X. Liu, Q. Yu: Some Improvements on Digital Fringe-Multiplication Methods, Exp Tech. **17**, 26–29 (1993)
34. T. Y. Chen: Digital Fringe Multiplication of Photoelastic Images - A New Approach, Exp. Tech. **18**, 15–18 (1994)
35. T. Y. Chen, T. F. Chen: Whole-Field Automatic Measurements of Isochromatics and Isoclinics in Photoelastic Coatings, SPIE Proc. **2921**, 332–337 (1996)
36. A. S. Redner: Photoelastic Measurements by Means of Computer-Assisted Spectral-Contents Analysis, Exp. Mech. **25**, 148–153 (1985)
37. A. S. Voloshin, A. S. Redner: Automated Measurement of Birefringence: Development and Experimental Evaluation of the Technique, Exp. Mech. **29**, 252–257 (1989)
38. S. J. Haake, E. A. Patterson: Photoelastic Analysis of Frozen Stressed Specimens Using Spectral-Contents Analysis, Exp. Mech. **32**, 266–272 (1992)
39. A. S. Voloshin, C. P. Burger: Half-Fringe Photoelasticity - A New Approach to Whole Field Stress Analysis, Exp. Mech. **23**, 304–314 (1983)

40. A. V. S. S. S.R Sarma, S. A. Pillai, G. Subramanian, T. K. Varadan: Computerized Image Processing for Whole Field Determination of Isoclinics and Isochromatics, Exp. Mech. **32**, 24–29 (1992)
41. A. Asundi: Phase Shifting in Photoelasticity, Exp. Tech. **17**, 19–23 (1993)
42. W. C. Wang, T. L. Chen: Half-Fringe Photoelastic Determination of Opening Mode Stress Intensity Factor for Edge Cracked Stripes, Eng. Fract. Mech. **32**, 111–122 (1989)
43. P. K. Rastogi: *Holographic interferometry*, Springer, Heidelberg, Berlin (1994)
44. A. S. Redner: A New Automatic Polariscope System, Exp. Mech. **14**, 486–491 (1974)
45. J. Y. Yao: Digital Image Processing and Isoclinics, Exp. Mech. **30**, 264–269 (1990)
46. T. Y. Chen, C. H. Lin: An Improved Method for Whole-Field Automatic Measurement of Principal Stress Directions, Abst. Proc VIII Intl. Cong. Exp. Mech., (1996) pp 178–179
47. K. Ramesh, V. Ganapathy: Phase-Shifting Methodologies in Photoelastic Analysis - The Application of Jones Calculus, J. Strain Anal. **31**, 423–432 (1996)
48. T. Y. Chen, J. S. Lin: Computer Aided Photoelastic Stress Analysis of a Variable-Pitch Lead Screw, J. Strain Anal. **32**, 157–164 (1997)
49. W. C. Wang, T. L. Chen, S. H. Lin: Digital Photoelastic Investigation of Transient Thermal Stresses of Two Interacting Defects, J. Strain Anal. **25**, 215–228 (1990)
50. W. C. Wang, J. T. Chen: Theoretical and Experimental Re-examination of a Crack Perpendicular to and Terminating at the Bimaterial Interface, J. Strain Anal. **28**, 53–61 (1993)
51. J. T. Chen, W. C. Wang: Theoretical and Experimental Analysis of an Arbitrary Inclined Semi-infinite Crack Terminated at the Bimaterial Interface, J. Strain Anal. **30**, 117–128 (1995)
52. D. Mallik, C. P. Burger, A. S. Voloshin, E. Matsmoto: Stress Analysis of Adhesive Joints in Composite Structures Through HFP, Compos. Struct. **4**, 97–109 (1985)
53. S. J. Haake, E. A. Patterson, Z. F. Wang: 2D and 3D Separation of Stresses Using Automated Photoelasticity, Exp. Mech. **36**, 269–276 (1996)
54. C. H. Lin: Automated Analysis of Photoelastic Fringe Patterns. MS Thesis, National Cheng Kung University, Tainan (1995)
55. S. J. Haake, Z. F. Wang, E. A. Patterson: Evaluation of Full Field Automated Photoelastic Analysis Based on Phase Stepping, Exp. Tech. **17**, 19–25 (1993)

Optical Fiber Strain Sensing in Engineering Mechanics

James S. Sirkis

Smart Material and Structures Research Center, University of Maryland
College Park, MD 20742, USA
jere@eng.umd.edu

Abstract. Several fiber-optic sensors that are proving very useful for measuring the strain on the surface of structural components are described. The emphasis is on commercially available fiber-optic sensor configurations and instrumentation, as well as on many of the techniques required for successful measurements. Sensor fabrication, optical functionality, multiplexing, demodulation instrumentation, sensor packaging and sensor bonding techniques are all covered.

Embryonic optical fiber sensors were first introduced soon after low-loss single-mode fiber became commercially available in the late 1970s. Since then, optical fiber sensor technology has grown rapidly to the level where it is now making commercial inroads in field applications not well suited for traditional sensor technologies. Applications in environments with significant electromagnetic noise activity, in explosive environments where electrical sparks are intolerable, or in high temperature environments and applications requiring a large degree of design flexibility or passive sensor multiplexing[1] capabilities are all areas where fiber-optic strain sensors are making inroads. The primary agents responsible for the impressive level of development that has occurred in fiber-optic sensors over the past decade are lower costs and better performance of optical fiber components, and development of practical fiber-optic strain sensors that in many ways mimic the form and function of resistance strain-gage technology. This chapter purposefully excludes applications involving fiber-optic strain sensors embedded within structures. In doing so we avoid the Pandora's box of technical problems that are best faced once the basics of fiber-optic strain sensing are fully understood. The old adage "it is better to walk before you run" is the guiding principle for this chapter. Following this theme, this chapter describes using Intrinsic Fabry-Perot (IFP), Extrinsic Fabry-Perot (EFP), and Bragg grating sensors [1,2,3] in making strain measurements on the surface of structures. The topics covered include sensor fabrication, optical functionality, multiplexing, demodulation, instrumentation, sensor packaging and sensor bonding techniques. The reference list provided is sufficient for readers interested in other fiber-optic sensor technologies.

---

[1] Multiplexing refers to the capability of locating many optical fiber sensors along a single optical lead, and being capable of individually accessing the signal from each sensor.

1 Intrinsic Fabry-Perot Sensor

As the optical fiber sensor technology began to evolve beyond laboratory demonstrations, the need for sensors resembling the form and function of resistance strain-gages arose. One of the first sensors developed having these characteristics was the *intrinsic Fabry-Perot* (IFP) *sensor* [4,5,6,7,8]. The term *intrinsic* refers to the fact that the sensor uses the optical fiber itself as the transducer, as well as to carry the light to and from the transducer. In contrast, *extrinsic*, as in extrinsic Fabry-Perot, refers to the fact that the sensor uses an auxiliary optical media (air) as the transducer, with the optical fiber serving *only* to carry the light to and from the sensor.

1.1 Optical Arrangement

The optical arrangement for the IFP sensor is shown in Fig. 1. Light from a coherent source is launched into 2×1 coupler, which directs the light to an infiber Fabry-Perot cavity formed by two thin-film mirrors manufactured into the optical fiber. Part of the light is reflected from the first partial mirror and part of it is transmitted into the Fabry-Perot cavity. The light entering the Fabry-Perot cavity is reflected at the second internal mirror, and travels back through the first internal mirror back into the lead-in/out fiber. The two light components reflected back into the lead-in/out fiber interfere coherently, and the resulting intensity function is recorded by a photodetector at the output lead of the 2×1 coupler. The reflectivities of the two internal mirrors are often made such that the finesse[2] of the cavity is relatively low; therefore only a few internal reflections occur. Because of the low finesse, the resulting optical interference looks very much like dual-beam interference with an Optical Path length Difference (OPD) proportional to twice the cavity length.

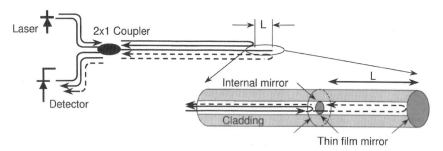

Fig. 1. Schematic of an intrinsic Fabry-Perot optical fiber sensor

The most difficult part of making IFP sensors is producing the thin film internal partial mirrors used to form the cavity. This is most often done using

---

[2] Finesse is a measure of the reflectivities of the mirrors that form the Fabry-Perot cavity. Finesse is related to the reflectivity, R, by $F = 4R^2/(1-R^2)^2$.

sputtering or vacuum deposition techniques. The process of using sputtering to make intrinsic Fabry-Perot mirrors is illustrated in Fig. 2 and enumerated below.

- The protective coating is removed, and the optical fiber is cleaved or polished, and cleaned following usual practices (Fig. 2a).
- Sputtering is used to fabricate a thin-film mirror on the end of the fiber (Fig. 2b). The materials most commonly used are titanium dioxide (TiO_2), multilayer combinations of TiO_2 and SiO_2, gold, and sometimes copper. TiO_2 is preferred because of its material compatibility with optical fiber material (SiO_2). The mirror reflectivity at this point in the process can be as much as 99%, and the thickness is as much as 100 nm. Many fibers can be sputtered simultaneously.
- The coated fiber is then fused to a cleaved and cleaned fiber, thereby producing an internal mirror (Fig. 2c). Because IFPs use low reflectivity internal mirrors, the quality of the internal mirror is primarily determined by the strength of the splice. The best splice strength can be obtained using multiple low-arc pulses. The reflectivity is controlled by adjusting the splicing time and arc current, with the reflectivity decreasing with each splice arc. Reflectivities of several percent are common at this point in the process, but reflectivities as high as 87% have been produced [4].

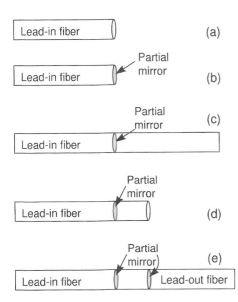

Fig. 2. Steps required to fabricate intrinsic Fabry-Perot sensors (**a**) strip, cleave, and clean the lead-in/out fiber, (**b**) deposit a thin film reflector on the cleaved fiber end, (**c**) fuse on a cleaved fiber using a series of low arc current pulses, (**d**) cleave the newly fused fiber to the desired cavity length, (**e**) fuse on a second coated fiber

- The fused fiber is then cleaved to the desired length (Fig. 2d) and fused with another coated fiber (Fig. 2e) to form the in-fiber Fabry-Perot cavity.

The strength degradation caused by the internal partial mirrors manufactured using the process described above has long been cited as the major drawback of IFP sensor technologies. The strength degradation arises because the thin films used to fabricate the mirrors inhibit strong glass-on-glass fusion splices. Fortunately, IFP sensor vendors have developed proprietary coatings and processes that seem to make strong IFP sensors.

1.2 Response to Strain and Temperature

The optical interference in IFP sensors occurs because the light entering the Fabry-Perot cavity travels further than the light immediately reflected at the first internal mirror. The optical phase change is, therefore, dependent only on the strain and temperature field in the cavity region, thereby producing a truly localized optical fiber sensor. In general, the phase response of IFP sensors depends on the entire thermomechanical strain state tensor [9,10]. Fortunately, sensors bonded to the surface of structures have a very unique and quantifiable strain state. Consider the photoelastic model of an optical fiber bonded to the surface of a loaded structure shown in Fig. 3[3]. Counting the isochromatic fringes in this figure shows that the far-field fringe order is roughly 2.5, while the maximum fringe order in the surface-mounted optical fiber is approximately 0.3 (a factor of 6.7 smaller) [11]. Equally important is the fact that this fringe is localized near the bottom of the optical fiber and that the fringe order in the core region of the fiber (where the light

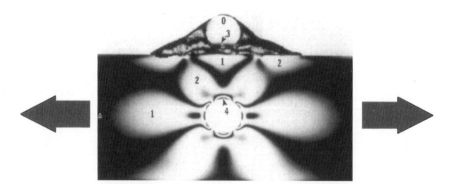

Fig. 3. Photoelastic model of an optical fiber sensor glued to the surface of a loaded structure. A hole is added to illustrate the structural load state

---

[3] The hole is added to the model of the structure to better illustrate the stress state.

propagates) is zero. This simple model shows that a negligible amount of the transverse strain in the structure is transferred directly to the optical fiber. Conversely, experiments have shown that the strain component parallel to the longitudinal axis of the fiber is fully transferred from the structure to the fiber and that the only non-zero transverse strains present in the fiber core are Poisson contraction strains associated with the longitudinal strain component. As a result, the equation relating temperature ΔT and axial strain ε_1 in the IFP sensor cavity to the coherent optical phase change $\Delta \phi$ is given by [9,12]

$$\Delta \phi = \frac{4\pi n L}{\lambda} \{P_\varepsilon[\varepsilon_1 + (\alpha_s - \alpha_f)\Delta T] + \zeta \Delta T\}, \tag{1}$$

where α_s and $\alpha_f (= 0.5\,\mu\varepsilon/^\circ\text{C})$ are the coefficients of thermal expansion of the structural material and fiber, respectively, and $\zeta (= 6.8\,\mu\varepsilon/^\circ\text{C})$ is the thermal-optic coefficient, $P_\varepsilon (= 0.79)$ is the strain-optic coefficient for the surface-mounted IFP sensor, L is the cavity length and λ is the wavelength of the light source. This expression applies only to the case where the sensor is bonded to the surface of a structure (not embedded) and where the strain and temperature field is uniform over the gage length of the sensor. Rearranging this expression leads to the following equation for strain

$$\varepsilon_1 = \frac{1}{P_\varepsilon}\left(\frac{\lambda \Delta \phi}{4\pi n L} - \zeta \Delta T\right) - (\alpha_s - \alpha_f)\Delta T. \tag{2}$$

The relative magnitude of the coefficients in front of strain and temperature are roughly the same, suggesting that the contribution of temperature to the phase change is of the same order as the contribution of the strain. This fact suggests that IFPs are subject to high thermal apparent strains.

Recall that thermal apparent strain is nothing more than the interpretation of thermally-induced sensor signals as resulting only from the strain field. With this in mind, the thermal apparent strain in IFP sensors can be quantified by letting the strain in (1) be zero, but then interpreting the temperature-induced phase change as resulting from strain. Following this approach, the thermal apparent strain sensitivity of IFP sensors is given by

$$\varepsilon_{\text{apparent}} = -\frac{\zeta \Delta T}{P_\varepsilon} - (\alpha_s - \alpha_f)\Delta T, \tag{3}$$

Using the above expression, the thermal apparent strain sensitivity is plotted in Fig. 4 as a function the structure's coefficient of thermal expansion CTE[4]. As seen in this figure, the thermal apparent strain is linear with increasing

---
[4] Only the magnitude of the thermal apparent strain sensitivity is given in this figure. We leave it to the reader to remember that the negative signs in (3) mean the apparent strain is compressive.

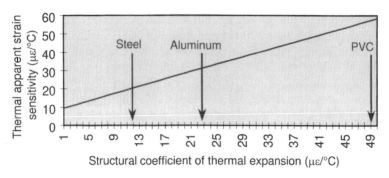

Fig. 4. Thermal apparent strain for an intrinsic Fabry-Perot (IFP) sensor plotted as a function of the structural coefficient of thermal expansion (CTE)

structural CTE[5]. The CTEs of a few common materials are identified in the graph for convenience. For example, the thermal apparent strain for aluminum is $\approx 20\,\mu\varepsilon/°$ C, which means that the equivalent of $\approx 20\,\mu\varepsilon$ occurs for every $1°$ C. This level of thermal apparent strain is much higher than most resistance strain-gage users would find acceptable. Because of this high thermal apparent strain, care must be exercised when using IFP strain sensors in thermally active structures. One obvious solution is to use dummy gages for temperature compensation. In the special case where the strain signals of interest have a higher frequency content than the temperature field, sensor signals can be high-pass filtered to eliminate the effects of the strain field. In fact, this is a standard option on commercial IFP sensor instrumentation. One final approach uses advanced dual-parameter strain/temperature sensors to overcome the high thermal apparent strain sensitivity. While these advanced sensor designs are the subject of intense investigation, no commercial products have yet been introduced.

High thermal apparent strain sensitivity is not the only optical issue that must be addressed when using IFP strain sensors. Intrinsic Fabry-Perot sensors are subject to an effect known as polarization fading [13,14,15,16,17]. Polarization fading occurs when unwanted transverse strains present in the optical fiber core cause stress-induced birefringence. The result of this birefringence is a strain-dependent coherent interference visibility function. Fortunately, stress-induced birefringence is rare in surface-mounted fiber-optic sensor applications. However, polarization fading can also be caused by asymmetries in the fiber core and/or intrinsic birefringence in the optical fiber components, which we have little control over. Polarization fading and its deleterious impact on the signal-to-noise characteristics of the sensor system

---

[5] While Fig. 4 plots the thermal apparent strain sensitivity as a function of only positive coefficients of thermal expansion, there are structural materials with negative coefficients of thermal expansion. Carbon-carbon composite materials are just one important example.

is usually defeated using automatic gain control concepts or a specialty fiber known as *polarization-maintaining* (PM) *fiber*[6] to maintain the state of polarization in all fiber components. Using PM fiber prevents the component of the coherent mixing that leads to the reduced fringe visibility.

1.3 Sensor Multiplexing

Sensor multiplexing refers to the ability to distribute sensors spatially through a network of optical fibers, and to demodulate all sensors using the same demodulation instrumentation. This requires unambiguously addressing intensity signals produced by the individual sensors in the network. One of the primary reasons for developing sensor multiplexing techniques is reducing the cost per sensor channel. The primary cost driver of fiber optic sensor systems is in the demodulator, so the ability to use more sensors with the same demodulation instrumentation becomes increasingly important as fiber-optic sensors move from laboratories to field applications [18,19]. Multiplexing becomes particularly important in applications involving many sensors such as automated factories, offshore platforms, medical diagnostics, bridges and dams, space structures, and the like. The multiplexing scheme used by commercial IFP instrumentation vendors is called spatial-division multiplexing (SMD), and is illustrated in Fig. 5. In this particular incarnation of SDM, each sensor is accessed by switching from one optical lead to another using an optical switch. An alternative form of SDM is to use a photodetector for each fiber to convert the optical signals to electrical signals, and then to use electrical switching schemes to address the individual sensors. SDM methods are appealing because they can be used with almost any type of optical fiber sensor configuration, including EFP and Bragg grating sensors. While switching SDM is certainly simple, having as many optical fiber leads as sensors may not be as desirable as having all sensors on a single optical fiber lead. It is also important to recognize that the measurement bandwidth of the spatial-division-multiplexed sensor systems is dictated by switching speed, instead of the demodulator bandwidth. Switching rates for multi-fiber optical switches are of the order of 2–10 Hz, and can therefore severely limit sensor signal bandwidth. Electrical switching rates are much faster, and therefore usually do not impact measurement bandwidth.

2 Extrinsic Fabry-Perot Strain Sensor

Extrinsic Fabry-Perot (EFP) sensors operate in much the same way as intrinsic Fabry-Perot sensors except that the cavity is formed in air instead of in an optical fiber. This simple fact offers some very important practical advantages, particularly in applications involving long-term measurements

---

[6] Polarization-maintaining fiber is also known as high-birefringence (Hi-Bi) fiber.

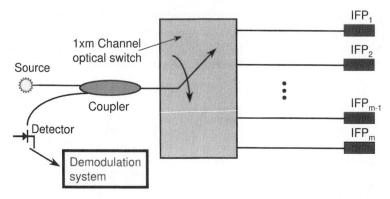

Fig. 5. Schematic of spatial division multiplexing intrinsic Fabry-Perot sensors using an optical switch

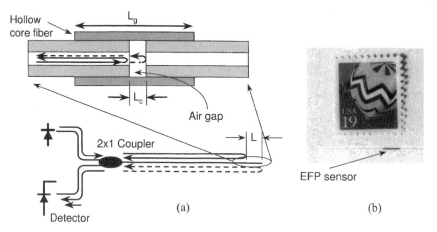

Fig. 6. (a) Schematic illustration of extrinsic Fabry-Perot (EFP) sensor operational principles (b) photograph illustrating the size of a typical EFP sensor (courtesy of Fiber and Sensor Technologies, Inc.)

and thermal-stress analysis. Combined with the sensor read-out instrumentation described in Sect. 7.2 and 7.4, EFP sensors are self-referencing. Self-referencing refers to the fact that measurements can be made relative to the time the sensor is manufactured, and are not interrupted if the electronic instrumentation is turned off. This feature is invaluable in applications involving measurement over days, months, or years. However, the "air-gap" configuration of EFP sensors leads to some unique limitations as well, as will be explained in the section devoted to the operating principles of EFP sensors. The remainder of this section discusses self-temperature-compensated strain-sensor configurations, multiplexing, and some interesting variations of the basic EFP sensor design.

2.1 Optical Arrangement

The basic configuration of extrinsic Fabry-Perot (EFP) sensors illustrated in Fig. 6a. EFPs are fabricated by gluing or fusion-welding two standard optical fibers into a hollow core fiber used for alignment [20,21,22]. The light carried by the lead-in fiber is partially reflected by the two air/glass interfaces to yield two independent electric field components, which ultimately recombine in the lead-in fiber in much the same way that occurs with intrinsic Fiber-Perot sensors. The two reflected electric field components propagate back through the 2×1 coupler and to the detector, which is presumably connected with appropriate sensor signal demodulation electronics. The reflectivity at air-glass interfaces is called Fresnel reflection, and is 4%. Because of this low reflectivity, the light reflected from the two air/glass interfaces produces *dual-beam* interference with an intensity function whose optical phase is proportional to twice the cavity length of the EFP sensor. Compared with fiber-optic sensors in general, EFP sensors are relatively easy to fabricate and can be made with gage lengths ranging from a few millimeters to a few centimeters. As an example, Fig. 6b shows a photograph of a 3 mm-long EFP sensor next to a postage stamp[7]. As a way of exploring some of their idiosyncrasies, consider the typical plot in Fig. 7 of EFP sensor intensity as a function of cavity length. Note how the fringe visibility becomes smaller as the cavity length increases. This behavior results because the light in the cavity is not guided and is therefore free to diverge as it propagates through the air gap. Beam divergence reduces the amount of light from the second air/glass interface coupled back into the lead-in/out fiber, but light reflected from the first air-glass interface is unaltered because it remains guided in the lead-in/out fiber. As a result, the amplitudes of the two interfering electric fields depend on the length of the air gap, and because fringe visibility is functionally dependent on the amplitudes of the interfering electric fields, it too is a function of the air gap-cavity length. In addition to reducing fringe visibility, beam expansion also influences the total optical *power budget*[8] of the sensor system. This is because the usable interference is produced by only a few percent of the total light injected into the sensor, which means that EFP sensors have very high *insertion losses*[9]. The resulting impact on the optical power budget limits the numbers of sensors that can be optically multiplexed.

---

[7] It is worth pointing out that all of the fiber-optic strain sensors described in this chapter can have gage lengths similar to the EFP sensor shown in Fig. 6b.
[8] Power budget refers to how the losses in the optical system reduce the total power reaching the detector.
[9] Insertion loss quantifies the optical loss of a specific fiber-optic component.

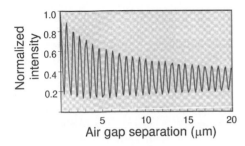

Fig. 7. Intensity as a function of extrinsic Fabry-Perot cavity length, illustrating the reduction in fringe visibility as the cavity length increase

2.2 Response to Strain and Temperature

Some of the more interesting features of EFP sensors can be illustrated by examining the expression relating the strain and temperature to the measured optical phase change:

$$\Delta \phi = \frac{4\pi}{\lambda} \{[\varepsilon + (\alpha_s - \alpha_f)\Delta T]L_g + \zeta_{air}\Delta T L_c\} \ , \tag{4}$$

where $\zeta_{air}(= 0.1\,\mu\varepsilon/^\circ C)$ is the thermo-optic coefficient of air, L_g is the distance between the hollow core fiber attachment points, L_c is the Fabry-Perot cavity length, and the other terms are the same as previously defined. This expression can be rearranged to solve for strain as follows

$$\varepsilon = \frac{\lambda \Delta \phi}{4\pi} \left[\zeta_{air}\left(\frac{L_c}{L_g}\right) + (\alpha_s - \alpha_f)\right]\Delta T \ . \tag{5}$$

The above expression does not include the strain-optic coefficient, P_ε, because EFP sensors are intrinsically independent of transverse strains. In other words, the optical phase change is solely dependent on the physical separation of the two air/glass interfaces forming the Fabry-Perot cavity. A positive by-product of being independent of transverse strains is that the sensor intensity is immune to polarization fading. Another interesting feature of (5) is that two characteristic lengths (L_g and L_c) govern the response to temperature and strain. This characteristic becomes invaluable in the design of self-temperature compensated EFP sensors. The final interesting feature of (5) is that the thermo-optic coefficient of air is more than a factor of 60 smaller than silica glass, resulting in an intrinsically smaller thermal responsivity of EFP sensors compared to IFP sensors. Even so, thermal apparent strain can be significant because the structural CTE is still included in the EFP sensor phase dependence. The relative magnitude of the thermal apparent strain is found following the approach described in Sect. 1.2, leading to the following expression in terms of the relevant optical and thermal parameters:

$$\varepsilon_{apparent} = -\left[\zeta_{air}\left(\frac{L_c}{L_g}\right) + (\alpha_s - \alpha_f)\right]\Delta T \ . \tag{6}$$

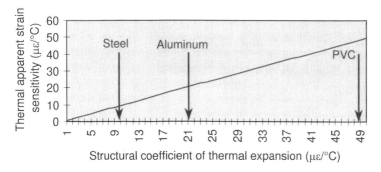

Fig. 8. Thermal apparent strain for an extrinsic Fabry-Perot (EFP) sensor plotted as a function of the structural coefficient of thermal expansion (CTE)

The thermal apparent strain is plotted in Fig. 8 as a function of structural CTE, and shares the same general characteristics as Fig. 4 for IFPs, in spite of their low intrinsic thermal sensitivity. The data in Fig. 8 is calculated assuming $L_c/L_g = 0.01$.

2.3 Self-temperature Compensation

One enviable option possessed only by EFP sensors is the ability to devise *self-temperature-compensated* (STC) *sensors* that completely eliminate the free thermal expansion effects from the sensor response. Temperature compensation is not the process of completely eliminating thermal effects from sensor responses; rather, it is the process of removing thermal-strain components that are not the result of mechanical stresses [23]. Self-temperature-compensated sensors involve replacing the second reflective fiber in the EFP sensor with a metallic wire, as illustrated in Fig. 9 [24]. The material and length of wire are chosen so that its elongation due to temperature exactly offsets the change in the air gap due to the free thermal expansion of the substrate plus the thermo-optic contributions from the air. Designing self-temperature compensating EFP sensors starts by examining the phase-strain-temperature equation that includes the thermal expansion of the wire. It is best, for the present purpose, to rearrange this expression so that the temperature-dependent terms are lumped together so that we may determine the conditions necessary to eliminate the temperature sensitivity. The resulting expression is

$$\Delta \phi = \frac{4\pi}{\lambda}\{\varepsilon L_g + [(\alpha_s - \alpha_f)L_g - \alpha_{mf}L_{mf} + \zeta_{air}L_c]\Delta T\}, \qquad (7)$$

where α_{mf} and L_{mf} are the coefficients of thermal expansion and length of the metal fiber (wire), respectively. Assume that the structure is undergoing only free thermal expansion (no thermal stresses), then forcing the terms in

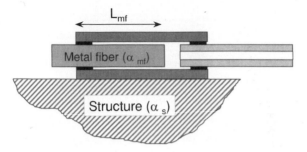

Fig. 9. Temperature-compensated extrinsic Fabry-Perot fiber-optic sensor

front of ΔT to be equal to zero leads to

$$\alpha_{\mathrm{mf}} L_{\mathrm{mf}} = (\alpha_{\mathrm{s}} - \alpha_{\mathrm{f}}) L_{\mathrm{g}} + \zeta_{\mathrm{air}} L_{\mathrm{c}} \; . \tag{8}$$

This expression allows one to select the combination of wire CTE and length to offset the effects of CTE and the thermo-optic effects of all other constituents. Note that different combinations of wire CTE and length are needed when the sensor is glued to different materials[10]. Because thermal strains are always composed of the superposition of free thermal expansion (no stress) and thermally generated mechanical strains (stress), using STC EFP sensors like the one described above automatically removes all strain components that do not lead to corresponding stress components. The ability to easily adjust the length of the metal wire at the time of sensor manufacture makes this temperature-compensation technique a practical option.

2.4 EFP Sensor Variants

While ease of fabrication is one of the chief advantages of EFP sensors, it is difficult to repeatably fabricate sensors with predefined gage lengths. Recall that the gage length of EFP sensors is defined by the attachment points between the alignment tube and the lead-in/out optical fibers, not the Fabry-Perot cavity. Uncontrollable fusion-weld geometry and/or the wicking of glue between the hollow core and optical fibers are sometimes responsible for the difficulties in defining the gage length. Furthermore, fatigue cracks can grow at the attachment points, leading to load-induced drift in gage length. Commercial EFP sensor vendors continue to develop proprietary manufacturing techniques that mitigate these issues such that sensor performance and repeatability have become similar to those normally expected of resistance strain gages. There have also been a few variations of the standard EFP develop to specifically address some of the concerns regarding gage length reliability. While none are commercially available, it is still interesting to briefly

---

[10] The need to design self-temperature compensated sensors for different substrates is common practice with virtually all strain sensor technology.

describe two EFP variants so that the reader can understand the range of possibilities that microfabrication can offer.

The variant of the EFP shown in Fig. 10 was developed in an effort to solve the problem of uncontrollable gage lengths. For obvious reasons, this EFP is called the uni-diameter tailed tapered Fabry-Perot interferometer [25]. It is fabricated by first fusing a standard diameter and small diameter (50 μm) optical fiber together to form a junction with a step jump in outer diameter but a uniform core diameter. A hollow core fiber with an outer diameter equal to the outer diameter of a standard fiber is slipped over the 50 μm diameter fiber and fused in place, as illustrated in Fig. 10. The length of the hollow core and 50 μm diameter fibers are chosen so that a small air gap is formed, thus providing an EFP sensor with a well-defined and robust gage length. While this is an interesting design, the fabrication of the sensor is so laborious that no commercial vendors are presently offering this technology.

A second variation of the EFP sensor that has emerged is called the in-line fiber etalon (ILFE) sensor [26,27]. The ILFE sensor is formed by fusing a hollow core fiber between two fiber-optic leads, as illustrated in Fig. 11. The hollow core fiber is manufactured to have the same outer diameter as the lead-in/out fibers in order to provide a well-defined gage length and a robust fusion-weld. The ILFE cavity length shown in Fig. 11 is ≈ 90 μm, but it may range from about 20 μm to 1000 μm, depending on the coherence length and

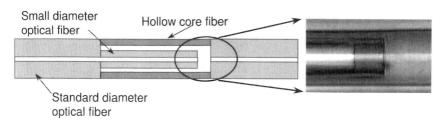

Fig. 10. Uni-diameter tailed tapered Fabry-Perot interferometer variation of the EFP sensor (courtesy of Anders Henriksson, FFA, Sweden)

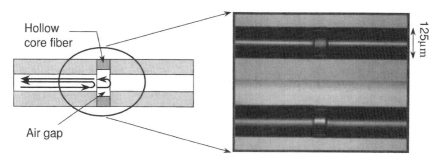

Fig. 11. In-line fiber etalon (ILFE) variant of the EFP sensor

power of the optical source. This variant of the EFP is very useful in high-strain-gradient applications because the gage length is so small, but cannot be made in a self-temperature-compensated format. Like the uni-diameter tailed tapered Fabry-Perot interferometer, the fabrication complexities have inhibited introduction of this technology into the marketplace.

2.5 Sensor Multiplexing

Extrinsic Fabry-Perot sensors can be multiplexed using the same range of techniques that work with IPF sensors. However, the beam expansion in the air-gap limits the sensor-system power budget and increases the cost and the sensor signal cross-talk. These issues have thus far limited commercial vendors to the simplest of multiplexing concepts–spatial-division multiplexing. As was discussed in Sect. 1.3 for IFP sensors, SDM is straightforward to implement but necessitates that each sensor be addressed though its own lead-in/out optical fiber.

3 Bragg Grating Strain Sensor

In-fiber Bragg grating sensors are very different from the EFP and IFP sensors described in the preceding sections. The most important differences come from the fact that the strain information is encoded by wavelength, instead of optical phase. The operating principles of ideal Bragg grating sensors are illustrated in Fig. 12, where light from a broadband source[11] interacts with the grating in such a way as to reflect a single wavelength, called the *Bragg wavelength*. In keeping with the conservation of energy, the transmitted spectrum shows a dip at the Bragg wavelength [28,29,30]. Whether viewing it in reflection or transmission, the Bragg wavelength is related to the grating pitch, Λ, and the mean refractive index of the core, n, by

$$\lambda_B = 2\Lambda n . \tag{9}$$

Examining this equation reveals that the utility of Bragg gratings as sensors arises from the fact that both the fiber refractive index and the grating pitch vary with strain and temperature. The above expression also provides some insight into Bragg grating manufacturing issues in that the refractive index is determined by the choice of fiber. In contrast, the grating pitch is controlled during the manufacturing process thereby allowing Bragg wavelengths to be generated that fall within almost any spectral band of interest.

Perhaps the most important features of Bragg grating sensors are *multiplexing* ability and *self-referencing*. Bragg gratings stand out above all other

---

[11] Common broadband sources used with Bragg grating sensors are light emitting diodes (LEDs), superluminescent diodes (SLDs) and amplified spontaneous emission (ASE) from eribuim doped fiber amplifiers.

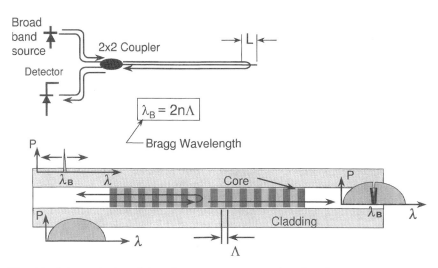

Fig. 12. Schematic describing the operating principles of in-fiber Bragg gratings

fiber-optic sensors in terms of the numbers of sensors that can be optically multiplexed. Bragg gratings are also unique in that they are finding a vast array of applications in telecommunications; therefore, more is known about in-fiber Bragg gratings, their fabrication, and their reliability than any other sensor configuration.

Do not be lead too quickly down the golden path by the apparent advantageous characteristics of Bragg grating sensors - they have drawbacks. First, they are intrinsic devices, and are therefore susceptible to high thermal sensitivity and birefringence. Additionally, Bragg grating sensors have a unique, but not always desirable, response to strain gradients along the grating length. All of these issues will be discussed in the sections that follow.

3.1 Bragg Grating Fabrication

The oldest and most flexible in-fiber Bragg grating fabrication technique is called the *holographic technique* (also known as the *interferometric technique*) [31], and is illustrated in Fig. 13. Two coherent optical waves from an ultraviolet (UV) laser with a known angle between them are launched onto a fiber from the side. The fiber is then adjusted such that it is aligned perpendicular to the plane of constructive and destructive fringes. This configuration is maintained for a pre-determined exposure time, at the end of which the *photorefractive effect* in the germanium doped optical fiber produces a periodic refractive-index variation matching the pitch of the interference pattern. The holographic technique making of Bragg gratings offers a large degree of flexibility because the pitch of the interference pattern, and therefore the Bragg grating pitch, can easily be controlled by simply adjusting the angle, θ, between the two interfering beams. The downside is that it

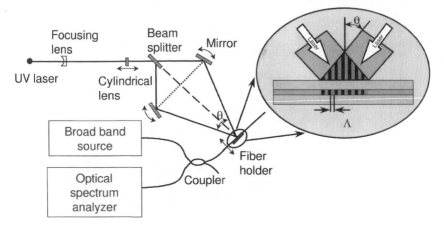

Fig. 13. Typical holographic procedure of Bragg grating fabrication

requires careful vibration isolation. Additionally, the lasers used to produce the dual-beam interference must have sufficient coherence and beam stability.

The photorefractive effect used to produce Bragg gratings is initiated when light from an ultraviolet (UV) laser interacts with germanium-doped fiber to produce defect centers in the glass. Conventional telecommunications-grade Ge-doped optical fibers exhibit a relatively weak photosensitive response so that hydrogen loading[12] is generally used to reduce the exposure times. Vendors are presently developing proprietary optical fibers with dramatically enhanced photosensitvity, but these fibers were not available for purchase at the time of writing. The good news is that the UV light required to write gratings is produced by several common lasers [32]. The bad news is that all of these lasers are very expensive. Kryton Fluoride (KrF) excimer lasers are by far the most popular lasers for writing gratings because they deliver significant energy density pulses to the fiber at repetition rates exceeding 100 Hz. KrF Excimer lasers generate light at wavelength of 248 nm, and are known for their poor beam quality and very short coherence lengths. Because of their low coherence, KrF lasers are better suited for phase mask (discussed in the next paragraph) techniques. Frequency-doubled argon-ion lasers[13] and frequency-doubled dye-lasers are second in popularity. These lasers generally have good coherence, making them suitable for holographic fabrication techniques. The relatively low power of these CW lasers (roughly 100–200 mW), leads to slow grating writing times. A 100% reflectivity Bragg

---

[12] Hydrogen loading is the process of exposing the optical fiber to hydrogen at elevated pressures for an extended period of time. This process is known to significantly enhance the photosensitivity of germanium-doped optical fiber, although the precise reasons for this enhancement is not all that clear.

[13] Frequency-doubled argon-ion lasers are called FReD lasers by their manufacturer, Coherent Laser Group.

grating written with a KrF laser takes about 30 s but requires roughly 5 min with a FReD laser.

An alternative grating fabrication technique, which has the same basic principle as the one just discussed above, uses what is known as a phase-mask [33]. A commercially available transparent mask with phase corrugations is used as a diffracting element to produce the two coherent UV beams[14] for the formation of Bragg gratings. As shown in Fig. 14, an uncoated optical fiber is placed beside the phase-mask, and light from a UV laser is incident from the other side. Intrinsic optical stability and the ability to write gratings using lasers with almost negligible coherence are the two most important advantages of phase-mask fabrication techniques. The biggest disadvantage is the difficulty of fabricating gratings with different pitches without resorting to different phase masks, which is an expensive proposition. *Strain tuning* is one alternative that enables the Bragg wavelength to be adjusted to a small degree (no more than ≈ 3 nm). In this process, the fiber is pre-strained during writing and released afterwards. Releasing the fiber causes it to contract. This, in turn, reduces the grating pitch and therefore the wavelength. While not as versatile as the holographic method of fabricating gratings, phase-mask fabrication methods are generally more robust, require a smaller set-up time, and yield higher quality gratings. Properly done, the holographic or phase-mask methods can produce gratings with insertion losses less than ≈ 0.1 dB, spectral bandwidths less than 0.2 nm, and reflectivities greater than 99%[15].

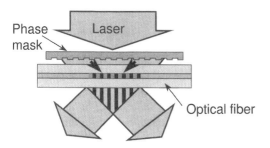

Fig. 14. Phase mask method of writing in fiber Bragg gratings

---

[14] Other diffraction orders, including the zero order, are suppressed by controlling the geometry of the corrugations comprising the phase mask.

[15] The reflectivity and spectral bandwidth are critical parameters in telecommunications applications, and therefore grating manufacturers go to great lengths to determine reflectivity and bandwidth requirements of end users. This is not the case when it comes to sensors. Generally speaking, grating reflectivities ranging from 85% to 100% and spectral bandwidths ranging from 0.2 nm to 0.5 nm are sufficient for sensor applications.

3.2 Strain and Temperature Response

Bragg grating sensors encode these measurands in the Bragg wavelength (instead of the optical phase) through the intrinsic response of the optical fiber. Therefore, the functional dependence of these sensors bears resemblance to the equations used for the intrinsic Fabry-Perot sensors described earlier in this chapter. Referring to (9), the strain and temperature response arises in Bragg gratings due to both the physical elongation of the sensor (and corresponding fractional change in grating pitch, Λ) and the change in fiber index, n, due to photoelastic effects. For a Bragg grating sensor bonded to the surface of a structure, the strain and temperature are related to the change in the Bragg wavelength by

$$\frac{\Delta \lambda_B}{\lambda_B} = P_\varepsilon[\varepsilon + (\alpha_s - \alpha_f)\Delta T] + \zeta \Delta T, \tag{10}$$

where λ_B is the unstrained Bragg wavelength given by (9), $\Delta \lambda_B$ is the strain- and temperature-induced change in the Bragg wavelength, and all of the other parameters are the same as defined in (1). Rearranging this equation for strain yields

$$\varepsilon = \frac{1}{P_\varepsilon}\left(\frac{\Delta \lambda_B}{\lambda_B} - \zeta \Delta T\right) - (\alpha_s - \alpha_f)\Delta T. \tag{11}$$

This expression is very similar to the corresponding expression for IFP sensors (1), and therefore also has a significant response to temperature. In fact, zeroing the strain term in (11) leads to an identical expression to (3) for the thermal apparent strain. As a result, the thermal apparent strain plot given in Fig. 4 also holds for Bragg gratings.

Birefringence affects interferometric sensor signals by introducing strain-dependent polarization fading into the intensity signal, which can ultimately reduce the signal-to-noise ratio of measured phase signals. In the case of Bragg gratings, birefringence works to produce two refractive indices with magnitudes dependent on the strain and temperature state, and these two refractive indices in turn produce two independent Bragg wavelengths [34] with orthogonal polarizations. Fortunately, this troublesome effect occurs rarely in surface-mounted applications so long as the adhesive layer is thin and the temperature is not extremely low, thus avoiding transverse compressive stresses in the fiber.

3.3 Effects of Gradients

We were careful in the preceding subsections to examine only those cases where the strain field or temperature were uniform over the grating length. This was done to clarify the physical nature of grating transduction mechanisms without overly confusing the relevant issues. When gradients are applied to Bragg gratings, the concept of a single Bragg wavelength becomes obscured both figuratively and literally. When a strain gradient is applied to a grating, each section of the grating experiences a different strain, and therefore the grating pitch is no longer uniform. In essence, we now have a series of cascaded gratings with slightly different pitches and therefore slightly different Bragg wavelengths [35]. In extreme cases, the superstition of all of the resulting Bragg wavelengths can completely smear out the reflected spectrum to the point where no spectral feature associated with the grating can be identified.

3.4 Bragg Grating Reliability

Telecommunications systems are expected to last upwards of 35 years, and so it should come as no surprise to find that before any fiber-optic component is installed it undergoes rigorous optical and mechanical reliability testing. Bragg gratings are no different in this regard, and have been subjected to a battery of humidity, temperature and strain cycling tests. Environmental tests intended to investigate the temperature and humidity influence on the optical characteristics of Bragg gratings showed that they had limited impact. No change in grating properties were observed after 1000 h of 85° C at 85% relative humidity, nor were any changes in grating properties observed after 1000 thermal cycles from −40° C to 85° C or 512 cycles from 21° C to 427° C. However, in order to achieve thermal stability at the higher temperatures, the gratings must be annealed at temperatures exceeding the intended operating temperature [36].

Mechanical testing performed to date also indicates no change in grating properties, even after 1.4 million strain cycles from 0 to 2500 $\mu\varepsilon$. Additionally, Bragg gratings fabricated from high-strength fiber and properly handled during the writing process have exhibited mean strengths in excess of 4.8 GPa provided that proper fiber-stripping techniques are used [37].

3.5 Sensor Multiplexing

Bragg grating sensors encode measurement fields in the Bragg wavelength, which offers unique opportunities to use *Wavelength-Division Multiplexing* (WDM) topologies to multiplex sensors. In fact, Bragg grating sensors have unparalleled multiplexing potential. Multiplexing is achieved by producing an optical fiber with a sequence of spatially separated gratings, each with a different pitch. The output of the multiplexed sensor is processed through wavelength-selective instrumentation such as an optical spectrum analyzer or a scanning optical-bandpass-filter. The resulting optical signal associated with a given pitch, Λ_n, where $n = 1, 2, \ldots, N$, is centered at the associated Bragg wavelength, λ_{Bn}, in the spectral domain (see Fig. 15) [38,39,40]. A measurement field at grating n is uniquely encoded as a perturbation Bragg wavelength, λ_{Bn}. For example, Fig. 15 shows the spectrum produced by eight gratings serialized along a 12 m length of optical fiber. A thermomechanical

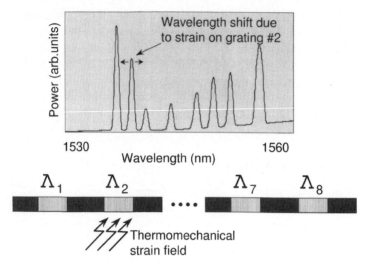

Fig. 15. Multiplexing with Bragg grating optical fiber sensor

strain field applied to the second grating produces a shift in the second Bragg wavelength. Having all of the sensors located along an optical lead like the one in Fig. 15 is one of the most important advantages of wavelength-division-multiplexed Bragg grating sensors.

There is, of course, a limit to the number of sensors that can be multiplexed. Each sensor is assigned to a different region of the available source spectrum, and therefore the spectral width of the source plays a critical role in determining the number of sensors that can be multiplexed. So too does the amount of strain and temperature-induced shift in the Bragg wavelength. Bragg wavelengths that overlap, even partially, become indistinguishable from each other and therefore render both sensors unusable. As a result, the spectral separation of the unstrained Bragg wavelength is an important consideration when designing a sensor system. Determining the number of sensors in a multiplexed system, and the requried spectral separation, starts by using (10) to estimate the maximum change in Bragg wavelength, called $\Delta\lambda_{max}$, that can be expected from the load field. This requires at least an approximate estimate of the maximum levels of strain and temperature that can be expected from the structure during the duration of the measurement. Assuming a worst case scenario where any two adjacent gratings see the same maximum wavelength shift, but in opposite directions, the required spectral separation between any two adjacent Bragg wavelengths is $\Delta\lambda_{strain} = \Delta\lambda_{max} - (-\Delta\lambda_{max}) = 2\Delta\lambda_{max}$, as illustrated in Fig. 16. If the spectral width of the source is given by $\Delta\lambda_{source}$, then the maximum number of Bragg gratings that can be multiplexed using wavelength-division techniques is $N_{max} = \Delta\lambda_{source}/2\Delta\lambda_{max}$. As an example, a typical source with a central wavelength of ≈ 1550 nm will have a spectral bandwidth of

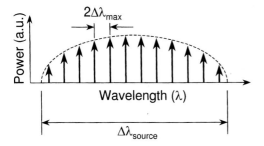

Fig. 16. Schematic showing Bragg grating wavelength spacing relative to the optical source bandwidth

40 nm. Assuming an isothermal strain field where the maximum strains are 2000 $\mu\varepsilon$, then $\Delta\lambda_{max} = 2.45$ and $N_{max} = 8$.

It is noteworthy that there are other multiplexing schemes that are not as limited by the source spectrum as the traditional WDM described above. One method gaining in popularity is based on Fourier transforms [41]. While comparatively slow because it requires intense numerical processing, this technique is capable of multiplexing roughly four times as many sensors for the same strain field. Unfortunately, Fourier-transform-based demodulation instrumentation is not yet commercially available at the time of writing. One other variation of WDM Bragg grating multiplexing that yields more sensors and is commercially available uses a combination of the optical switch-based spatial-division multiplexing described in Sect. 1.3 with the wavelength-division multiplexing described in this section.

4 Fiber-Optic Coatings and Cables

Optical-fiber sensor reliability has already been mentioned on several occasions. It is clear that the differences in construction of extrinsic Fabry-Perot, intrinsic Fabry-Perot, and Bragg grating sensors lead to unique concerns with regard to reliability. The internal mirrors used by the IFP sensors, the attachment points used in EFP sensors, and the high laser energies used in writing Bragg grating sensors are examples of the unique details influencing reliability. However, there are two components influencing reliability and measurement performance that all fiber-optic strain sensors have in common - the lead-in fiber and coatings. Obviously, strain measurements will not be possible if fractures occur in the fiber-optic sensors or the optical fiber carrying the light to the sensors. The coatings used to protect the optical fiber in the sensor system therefore has considerable influence on the overall measurement reliability. The coatings used on the sensors themselves must also provide for effective strain transfer between the structure and sensor.

Optical fibers are made from amorphous (glassy) quartz, which is an intrinsically low fracture toughness material. Low fracture toughness often implies low mechanical strength. However, optical fibers have very high mechanical strength, approaching a factor of three higher than the strongest carbon steels. These apparently contradictory statements (low fracture toughness and high strength) arise from the fact that optical fiber manufacturing conditions are so well controlled that the initial surface flaws are on the atomic scale. When the fiber is properly coated, these very small flaws rarely exceed the critical length required for unstable crack growth. The highest strength fiber (proof-tested above 200 ksi) is fabricated in Class 1 laminar-flow hoods and is immediately coated with a high toughness material to protect the fiber from handling-induced surface flaws. The implication of this discussion is that coatings play an important role in determining the strength of optical fiber sensors. Coatings also play an important role in transferring strain for the structure to the optical fiber sensor.

There is a surprisingly large number of standard coatings offered by optical fiber vendors and fiber-optic sensor vendors. Ultraviolet curing acrylate polymers are unquestionably the most common material used to coat optical fibers because the are cost effective, offer good protection, and can be easily stripped from optical fibers. However, these coatings are limited to operating in a relatively low temperature range (less than $\approx 125°$ C) [42]. In addition, they are very compliant, are susceptible to creep, and are relatively thick ($\approx 63\,\mu m$)[16]. As a result, acrylate-coated sensors can be ineffective in transferring strain from the structure or sensor package. For these reasons, commercial fiber-optic sensors are rarely supplied with acrylate coatings. Some vendors will supply sensors with no coating at all, but most will use polyimide coatings. Polyimide has a higher stiffness than acrylate and can operated in temperature environments reaching $\approx 500°$ C [42]. In addition, the thickness of polyimide coatings is generally less than $25\,\mu m$. The combination of these factors leads to a better strain transfer characteristic.

While polyimide is the preferred coating for many applications, there are environments that are better served using more specialized materials. High temperature, high humidity, and high alkalinity are three environments worth discussing. The high softening point of amorphous silica ($\approx 1300°$ C) means that it is possible to use optical fiber sensors in applications involving very high temperature fields. However, this potential for high-temperature operation can only be realized if high-temperature, inert coatings are used to prevent oxidation of the fiber surface. Oxidation causes the optical fiber to develop surface flaws, which have the effect of making the optical fiber fragile. Polyimide coatings are typically used for applications below 500° C, but gold alloys are commonly used for applications near 1000° C [43,44]. Gold coatings were originally developed to prevent moisture from degrading the strength of optical fibers, as will be described next. Other more exotic high-

---

[16] This comment regarding the thickness is made relative to the optical fiber diameter. The diameter of optical fibers is typically $125\,\mu m$, and so $63\,\mu m$ is indeed large on this relative scale.

temperature coating materials continue to be the focus of research, but none have yet reached the commercial market place.

Moisture absorption is known to break down the covalent bonds in amorphous silica dioxide optical fibers [45]. This breakdown is known as stress corrosion because it is accelerated in the presence of non-zero stress states. Stress corrosion is well known to the optical fiber community and has been effectively defeated through the use of hermetic coatings. Off-the-shelf hermetic coatings are offered by many fiber vendors [45]. One popular hermetic coating incorporates a 50 nm amorphous carbon coating [46,47]. This is done by depositing a carbon outer layer onto the fiber preform before manufacturing the fiber. Hermetic seals are also produced using metallic coatings (aluminum or gold are common) [48,49].

The issue of coating selection is more complex when the strain measurements involve concrete structures because of their high alkalinity. Mixing cement with water leads to a hydration reaction, which produces calcium hydroxide as one of its by-products. It is this calcium hydroxide that is responsible for the high alkalinity of concrete. Coatings capable of fully protecting optical fibers from high alkali have not yet been identified, although there are ongoing tests designed to do so [50]. Acrylate, polyimide, fluorine thermoplastic, and Tefzel coatings are among those that have been investigated, and all experienced some degradation (cracks, crazing, etc.) as a result of alkalines and hydration [50,51]. The issues here are two-fold. The coating must not only prevent surface flaws and water absorption into the fiber, it must also provide sufficient strain transfer [51,52]. As is often the case, these requirements can sometimes be competing. The tests conducted so far rule out only acrylate coatings, but most of the data published covers relatively short exposure times (of the order of months) and does not include hermetic coatings. Clearly, further work in this regard is needed if high-reliability fiber-optic sensor systems intended for use with concrete are to be developed.

The choice of coating does not fully protect optical fibers from failure due to unexpected loads that commonly occur when making strain measurements. For example, there is always a potential for loads from accidental impacts (e.g. dropped tools and other similar episodes). Fortunately, a vast array of small-diameter fiber-optic cable structures have been developed by the telecommunications industry to protect against these types of impacts. These cables usually include a combination of a $\approx$ 3 mm diameter hard-polymer coating extruded over standard-coated optical fiber. The fiber assembly is then usually encased in a structurally reinforced loose-fitting tertiary tube [53,54]. The optical fiber leading to the commercial sensor packages shown in Fig. 17 all use a form of fiber-optic cable structure.

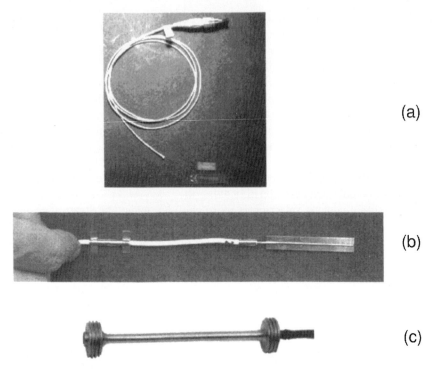

Fig. 17. Different packaging options for fiber-optic sensors: (**a**) strain gage on a polyimide backing (courtesy of Fiber and Sensor Technologies) (**b**) weldable strain gage (courtesy of Fiber Dynamics, Inc.) (**c**) strain gage package design for embedding in concrete (courtesy of FISO Technologies, Inc.)

5 Commercial Packaging

While most vendors will simply sell coated Bragg gratings as sensors, there has been some recent emphasis on developing packaging options for a few specific applications. There are not nearly as many packaging options available as there are with resistance strain-gages, but this is to be expected given the relative immaturity of the fiber-optic strain sensor technology. A wider variety of packaging formats will certainly become available as the market for fiber optic strain sensors grows. Nevertheless, Fig. 17 shows a photograph of a few packaged fiber optic sensors. Figure 17a shows an EFP sensor sandwiched between polyimide films much as is done with resistance strain gages. The second packaging option, seen in Fig. 17b, is a weldable strain gage fabricated by bonding a Bragg grating inside a metallic tube, which itself is welded to a metallic strip. This metallic strip is then welded to the structure of interest. The final option seen in Fig. 17 (17c) is a Bragg grating placed in the center of a "bar-bell" package for concrete embedding applications.

6 Sensor Bonding Techniques

Fiber-optic strain sensor bonding techniques borrow techniques and products from resistance strain gage bonding techniques. Gluing, metal arc spraying, welding, as well as a host of other techniques can be used. As with standard resistance strain gages, care must be taken to properly prepare the surface, not to damage the sensors, and to avoid residual stresses caused by adhesives[17] Because fiber-optic strain gages are in relative infancy compared to resistance strain gages, there are no industry-wide standards for bonding procedures. The EFP sensor vendors have made the most progress towards standardization [55,56,57]. Therefore, this section outlines the steps used to bond EFP strain sensors to the surface of structural materials. This process can easily be adapted for use with Bragg grating and IFP sensors.

Step 1: Degrease the surface using a degreaser or isopropyl alcohol.

Step 2: Abrade the specimen surface using dry 220 or 320 grit SiC sand paper.

Step 3: Clean the surface with Measurements Group, M-Prep Conditioner A (or equivalent product). Repeatedly Apply M-Prep Conditioner A (or equivalent) and wipe with clean Kimwipes.

Step 4: Liberally apply M-Prep neutralizer (or equivalent) and scrub with cotton swabs (This makes the surface alkalinity ≈ 7.0).

Step 5: Align the sensor along the desired direction and tape to the specimen using electrical or Scotch tape as indicated in Fig. 18a.

Step 6: Apply a dab of partially cured five minute epoxy ≈ 3 mm from the sensor (Fig. 18b). This maintains sensor alignment for the adhesion process.

Step 7: Gently remove the tape from the fiber tip.

Step 8: Mix Measurements Group AE-10 epoxy adhesive (or equivalent) slowly to prevent entraining air bubbles in the adhesive.

Step 9: Apply the adhesive slowly using a linear motion parallel to the sensor orientation. Make sure the entire sensor is covered from the tip to the five minute epoxy, and make sure not to produce an overly thick adhesive layer.

Step 10: Use Scotch tape over the adhesive to hold the sensor in contact with the surface (Fig. 18c). Do not apply excessive pressure to the sensor.

Step 11: Allow AE-10 to cure at room temperature for 6 - 10 hours. Elevated temperature curing can induce residual compressive strains.

---

[17] Residual stresses can cause birefringence effects or cause strain gradients that can smear the spectrum of Bragg gratings. Residual stresses can be mitigated by using room curing adhesives.

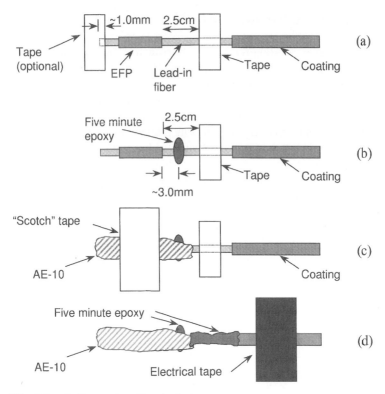

Fig. 18a–d. Sensor bonding technique

Step 12: Remove all tape, and then apply five minute epoxy to the uncoated optical fiber. Then use electrical tape to attach the coated section of the fiber to the structure (Fig. 18d). It is also advisable to cover the entire sensor with electrical tape if the sensor will be exposed to UV light (e.g. sunlight) for extended periods of time. UV light tends to degrade adhesives.

7 Demodulation

Demodulation is a generic term describing the electro-optic signal processing that converts electrical signals, proportional to the modulated optical intensity, to signals proportional to the applied strain. At first glance, one might think that demodulators are simply some form of amplifier because the optical signal at the photodetector definitely changes when the sensor experiences a strain field. However, the detector records the optical intensity, which is almost always a heavily non-linear (sometimes periodic) function of strain. It is the job of the demodulator to convert this non-linear function to a linear one [58,59,60]. There is no end to the number of demodulation techniques that have been developed over the years for IFP, EFP, and Bragg grating sensors. Discussing them all is beyond the scope of this chapter. Instead, this chapter focuses on the demodulation concepts that form the foundation of commercially available demodulation products.

7.1 Serrodyne Fringe Counting (IFP Sensors)

Intrinsic Fabry-Perot fiber-optic pressure sensors are making strong inroads into the automotive industry, and this success is helping to transition the sensors into the strain sensors market. The only commercial demodulator for IFP sensors available at the time of this writing is based on an adaptation of the "serrodyne" demodulation scheme that is so popular in the fiber-optic sensor community [61]. A schematic of this approach is shown in Fig. 19, where light from a distributed feedback laser is coupled into a $1 \times N$ coupler that distributes the light to N optical fiber sensors. A Faraday isolator is located just after the DFP laser to prevent downstream reflectors from coupling light back into the laser. Without an isolator, the feedback causes instability in the laser's center wavelength, with adverse consequences for sensor performance. The light reflected from each IFP sensor travels back though a 2×1 directional coupler to a photodetector (labeled PD in Fig. 19). In this architecture, each sensor has its own 2×1 coupler and photodetector. To realize sensor signal demodulation, the laser is driven with a periodic, slightly nonlinear, sawtooth current waveform to produce a linear chirp[18] during each modulation period [62,63]. The principle of operation is to use the chirp to drive the optical phase change over 2π radians every modulation period, even if no strain is present. The resulting interference function will result in a periodic waveform whose relative phase depends on both the known linear chirp and the additional unknown strain. Strain-induced deviations from the known chirp-induced phase change are measured using a fringe counting scheme implemented in a microprocessor. The microprocessor operates on each photodetector signal in sequence, which means that the measurement bandwidth per sensor depends on the number of sensors being interrogated.

A commercial demodulator utilizing the serrodyne fringe counter is shown in Fig. 20. This IFP demodulator can monitor up to 24 IFP sensor channels, with each sensor having a 2.4 Ksamp/sec bandwidth. This system can also be configured to monitor 12 channels at 4.8 Ksamp/sec or four channels at 7.2 Ksamp/sec. All sensors are connected through the back panel, and a handheld touch pad is used to control the instrument. Serrodyne fringe counters are linear and have a resolution of $\approx 0.05\,\mu\varepsilon$ for $\approx 1\,\mathrm{cm}$ gage length sensors. Pairs of channels can be used to perform temperature compensation by using

---

[18] A linear chirp is a linear variation of the optical frequency with time.

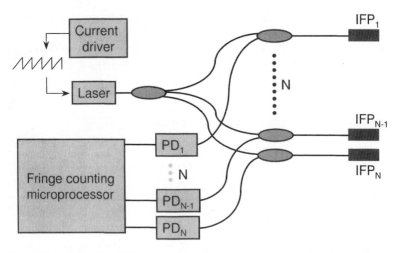

Fig. 19. Schematic of the serrodyne fringe counter demodulator

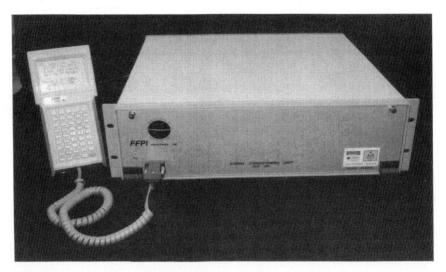

Fig. 20. Photograph of a commercial IFP sensor demodulator (SCU-100) that uses the serrodyne fringe counting technique (courtesy of Fiber Dynamics, Inc.)

one sensor to measure strain and one to measure temperature, and then using software to remove free thermal expansion effects.

7.2 White Light Cross-Correlator (EFP Sensors)

The key to using EFP strain sensors in stress-analysis applications is to find a practical way of obtaining precise and reliable Fabry-Pérot cavity length measurements. The white-light cross-correlator offers [64,65] a unique and

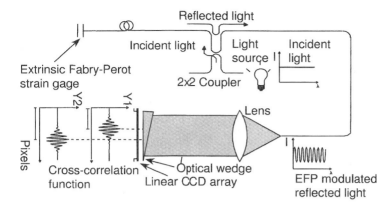

Fig. 21. Schematic of a white light cross correlator demodulator for EFP sensors

powerful way to make absolute Fabry-Perot cavity length measurements with good accuracy, linearity, and repeatability. Figure 21 provides a schematic diagram describing the operationing principle of the cross-correlator interrogation technique. Light from a broadband source is launched into one arm of a 2×2 coupler and directed toward the EFP strain sensor. The light signal is reflected back towards the cross-correlator portion of the instrument where it is expanded by a lens, transmitted through a thin optical wedge[19] and then detected by a linear CCD array. Prior to entering the lens, the light is a sinusoidal function of wavelength corresponding to the modulation transfer function of a low-finesse Fabry-Perot interferometer. In essence, the optical wedge is also a Fabry-Perot interferometer formed between two air-glass interfaces. As a result, the optical path length of this Fabry-Perot interferometer depends upon the spatial location along the wedge. In the absence of a sensor, light passing through the wedge results in another sinusoidal function, but this one is a function of position along the wedge. The combination of the wedge and the sensor yields the cross-correlation for which the instrument is named. Consider an EFP with a cavity length of d. The light reflected by this sensor will be transmitted maximally at the pixel location of the CCD array where the optical path length of the wedge is equal to d, as is shown in Fig. 21. A change in the EFP sensor cavity length due to strain will cause the optical correlation function (light transmitted through the wedge) to shift back and forth along the CCD array. Signal-processing software is used to track the maximum-intensity peak, and to relate its displacement to strain.

Whitelight cross-correlation techniques can make absolute measurements, meaning that the user can disconnect the gauge, shut the system down, etc., and not lose track of the strain value when the system is turned back on

---

[19] The wedge in this instrument is very shallow, thereby necessitating dielectric thin-film fabrication techniques. Fortunately, this type of thin-film fabrication is amenable to batch fabrication and is therefore very inexpensive.

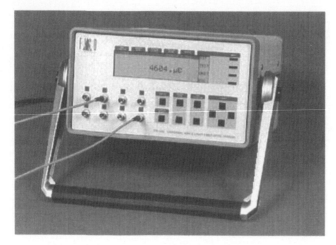

Fig. 22. Photograph a commercial EFP sensor demodulator (FTI-100) that uses the white light cross correlator technique (courtesy of FISO Technologies)

(i.e. no "bridge balancing" is required). Figure 22 shows a photograph of a commercially available EFP demodulator based on the white-light cross-correlator concept. Typical characteristics of this system are a dynamic range of 1:15000, precision of better than 0.025%, and resolution better than 0.01%. This particular instrument is designed to multiplex eight sensors. Multiplexing in this case is accomplished using an electrically-based spatial-division technique in which each channel is sequentially accessed by turning on and off the optical sources associated with each sensor. All sensors are interrogated using the same cross-correlator, although individual cross-correlators can be used for each sensor.

7.3 Scanning Fabry-Perot Filter (Bragg Grating Sensors)

A Bragg grating demodulation system that has recently become popular is based on a tunable spectral filter for wavelength discrimination [66,67]. The tunable wavelength filter used in this Bragg grating read-out instrument is based on the spectral transfer function of high-finesse optical fiber Fabry-Perot interferometers. These interferometers, in turn, are based on collimated air-gap configurations that use piezoelectric materials to control the separation of the reflective surfaces forming the cavity. The way these Fabry-Perot cavities are used as tunable filters is illustrated in Fig. 23. The spectral transfer function is a series of evenly spaced spectral peaks, which are usually separated by $\approx 50\,\text{nm}$[20]. A voltage signal applied to the piezoelectric device expands and contracts the Fabry-Perot cavity, which has the result of shifting

---

[20] The separation of two adjacent peaks in the spectral transfer function of a Fabry-Perot interferometer is called the Free Spectral Range (FSR).

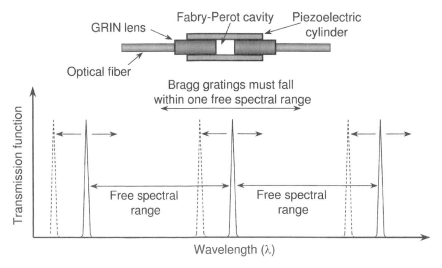

Fig. 23. Transmission function of a tunable-fiber Fabry-Perot filter

the spectral location of the Fabry-Perot bandpass peaks in unison. Only one of the bandpass peaks is used to make the Bragg grating read-out system by designing the filter such that only one of the spectral peaks falls in the wavelength range where Bragg gratings will reside. The fiber Fabry-Perot tunable filter scans the entire wavelength range containing the Bragg gratings, which has the result of producing an electrical signal proportional to the Bragg grating spectra (much like the one shown in Fig. 15). This spectrum is then differentiated numerically, and the zero crossings are tracked to measure the respective Bragg wavelengths. High accuracy and resolution are achieved using temperature-stabilized wavelength references.

Figure 24 shows a photograph of a commercially available Bragg grating demodulator that uses the scanning Fabry-Perot filter technique (FBG-IS Micron Optics, Inc.). The FBG-IS is capable of accommodating 30 gratings, depending on the strain range (as described in Sect. 3.5). The measurement bandwidth of the FBG-IS is $\approx 5\,\text{Hz}$ and the strain resolution is $\approx 5\,\mu\epsilon$. The bandwidth of this instrument is limited by the software interface between the FBG-IS and the controlling computer.

7.4 Spectral Interrogation (EFP Sensors)

It should come as no surprise that more than one supplier is marketing EFP sensor demodulation systems, since EFPs have been commercially available the longest. The demodulation technique described in this section is similar in some respects to the white-light cross-correlation technique described in Sect. 7.2. Like white-light cross-correlation, the technique shown in Fig. 25 uses a broadband optical source and and a linear CCD array. In

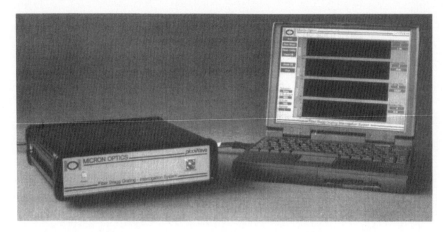

Fig. 24. Commercial fiber-optic Bragg grating instrumentation system interfaced with a laptop personal computer (courtesy of Micron Optics, Inc.)

this case, however, a diffraction grating is used instead of an optical wedge. The diffraction grating used in combination with the CCD array produces a mini-spectrometer, which in this case is used to determine the strain by examining the wavelength versus power relationship of the signal returned from the sensor. Because this method uses spectral information, it is known as spectral interrogation. In general, the EFP wavelength versus optical output power signal seen by the spectrometer, shown by the insert in Fig. 25, may be characterized as a Gaussian amplitude from the spectral profile of the broadband source modulated by the sinusoidal function produced by the sensor [68]. The separation of the maxima in the sinusoidal component of the spectrum is related to the separation of the two fiber ends, S, in an EFP sensor by

$$S = \frac{1}{2}\lambda_1\lambda_2/(\lambda_2 - \lambda_1) \,, \tag{12}$$

where λ_1 and λ_2 are the wavelengths at the center of the two adjacent nulls (or peaks). Once S is determined, the strain is given by $\varepsilon = \Delta S/L_g$, where ΔS is the change in EFP gap from one load state to the next and L_g is the gage length of the sensor. There are many methods of determining the peak separation. One method uses the fast Fourier transform (FFT) combined with zero padding and interpolation [69]. Simplicity and self-referencing (absolute measurement) are the two most attractive features of spectral interrogation. Recall that this self-referencing capability is particularly useful for strain monitoring applications where it is not practical or possible to keep a sensor continuously connected to the read-out instrumentation.

Figure 26 shows a photograph of a commercially available EFP sensor demodulator that uses the spectral processing concept. This four-channel demodulator (Fiber and Sensor Technologies Fiberscan 2000-4) performs all

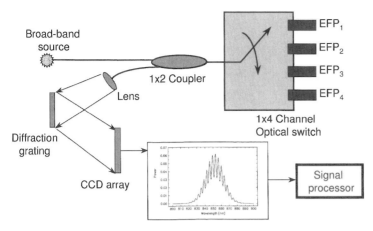

Fig. 25. Schematic of a spectral interrogation demodulator for EFP sensors

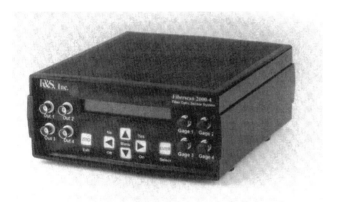

Fig. 26. Commercial four channel extrinsic Fabry-Perot sensor read-out instrument based on spectral interrogation (courtesy of Fiber and Sensor Technologies, Inc.)

calculations using an on-board microprocessor to achieve a displacement resolution of 1 nm, which correlates to sub-microstrain resolution for typical EFP sensor gage lengths ($\approx 2 - 8$ mm). The system has a displacement range of hundreds of micrometers, providing a strain range exceeding $50000\,\mu\varepsilon$. The Fiberscan 2000-4 system interrogates EFP sensors at a rate of 800 scans per second, which allows the system to accommodate moderate strain/displacement rate measurements with low measurement error. The output refresh rate of the system is user-configurable, which allows the user to make performance trade-offs between refresh rate, measurement resolution and measurement range. Multiplexing is accomplished using optical switching technology.

7.5 Other Demodulators

There continues to be interest in developing new demodulation technologies to fill operational gaps left by existing commercial demodulators. This is particularly true with Bragg gratings, where higher sensitivity, higher measurement bandwidth, greater multiplexing capacity, and lower costs are all desired by the strain measurement practitioners. Ratiometric demodulation is a simple demodulation scheme developed specifically to reduce complexity and increase measurement bandwidth [70,71]. This demodulation technique passes the light reflected from the grating through an optical filter with an attenuation function that is linear with wavelength. This linear optical filter is used to convert strain-induced changes in Bragg wavelength to proportional changes in optical intensity, which can easily be measured using low-cost photodetectors. The measurement bandwidth of ratiometric demodulators is limited only by the photodetector bandwidth, which can be as high as several GHz, but it sacrifices multiplexibility for measurement bandwidth. Even so, demodulation products based on ratiometric principles are commercially available.

Another Bragg grating demodulation concept that shows promise treats the Bragg wavelength as a laser source, and load-induced changes in Bragg wavelength as a current-induced modulation in the wavelength source. These changes in Bragg wavelength can be converted to a measurable optical phase by passing the light through a interferometer with a non-zero optical pathlength difference [72]. Interferometric demodulation is highly sensitive and has good measurement-bandwidth characteristics, but it is relatively complicated and has limited multiplexing capability. A recent entry to the interferometric family of demodulators uses serially multiplexed gratings with identical Bragg wavelengths and Fourier transform processing [73]. This Fourier-transform interferometric demodulator has exactly the opposite characteristics to that of the previous two demodulation techniques. It can multiplex literally hundreds of Bragg grating sensors, but has a measurement bandwidth of less than a few Hz and has relatively low strain sensitivity.

The final demodulation technology worth mentioning essentially uses inexpensive one-dimensional and two-dimensional spectrometers [74,75,76]. The two-dimensional version of this technique uses a diffraction grating with two-dimensional CCD imager to produce many spectrometers operating in parallel. In this way, arrays of serially multiplexed Bragg gratings in different fibers can be interrogated simultaneously. This demodulation technology is relatively new, but has the potential to preserve the balance between multiplexing capacity, measurement bandwidth, and strain sensitivity.

8 Sensor Comparison

This section provides a brief comparison of the sensors discussed in this review. The comparison is purposefully restricted to a commercial fiber-optic strain gage perspective. As a result, it does not include all possible configurations and options, rather it includes only configurations and options that are available for purchase. Obviously new products will become available as time goes on, so the discussion in this section will quickly become obsolete. However, it will provide a reasonable means for fiber-optic sensor users to judge the relevant merits of the IFP, EFP and Bragg grating sensors. Table 1 compares these sensors from the perspective of thermal sensitivity, multiplexibility, gage length, polarization fading, fabrication ease, and maturity. In strain sensing applications, all three sensors have intrinsically high thermal apparent strain sensitivity, primarily because of the coefficient of thermal expansion of the structure to which the sensors are bonded. Extrinsic Fabry-Perot sensors are unique in that they can be made in self-temperature-compensated (STC) formats. As a result, EFP sensors are the only sensors that can be classified as having a low thermal apparent strain sensitivity. The high thermal apparent strain sensitivity of the other two sensors can be overcome by using (1) dummy gages, (2) an auxiliary fiber-optic temperature sensor, or (3) high-pass filtering if the strain signal is dynamic and the temperature signal is not. While effective, none of these options is as attractive as using a STC EFP gage.

Table 1. Comparison of Fiber-optic Strain Sensors

Sensor	Thermal Apparent Strain Sensitivity	Multi-plexing	Gage Length	Polarization Fading	Fabrication Ease	Maturity
Intrinsic Fabry-Perot	High	SDM	$\sim 20\,\mu m$ to ~ 2 m	Yes	Moderate	Moderate
Extrinsic Fabry-Perot	Low	SDM	~ 2 mm to ~ 3 cm	No	Moderate	Moderate
Bragg-grating	High	WDM	~ 1 mm to ~ 3 cm	Yes	Moderate	Moderate

As far as multiplexibility is concerned, the issue is not whether the optical fiber strain gages can be multiplexed, because they all possess this potential in one form or another. At issue is the number of sensors that can be multiplexed, the cross-talk between sensors, and the complexity of potential multiplexing schemes. The optical switch-based spatial-division multiplexing (SDM) used with EFP and IFP sensors is very simple, but it limits the measurement bandwidth and requires one lead-in/out fiber per sensor. Whether or not these characteristics are problematic depends on the application at hand. There is also the option of using electrical-switching concepts where each optical fiber has its own photodetector. In terms of multiplexing fiber-optic strain sensors on a single optical fiber, wavelength-division multiplexing (WDM) Bragg grating sensors are the undisputed champion. Of course, the number of sensors that can be multiplexed is limited by the spectral bandwidth of the optical source and by the anticipated wavelength shifts induced by the measurement field. It is worth noting that it is possible to use other kinds of multiplexing concept with IFP, EFP, and Bragg grating sensors. Coherence division, frequency division, time division and others have all been demonstrated, but they are not available as commercial off-the-self products.

Comparison of gage length is straight forward. By their designs, commercial EFP and Bragg grating sensor gage lengths are always less than a few centimeters. In contrast, the guiding nature of IFP sensors makes gage lengths as long as a few meters possible[21]. The sensitivity of IFP sensors scales with gage length, so longer gage lengths mean highly sensitive strain sensors. However, the potential for polarization fading also increases with gage length, making design trade-offs between sensitivity and polarization fading necessary. Bragg gratings are sensitive to polarization effects as well, although not in a way that usually affects the signal-to-noise characteristics of the sensor system. EFPs are unique in that they are immune to polarization fading.

Fabrication ease and maturity are the last of the comparisons in Table 1. Intrinsic and extrinsic Fabry-Perot sensors can be considered moderately difficult to fabricate mainly because they require multi-step fabrication processes, which are still done by hand. Until automated techniques are developed, these sensors will remain relatively expensive compared to resistance strain gages. Bragg gratings also fall into the moderate range, not so much because they are all that difficult to manufacture (they are not), but because the fabrication facility requires a considerable capital investment. These sensors are also mostly made by hand, although the telecommunications market is pushing vendors towards complete automation. This automation is expected to significantly decrease the cost of Bragg gratings.

The final comparison in Table 1 is maturity. Intrinsic Fabry-Perot (IFP) Sensors, Extrinsic Fabry-Perot (EFP) sensors, and Bragg gratings have all reached roughly the same level of moderate maturity, even though Bragg gratings are a much younger technology. Several small companies market IFP and EFP sensors, and have put in a commensurate level of effort in solving engineering issues associated with sensor implementation. Bragg gratings, on the other hand, are maturing rapidly because of their many important uses in the telecommunications industry.

---

[21] The gage length of long IFP sensors is limited by the coherence length of the laser used by the instrumentation.

References

1. B. Culshaw, J. Dankin: *Optical Fiber Sensor: Systems and Applications*, 1 and 2, Artech House, Norwood, MA (1988)
2. E. Udd (Ed.): *Fiber Optic Sensors: An Introduction For Engineers and Scientists*, Wiley, New York (1990)
3. R. D. Turner, T. Valis, W. Hogg, R. Measures: Fiber-Optic Sensors of Smart Structures, J. Intell. Mater. Syst. Struct. **1**, 26 (1990)
4. C. E. Lee, W. N. Gibler, R. A. Atkins, H. F. Taylor: In-line Fiber Fabry-Perot Interferometer with High-Reflectivity Internal Mirrors, J. Lightwave Technol. **10**, 1376 (1992)
5. J. Stone, D. Marcuse: Ultrahigh Finesse Fiber Fabry-Perot Interferometers, J. Lightwave Technol. **4**, 382 (1992)
6. C. E. Lee, J. J. Alcoz, Y. Yeh, W. N. Gibler, R. A. Atkins, H. F. Taylor: Optical Fiber Fabry-Perot Sensors for Smart Structures, J. Smart Mater. Struct. **1**, 123 (1992)
7. C. E. Lee, H. F. Taylor: Interferometric Optical Fiber Sensors Using Internal Mirrors, Electron. Lett. **24**, 193 (1988)
8. L. F. Stokes, M. Chodorow, H. J. Shaw: All Single-Mode Fiber Resonator, Opt. Lett. **7**, 288 (1982)
9. J. S. Sirkis, H. W. Haslach: Complete Phase-Strain Model for Structurally Embedded Interferometric Optical Fiber Sensors, J. Intell. Mater. Syst. Struct. **2**, 3 (1991)
10. J. S. Sirkis: Unified Approach to Phase-Strain-Temperature Models for Smart Structure Interferometric Optical Fiber Sensors: Part I, Applications, Opt. Eng. **32**, 752 (1993)
11. J. S. Sirkis, H. W. Haslach: Interferometric Strain Measurement by Arbitrarily Configured, Surface-Mounted, Optical Fibers, J. Lightwave Technol. **8**, 1497 (1990)
12. T. Valis, D. Hogg, R. M. Measures: Thermal Apparent Strain Sensitivity of Surface Adhered, Fiber Optic Strain Gauges, Appl. Opt. **31**, 7178 (1992)
13. D. W. Stowe, D. R. Moore, R. G. Priest: Polarization Fading in Fiber Interferometric Sensors, IEEE Trans. Microw. Theory Tech. **MTT-3D**, 1632 (1982)
14. N. J. Frigo, A. Dandridge, A. B. Tveten: Technique for the Elimination of Polarization Fading in Fiber Interferometers, Electron. Lett. **20**, 319 (1984)
15. K. H. Wanser, N. H. Safar: Remote Polarization Control for Fiber Optic Interferometers, Opt. Lett. **12**, 217 (1987)
16. A. D. Kersey, M. J. Marrone, A. Dandridge, A. B. Tveten: Optimization and Stabilization of Visibility in Interferometric Fiber Optic Sensors Using Input Polarization Control, J. Lightwave Technol. **6**, 1599 (1988)
17. A. D. Kersey, M. Marrone, A. Dandridge: Observation of Input Polarization Induced Phase Noise in Interferometric Fiber Optic Sensors, Opt. Lett. **13**, 847 (1988)
18. R. Kist: Point Sensor Multiplexing Principles, in B. Culshaw, J. Dankin (Ed.): *Optical Fiber Sensors: Systems and Application*, Vol. 2, Artech House, Norwood, MA (1989)
19. Kersey: Distributed and Multiplexed Fiber Optic Sensors, in E. Udd (Ed.): *Fiber Optic Sensors: An Introduction for Engineers and Scientists*, Wiley, New York (1990)

20. A. Wang, M. S. Miller, D. Sun, K. A. Murphy, R. O. Claus: Advances in the Extrinsic Fabry-Perot Interferometric Optical Fiber Sensors, in: Fiber Optic Smart Structures and Skins V, Boston, Ma, Proc. SPIE **1798**, 32 (1992)
21. K. Murphy, M. F. Gunther, A. M. Vengsarkar, R. O. Claus: Quadrature Phase Shifted, Extrinsic Fabry-Perot Optical Fiber Sensors, Opt. Lett. **16**, 273 (1991)
22. R. O. Claus, A. Gunther, A. Wang, K. Murphy, D. Sun: Extrinsic Fabry-Perot Sensor for Structural Evaluations, in F. Ansari (Ed.): Applications of Fiber Optic Sensors in Engineering Mechnics, ASCE, 60 (1993)
23. J. W. Dally, W. F. Riley: *Experimental Stress Analyis*, 3rd edn., McGraw-Hill, New York (1991)
24. Fiber-Optic Strain Gauges, FISO Technologies Technical Note (1997)
25. Henriksson, F. Brandt: Design and Manufacture of and EFP Sensor for Embedding in Carbon/Epoxy Composites, in: Proc. 10th Conf. on Optical Fiber Sensors, Glasgow, 472 (1994)
26. J. S. Sirkis, T. A. Berkoff, R. T. Jones, H. Singh, E. J. Friebele, A. D. Kersey, M. A. Putnam: In-Line Fiber Etalon (ILFE) Strain Sensor Fiber Optic, J. Lightwave Technol. **13**, 1256 (1995)
27. H. Singh, J. S. Sirkis, J. Andrews, R. Pulfrey: Evaluation of Integrated Optic Modulator-Based Detection Schemes for In-Line Fiber Etalon Sensors, J. Lightwave Technol. **13**, 1772 (1995)
28. J. R. Dunphy, G. Meltz, W. W. Morey: Multi-Function, Distributed Optical Fiber Sensor for Composite Cure and Response Monitoring, in: Fiber Optic Smart Structures and Skins III, San Jose, Ca, Proc. SPIE **1370**, 116 (1990)
29. R. M. Measures, S. M. Melle, K. Liu: Wavelength Demodulated Bragg Grating Fiber Optic Sensing Systems for Addressing Smart Structure Critical Issues, J. Smart Mater. Struct. **1**, 36 (1992)
30. W. W. Morey, G. Ball, H. Singh: Applications of Fiber Grating Sensors, in: Fiber Optic and Laser Sensors XIV, Denver Co, Proc. SPIE **2839**, 2 (1996)
31. G. R. Meltz, W. W. Morey, W. H. Glen: Formation of Bragg Gratings in Optical Fibers by a Transverse Holographic Method, Opt. Lett. **14**, 823 (1989)
32. Malo, F. Bilodeau, J. Albert, D. C. Johnson, K. O. Hill, Y. Hibino, M. Abe: Photosensitivity in Optical Fibers and Silica-on-Substrate Waveguides, in: Photosensitivity and Self Organization in Optical Fibers and Wavegudies, Quebec, Canada, Proc. SPIE **2044**, 18 (1993)
33. K. O. Hill, B. Mallo, F. Bilodeau, D. C. Johnson, J. Albert: Bragg Gratings Fabricated in Monomode Photosensitive Optical Fiber by UV Exposure Through a Phase Mask, Appl. Phys. Lett. **62**, 1035 (1993)
34. R. B. Wagreich, W. A. Atia, H. Singh, J. S. Sirkis: Effects of Diametric Load on Fibre Bragg Gratings Fabricated in Low Birefringece Fibre, Electron. Lett. **32**, 1223 (1996)
35. R. Wagreich, J. S. Sirkis: Distinguishing Fiber Bragg Grating Strain Effects, in: Proc. 11th International Conf. Optical Fiber Sensors, Williamsburg, VA (1997) p. 20
36. H. Patrick, S. L. Gilbert, A. Lidgard, M. D. Gallagher: Annealing of Bragg Gratings in Hydrogen Loaded Optical Fiber, J. Appl. Phys. **78**, 2940 (1995)
37. 3M Fiber Bragg Grating Application Note, The Mechanical and Optical Reliability of Fiber Bragg Gratings (1996)
38. W. W. Morey, J. R. Dunphy, G. Meltz: Multiplexing Fiber Bragg Grating Sensors, in: Proc. Distributed and Multiplexed Fiber Sensors I, Boston MA, Proc. SPIE **1586**, 216 (1991)

39. K. T. V. Grattan, B. T. Meggit: *Optical Fiber Sensor Technology*, Chapman & Hall, London (1995)
40. D. Kersey, T. A. Berkoff, W. W. Morey: Multiplexed Fiber Bragg Grating Strain Sensor System with a Fiber Fabry Perot Wavelength Filter, Opt. Lett. **18**, 1370 (1993)
41. Melvin et al.: Integrate Vehicle Monitoring (IVHM) for Aerospace Vehciles, in F. K. Chang (Ed.): Structural Health Monitoring: Current Status and Perspectives, Technomic, Lancaster (1997) p. 705
42. S. S.J. Roberts, R. Davidson: Mechanical Properties of Composite Materils Containing Embedded Fibre Optic Sensors, Fiber Optics Smart Structures and Skins IV, Boston, MA (1991) p. 326
43. S. E. Baldini, E. Nowakowski, H. G. Smith, E. J. Friebele, M. A. Putnam, R. Rogowski, L. D. Melvin, R. O. Clause, T. Tran, M. Holben: Cooperative Implementation of a High Temperature Acoustic Sensors, Fiber Optics Smart Structures and Skins IV, Boston, MA (1991) p. 125
44. R. Claus, M. Gunther, A. Wang, K. Murphy: Extrinsic Fabry-Perot Sensor for Strain and COD Measurements from -200 to 900 °C, Smart Mater. Struct. **1**, 237 (1992)
45. R. Kurkjian, D. Inniss: Understanding Mechanical Properties of Lightguides: A Commentary, Opt. Eng. **30**, 681 (1991)
46. E. S. R. Sikora: Examination of the Strength Characteristics, Hydrogen Permeation and Electrical Resistance of the Carbon Coating of a Number of 'Hermetic' Optical Fibers, in: International Wire and Cabling Symposium, Atlanta, GA (1989) p. 663
47. M. G. Estep, G. S. Gleasemann: The Effect of Carbon Overcoating on the Mechanical Behavior of Large Flaws, Optical Materials Reliability and Testing: Benign and Adverse Environments, Boston MA, Proc. SPIE **1971**, 18 (1992)
48. R. Kurkjian: Hermetically Coated Fibers, in: Fiber-Optic Materials and Components, San Diego, CA, Proc. SPIE **2290**, 237 (1994)
49. J. S. Andreassen: Experimental Study on Reliability of Stress Free Aging Effects on Hermetically Coated Fibres, in: Fiber Optic Materials and Components, San Diego CA, Proc. SPIE **2290**, 229 (1994)
50. W. Habel, B. Hillemeier: Results in Monitoring and Assessment of Damages in Large Steel and Concrete Structures by Means of Fiber Optic Sensors, in: Proc. Smart Systems for Bridges, Structures, and Highways, San Diego CA, Proc. SPIE **2446**, 25 (1995)
51. W. Habel, M. Hopcke, F. Basedau, H. Polster: The Influence of Concrete and Alkaline Solutions on Different Surfaces of Optical Fibers Sensors, in: 2nd European Conference on Smart Structures and Materials, Institute of Physics, Bristol, Glasgow, UK (1994) p. 168
52. K. Y. Leung, D. Darmawangsa: Interfacial Changes for Optical Fibers in a Cementitious Environment, unpublished
53. J. Hayes: *Fiber Optics Technicians Manual*, Delmar, Albany, New York (1996)
54. S. Ungar: *Fibre Optics: Theory and Applications*, Wiley, New York (1989)
55. Fiber Optic Strain Gauge Installation Guide, FISO Technologies (1997)
56. Strain Gage Installation with M-Bond 200 and AE-10 Adhesive Systems, Measurements Group Tech. Note
57. EFP Pre-Installation Surface Preparation Procedures, F&S Technical Note

58. J. S. Sirkis, C. C. Chang: Embedded Fiber Optic Strain Sensors, in C. Jenkins (Ed.): *Manual on Experimental Methods for Mechanical Testing in Composites* SEM, Bethel, CT (1998)
59. D. A. Jackson: Monomode Optical Interferometers for Precision Measurement, J. Phys. **18**, 981 (1985)
60. K. Liu, R. M. Measures: Signal Processing Techniques for Localized Interferometric Fiber-Optic Strain Sensors, J. Intell. Mater. Syst. Struct. **3**, 432 (1992)
61. A. Jackson, A. D. Kersey, M. Corke, J. D. C. Jones: Pseudoheterodyne Detection Scheme for Optical Interferometers, Electron. Lett. **18**, 1081 (1982)
62. R. Sadkowski, C. E. Lee, H. F. Taylor: Multiplexed Interferometric Fiber-Optic Sensors with Digital Signal Processing, Appl. Opt. **34**, 5861 (1995)
63. D. Lee, A. F. Taylor: A Fiber-Optic Pressure Sensor for Internal Combustion Engines, Sensors, **20** (1998)
64. C. Belleville, G. Duplain: White-light Interferometric Multimode Fiber-optic Strain Sensor, Opt. Lett. **18**, 78 (1993)
65. Morin, S. Caron, R. Van Neste, M. H. Edgecombe: Field Monitoring of the Ice Load of an Icebreaker Propeller Blade Using Fibre Optic Strain Gauges, in: Smart Sensing, Processing, and Instrumentation, San Diego CA, Proc. SPIE **2718**, 427 (1996)
66. K. Kersey, T. A. Berkoff, W. W. Morey: Multiplexed Fiber Bragg Grating Strain Sensor System with a Fiber Fabry Perot Wavelength Filter, Opt. Lett. **18**, 370 (1993)
67. Miller, T. Li, J. Miller, K. Hsu: Multiplexed Fiber Gratings Enhance Mechanical Sensing, Laser Focus World, 119, March (1998)
68. V. Bhatia, K. Murphy, R. O. Claus: Optical Fiber Based Absolute Extrinsic Fabry-Perot Interferometric Sensing System, Meas. Sci. Technol. **7**, 58 (1996)
69. K. A. Shinpaugh, R. L. Simpson, A. L. Wicks, S. H. Ha, J. L. Fleming: Signal-processing Techniques for Low Signal-to-Noise Ratio Laser Doppler Velocimetry Signals, Exp. Fluids **12**, 319 (1992)
70. S. M. Melle, K. Liu, R. M. Measures: Practical Fiber-Optic Bragg Grating Strain Gauge System, Appl. Opt. **32**, 3601 (1993)
71. M. A. Davis, T. A. Berkoff, A. D. Kersey: Demodulator for Fiber Optic Bragg Grating Sensors Based on Fiber Wavelength Division Couplers, Proc. Smart Sensing, Processing and Instrumentation, Orlando FL, Proc. SPIE **2191**, 86 (1994)
72. A. D. Kersey, T. A. Berkoff, W. W. Morey: High Resolution Fiber Grating Based Strain Sensor with Interferometric Wavelength Shift Detection, Electron. Lett. **28**, 236 (1992)
73. L. Melvin: Integrated Vehicle Health Monitoring (IVHM) for Space Vehicles, in: Proceedings of International Workshop on Structure Health Monitoring, Stanford, CA (1997), p. 705
74. G. Askins, M. A. Putnam, E. J. Friebele: Instrumentation for Interrogating Many-element Fiber Bragg Grating Arrays, in: Smart Sensing, Processing, and Instrumentation, Proc. SPIE **2444**, 257 (1995)
75. S. Chen, Y. Hu, L. Zhang, I. Bennion: Digital Wavelength and Spatial Domain Multiplexing of Bragg Grating Optical Fiber Sensors, in: Proc. 11th International Conference Optical Fiber Sensors, Sapporo. Japan (1996) p. 100
76. Y. Hu, S. Chen: Multiplexing Bragg Gratings Using Combined Wavelength and Spatial Division Techniques with Digital Resolution Enhancement, Electro. Lett. **33**, 1973 (1997)

Long-Gage Fiber-Optic Sensors for Structural Monitoring

Daniele Inaudi

Smartec SA,
via al Molino 6, CH-6916 Grancia, Switzerland
inaudi@smartec.ch

Abstract. Long-gage and distributed sensors take advantage of the ability of optical fibers to guide light signals over large distances and following sinuous paths. This allows the measurement of deformations and temperatures inside or at the surface of structures. The sensors can be arranged into a network that mimics the nervous system of our body and monitors the health of the structure. Long-gage and distributed sensors are ideal for the global monitoring of large structures, where they allow a good coverage of the structure with reduced number of sensors and little a priori knowledge on its degradation modes. Long-gage sensors give an integrated or average measurement of strain or temperature over lengths of typically a few tens of centimeters to a few tens of meters. Interferometric and microbending sensors are typical examples of this category. On the other hand, distributed sensors allow the measurement of multiple points along a single fiber. These sensors are mostly based on different types of light scattering including Rayleigh, Raman and Brillouin. Both types of sensors have already found niche applications where their characteristics and performance surpass those of conventional sensors.

1 Introduction: Monitoring versus Measuring Structures

The word "structure" describes a large spectrum of engineering elements with a load-bearing or load-transferring function. You can think of bridges, buildings, airplane wings, frames, vehicles and so on. These structures undergo deformations that are either the results of the applied loads or are due to changes in the materials constituting the structure. Deformations can be an integral part of the structure's function (normal deformations) or indicate a degradation of the structure itself (abnormal deformations). Studying the structure's deformations is therefore a good means of assessing its performance and diagnosing possible degradations in its functions.

The study of structural deformations can be approached in two different ways: measuring and monitoring. Measuring refers to an action that is performed on the structure at pre-defined times and under pre-defined conditions to obtain data significant for the structure behavior. Typical examples of measurements include surveying with triangulation or leveling systems, photogrammetry, deformation analysis with whole-field methods (moiré, speckle interferometry) and visual inspection. Measurements usually extend over very short time-spans compared to the lifetime of the structure.

Monitoring, on the other end, usually follows the structure over its whole evolution. The monitoring system will automatically record data on the structure at programmed intervals, analyze, reduce and store the obtained values, and create a track-record able to diagnose the structure performance and allowing an extrapolation of its probable future evolution. Monitoring is usually realized by installing sensors that measure the relevant parameters relative to the structure and its environment. A monitoring system has the difficult task of identifying behavior and degradation that is not always predictable at the time of its design. In most cases, it is not possible nor reasonable to install a sensor network able to diagnose any possible (even if improbable) behavior of the structure. The aim of a monitoring system should be to detect deviations from the normal or predicted performance. If such deviations are detected, additional monitoring and measurements will be used to obtain a better understanding of the phenomena.

The diagnosis of human illness is an excellent example of effective combination of monitoring and measuring. Our nervous system constitutes the monitoring part of the diagnosis process. It is able to detect anomalies and to indicate them through the experiencing of pain. It is also able to give a rough indication of the nature, position and magnitude of the detected problem. It will therefore be possible to distinguish between a broken leg and a small headache. If the pain is important and persistent, we relay on measurements to pinpoint the exact cause and find an adequate cure. This can be done by a doctor taking blood samples, X-rays or carefully observing the patient. This dichotomy between monitoring and measurements is therefore natural but also very efficient. Living inside an NMR scanner would be an excellent means for early detection of any disease that might arise. This extreme monitoring approach is fortunately not common practice.

As its title implies, this review will concentrate on structural monitoring. Many of the sensor techniques presented here can, however, be used as measurement techniques as well. In many real cases the boundary between measurement and monitoring is not as sharp as presented in this introduction.

2 Long-Gage versus Short-Gage Sensors

The monitoring of a structure can be approached either from the material or from the structural point of view [1]. In the first case, monitoring will concentrate on the local properties of the materials used in the construction (e.g. concrete, steel, and timber) and observe their behavior under load or ageing. Short base-length strain sensors are the ideal transducers for this type of monitoring approach. If a very large number of these sensors are installed at different points in the structure, it is possible to extrapolate information about the behavior of the whole structure from these local measurements.

Since it is impossible to cover the whole structure with such sensors, some *a priori* knowledge about the most interesting and representative locations to be analyzed is required. It will than be assumed that the rest of the structure lying between the measurement points will behave in a similar or predictable way.

In the structural approach, the structure is observed from a geometrical point of view. By using long sensors with measurement bases of the same order of magnitude as the typical size of the structure (for example a few meters for a bridge), it is possible to gain information about the deformations of the structure as a whole and extrapolate the global behavior of the construction materials. The structural monitoring approach will detect material degradation like cracking or flow only if it has an impact on the form of the structure. Since it is reasonably possible to cover the whole structure with such sensors, no *a priori* knowledge about the position of the degradations to be observed (for example the exact position of a crack) is required. This approach usually requires a reduced number of sensors compared to the material monitoring approach.

The availability of reliable strain sensors like resistance strain gages or, more recently, of fiber-optic strain sensors as Fabry-Perot interferometers and fiber Bragg gratings [2,3] has historically concentrated most research efforts in the direction of material monitoring rather than structural monitoring. This latter has usually been applied using external means like triangulation, dial gages and invar wires. Fiber-Optic sensors [4,5,6,7,8,9,10,11,12,13,14,15,16] offer an interesting means of implementing structural monitoring with internal or embedded sensors.

3 Interferometric Sensors

Interferometric sensors use the coherence of light to produce highly sensitive measurements. The light travels through two distinct paths and is than recombined to produce an interference pattern. When the two paths are exposed to different perturbations, it is possible to obtain a measurement of their magnitude by observing the intensity variations at the output of the interferometer. In most configurations, the interferometer is made of a reference and a measurement arm. Both arms are exposed to the same perturbation fields except for the one to be measured, which is applied to the measurement arm only. In the case of deformation measurements, the measurement arm will be mechanically coupled to the structure, while the reference arm will be independent of it.

3.1 Optical Arrangements

Interferometric fiber-optic sensors come in a large variety of configurations. The most common are known under the names of their inventors[1]: Fabry-Perot, Michelson, Mach-Zehnder, and Sagnac. These set-ups are illustrated in Fig. 1. The Fabry-Perot interferometer has no proper reference arm and presents a path unbalance corresponding to twice the optical length of the measurement arm. The Michelson and Mach-Zehnder types are functionally equivalent, the first being a "folded" version of the second. The Sagnac interferometer is a special case because both arms are formed by the same section of fiber; the difference between them is given by the opposed propagating direction. This interferometer will therefore be sensitive only to time-dependent effects that affect the two propagation directions differently. Sagnac interferometers are particularly useful for the measurement of rotation rates and are often used as fiber-optic gyroscopes.

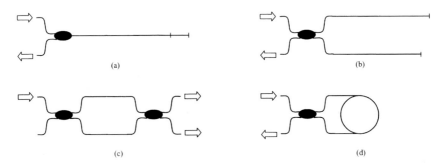

Fig. 1. Most common interferometric set-ups. (**a**) Fabry-Perot, (**b**) Michelson, (**c**) Mach-Zehnder, (**d**) Sagnac

3.2 Strain and Temperature Sensitivity

The strain and temperature sensitivity of all interferometric set-ups can be described by the following equations

$$\frac{\Delta \phi}{\phi} = P_\varepsilon D\varepsilon ,\qquad(1)$$

$$\frac{\Delta \phi}{\phi} = P_\mathrm{T} DT ,\qquad(2)$$

where

$$\phi = \frac{2\pi n L}{\lambda} .\qquad(3)$$

---

[1] All these interferometers were introduced in their bulk-optics form with light travelling in air rather than guided in a fiber. Since the fiber-optic equivalents are functionally identical, the same names are used.

Typical values for standard single-mode fibers are $P_\varepsilon = 0.78 \times 10^{-6}\ \mu\varepsilon^{-1}$ and $P_T = 6.67\ 10^{-6} T^{-1}$. These equations describe the phase variation induced by a strain or a temperature variation. The total phase change of the interferometer is given by the difference in integrated phase change between its two interfering paths. For standard single-mode fibers a temperature variation of $1\,°C$ roughly corresponds to a strain variation of $10\ \mu\varepsilon$. If pure integrated strain readings are required, it is therefore important to eliminate the parasitic temperature dependence. This can be done either by measuring the temperature separately and then correcting the strain readings, or more simply by exposing the reference fiber to the same temperature field. This will leave only a parasitic temperature dependence due to the thermal expansion coefficient of fused silica of about $0.5\ \mu\varepsilon/°C$.

3.3 Demodulation Techniques

The high sensitivity of interferometric set-ups comes with a major drawback: the output intensity is not only non-linear but even ambiguous. Since the interference of the two optical paths produces a sinusoidal modulation of the output signal, the same detected intensity can correspond to an infinity of input values. The demodulation of an interferometer therefore requires *ad hoc* techniques able to resolve the non-linearity and ambiguity of the signal. A complete discussion of the many available demodulation techniques is beyond the scope of this review. The following paragraphs will, however, give an overview of a few widely used demodulation techniques adapted for structural monitoring purposes.

Coherent Demodulation. If the interferometer is illuminated with a source having a coherence length larger than the path unbalance between the two interferometer arms, the detected intensity will have a cosinusoidal dependence on the phase difference between the arms. Stable laser diodes or gas lasers are the ideal light source for coherent demodulation. In order to demodulate the input signal, two main approaches can be followed: phase retrieval or fringe counting. In the first case the demodulator will produce an output proportional to the phase of the interference signal. To obtain this signal unambiguously it is necessary to add a known phase modulation to the unknown phase changes produced by the perturbation to be measured. This can be done with a phase modulator installed in one of the interferometer arms or by modulating the wavelength of the light source: an approach known as phase generated carrier [18]. If the phase is retrieved with a time-resolution compatible with the Shannon theorem (the sampling frequency should be at least twice the maximum signal frequency) it will be relatively easy to resolve the 2π ambiguity and obtain an unwrapped phase proportional to the input signal.

In the case of fringe counting, the periodic interference fringes are used as the ticks on a meter stick. By counting the passing fringes and multiplying by the sensitivity factor given by (1) and (2), one can obtain the magnitude of the perturbation applied to the measurement arm. Fringe counting is less precise than phase retrieval but can be implemented more simply. The main challenge resides in the sign ambiguity but can often be solved by using two signals in quadrature.

A more detailed discussion of the demodulation of coherent interferometers can be found in [17,19]

Low-Coherence Interferometers. If a broadband source is used instead of a coherent source to illuminate the interferometer, interference fringes will be visible only when the path unbalance between the interferometer arms is smaller than the coherence length of the source. Typical broadband sources, also called white light or low-coherence sources, are the light emitting diodes(LED) and the Superluminescent LED (SLED) and have a coherence length of a few tens of micrometers. Low-coherence interferometry can be used to demodulate interferometers by using a technique appropriately named path matching. It consists of cascading the measurement interferometer with a second one able to introduce an accurately known path unbalance [20]. Only when the path unbalance in the second interferometer matches the one in the first,will the interference fringes become visible.

The main interest of this technique consists in the elimination of the fringe ambiguity. Using low-coherence interferometry it is possible to switch off the demodulator without loosing information about the sensor's phase. This is a major advantage in the case of monitoring, since uninterrupted operation is difficult to be guaranteed in the case of long-term measurements.

An example of a commercially available low-coherence demodulator is the SOFO system [21] shown in Fig. 2. It was developed at the Swiss Federal Institute of Technology in Lausanne and is now commercialized by SMARTEC in Switzerland.

The measurement set-up is based on a double, all-fiber, Michelson interferometer in tandem configuration. The 1300 nm radiation of a superluminescent LED with a coherence length of 30 μm is launched into a single-mode fiber and sent to the sensor installed into the structure. Near the region to be measured, the light is split, by means of a coupler, into the reference and the measurement fibers. The measurement fiber is pre-tensioned and mechanically coupled to the structure at two anchorage points in order to follow its deformations, while the reference fiber is free and acts as a temperature reference. Both fibers are installed inside the same pipe. The light is reflected by mirrors at the end of the fibers, recombined by the coupler and sent back to the reading unit. The analyzer is a Michelson interferometer with one of the arms terminated with a mobile mirror. It allows the introduction of an accurately known path difference between its two arms. The signal detected by the photodiode is pre-amplified and demodulated by a band-pass filter and a digital envelope filter. For each scan, three equidistant peaks are observed. The central peak is obtained when the Michelson analyzer is balanced, whereas the side peaks correspond to the mirror positions where the path unbalance

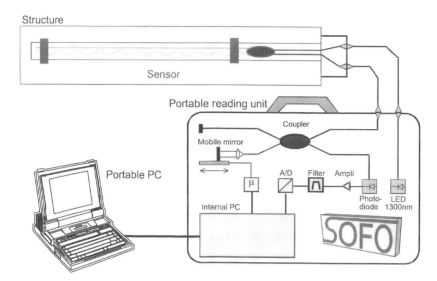

Fig. 2. Optical set-up of the SOFO Low-Coherence Interferometer

between the analyzer arms corresponds to the length difference between two twin partial reflectors or the end mirrors. By following the position of the side peaks it is possible to determine the total deformation undergone by the measurement fiber and therefore the deformation of the structure. This requires the knowledge of the relation between elongation of the fiber ΔL and the change in the time of flight Δt of the light guided in the fiber itself. This was experimentally determined to be (for standard single-mode fibers)

$$D L_{mm} = 0.128\, Dt(ps) . \tag{4}$$

The position of the peaks can be determined by computing the center of gravity of the peaks themselves. The resolution of the system is $2\,\mu m$ and its precision is 0.2% of the measured deformation. The resolution, precision and stability of this set-up are determined by the mechanical characteristics of the mirror scanning system. By using a high-precision displacement table it is therefore possible to guarantee the above-cited resolution even after years of operation.

The SOFO system has been successfully used to monitor more than 50 structures, including bridges, tunnels, piles, anchored walls, dams, historical monuments as well as laboratory models. An example of the application of this system is described in Sect. 3.5.

Modal Interferometers. Modal interferometers represent a special case, since both arms of the interferometer are contained in the same fiber. The two interfering paths are constituted by two different propagation modes of the fiber. Different set-ups have been proposed using different polarizations in a high-birefringence fiber [22] or higher order lateral modes [23]. These set-ups are particularly interesting when single-fiber operation is mandatory, as, for example, in the case of sensors imbedded in composite materials. However, these set-ups still have to prove their effectiveness for field applications.

3.4 Sensor Packaging

Since long-gage interferometric sensors can reach sizes of a few meters or tens of meters, their packaging assumes a particular importance. For some applications, it is possible to attach the fiber directly to the structure under test, while in other cases it is necessary to pre-package the fibers in order to protect them during installation and while in service. Sensor packaging is particularly important for application to civil structures where the demanding environment prevents fibers from being installed directly on construction materials [24].

Two approaches can be followed to couple the measurement fiber to the structure under test for deformation measurements: distributed (or full-length) and local coupling. In the case of distributed coupling the measurement fiber is attached to the structure along the whole active region. In this case the characteristics of the fiber coatings will have a strong influence on the sensitivity of the sensor. In local coupling sensors the measurement fiber is fixed to the host structure at two points and is free in-between. In order to follow both elongation and shortening, the fiber must be pre-strained. The two attachment points will define the limits of the measurement zone. The quality of the coupling between the fiber and the structure at the two attachment points will play an important role in the response of the sensor to deformations of the structure.

Distributed coupling offers the advantage an apparent simplicity, since the measurement fiber is identical over the whole active region and no special attachment points are needed. Furthermore, the fiber will respond to both elongation and shortening without the need for pre-stressing. The contact between the host material and the fiber over its whole length introduces two potential problems that should not be neglected: transversel strains and microbending. Since the fiber is surrounded by the host material or by the glue used to attach it to the structure, it is possible that it will be subject to transversel strain components perpendicular to the fiber axis. These components will alter the refractive index and the core diameter of the fiber and therefore induce an optical path variation. This variation will be incorrectly interpreted as a displacement in the fiber direction. The transmission of transversel strain from the structure to the fiber core depends strongly on the characteristics of the fiber coating. Soft coatings reduce these effects but also increase the possibility of creep effects in the axial direction. In civil structural monitoring it is often easy to isolate the fiber in a pipe so that no transversel strain will reach the fiber (local coupling). This is not the case in other areas of application, such as composite-material monitoring, where a pipe would alter significantly the structural behavior. In the distributed coupling approach it is possible to reduce the size of the sensing arm to that of a coated fiber. Distributed coupling is important in the case of metallic structures, where the measurement fiber can be simply glued on the surface or for mortars where a nylon-coated fiber can be embedded directly. As mentioned above, excessive microbending of the fiber should also be avoided. Microbending can be the result of the installation of a fiber into a granular material (like concrete or some composite materials). The micro-structure of the host material will induce local bending in the fiber that can lead to increased losses.

An example of distributed coupling is shown in Fig. 3. This sensor was designed for installation on membranes like those used in tensostructures to cover large areas. The measurement fiber is fused inside a plastic ribbon that can be attached to the membrane structure under test. The ribbon also includes a pipe containing the reference fiber.

In local coupling sensors, the measurement fiber is attached to the structure at the extremities of the active region and pre-stressed between these two points. The fiber is contained in a pipe or some other free space over its whole length. In this case the main problem to be solved is the creation of a reliable mechanical coupling between the fiber and the mechanical piece used to anchor it to the structure.

Since all the strains are transmitted from the structure to the fiber at the two anchorage points, the fiber has to be bonded to the anchorage points in a very rigid way. Besides the strains due to the structure deformations, these points have to react to the forces produced by the necessary pre-stressing of the measurement fiber.

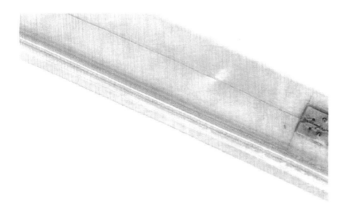

Fig. 3. Distributed coupling sensor for membrane monitoring (courtesy of Smartec SA, Grancia, Switzerland)

It was found that epoxy glues were well suited for obtaining the necessary rigidity and immunity from creep effects. Gluing onto the acrylate dual-layer coating was sufficient for short-term measurements with fiber elongation under 0.5% and for temperatures below 30 °C during the whole experiment. For long-term measurements, high tension or temperatures above 30 °C (such as that experienced during the setting process in concrete) it is necessary to locally remove the acrylate coating in order to glue directly onto the glass to avoid creep problems. The mechanical stripping of the acrylate coating induces, however, a brittleness of the fiber itself and increases the chances of sensor failure after a short period. Two solutions are possible to avoid this problem: accurate chemical removal of the coating or the use of polyimide coating. In the first case the acrylate coating is removed by attacking it with an acid (sulfuric acid) or a solvent (dichloromethane). A fiber stripped chemically maintains almost the same strength as a coated fiber. The use of polyimide-coated fibers makes the stripping process unnecessary, since the strains are easily transmitted through this thin and hard coating. No creeping is expected for this type of coating even at temperatures up to 300 °C. The main drawback of this type of coating is its greater cost.

The SOFO sensor shown in Fig. 4 constitutes an example of a local coupling sensor. This sensor contains both the pre-stressed measurement fiber and the free measurement fiber.

Fig. 4. Local coupling sensor for concrete embedding (courtesy of Smartec SA, Grancia Switerland)

3.5 Application Example: Monitoring the Versoix Bridge

The North and South Versoix bridges are two parallel twin bridges. Each one supports two lanes of the Swiss national highway A9 between Geneva and Lausanne. The bridges are classical bridges consisting of two parallel pre-stressed concrete beams supporting a 30 cm thick concrete deck and two overhangs.

In order to support a third traffic lane and a new emergency lane, the exterior beams were widened and the overhangs extended (see Fig. 5). The construction progressed in two phases: the interior and the exterior overhang extensions. The first one began with the demolition of the existing internal overhang followed by the reconstruction of a larger one. The second phase consisted of demolishing the old external overhang, widening the exterior web and rebuilding a larger overhang supported by metallic beams. Both phases were built in 14 m long stages.

Because of the added weight and pre-stressing, as well as the differential shrinkage between new and old concrete, the bridge bends (both horizontally and vertically) and twists during the construction phases. In order to increase knowledge of the bridge behavior, the engineers choose low-coherence SOFO sensors to measure the displacements of the fresh concrete during the setting phase [25] and to monitor long-term deformations of the bridge. The bridge was equipped with more than a hundred SOFO sensors. The sensors are 6 m long and placed parallel to the bridge length. The first two spans of the bridge were subdivided into 5 and 7 regions (called cells) respectively. In each cell, 8 sensors were installed at different positions in the cross-sections, as shown in Fig. 6.

Fig. 5. General view of the Versoix Bridge during the extension work (courtesy of IMAC-EPFL, Lausanne, Switzerland)

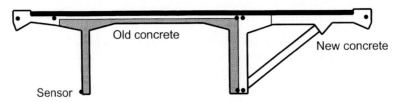

Fig. 6. Sensor positions in a typical cross-section. This installation is repeated 12 times in the first two spans of the bridge

Thanks to this sensor network it is possible to follow both the local and the global displacements of the bridge. The sensors were first used separately to quantify the concrete shrinkage and study the performance of different concrete-mix designs. Once the bridge was completed, the sensors were used in combination to calculate the horizontal and vertical curvature of each of the cells. By double-integration of these curvature measurements it was possible to calculate the horizontal and vertical displacement of the whole bridge [26,27].

Figure 7 shows the concrete deformations measured in one sensor during the first year. All the optical-fiber sensors for the same concrete-pouring stage show the same behavior. In the graph in Fig. 7, four phases are distinguishable: the first is the drying shrinkage (phase ①), followed by a stabilization phase (phase ②), and finally a zone of variation (phase ③) corresponding to the thermal elongation of the bridge. Phase ④ is due to the decrease of the bridge temperature during the month of November 1996. These variations are consistent with a temperature variation of about 10 °C that was actually observed.

During a load test, performed in May 1998 after the construction was terminated, the vertical displacement of the bridge was also monitored. Figure 8 hows the measurement with SOFO sensors (vertical displacement calculated) compared to that with mechanical gages (invar wires under the bridge) during the load test. The load was 6 trucks placed at the center of the second span of the bridge (position 73 in the graph).

The error curve is an estimate of the typical error of algorithm's (incertitude). The solid lines around the vertical calculated displacement correspond to this error estimate. The algorithm retrieves the position of the first pile and accurately reproduces the vertical displacement measured with mechanical gages.

4 Intensity-Based Sensors

An alternative fiber-optic sensor useful for the measurement of length variations is based on the principle of microbending. In this set-up, an optical fiber is twisted with one or more other fibers or with metallic wires [28] along its sensing length. When this fiber-optic twisted-pair is elongated the fibers

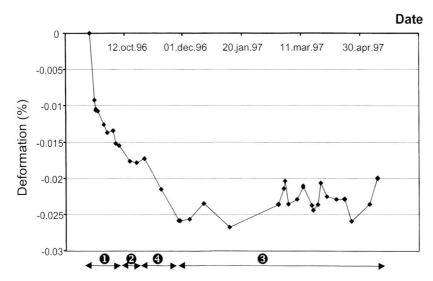

Fig. 7. Versoix Bridge: Concrete deformation over one year (Courtesy of IMAC-EPFL, Lausanne, Switzerland)

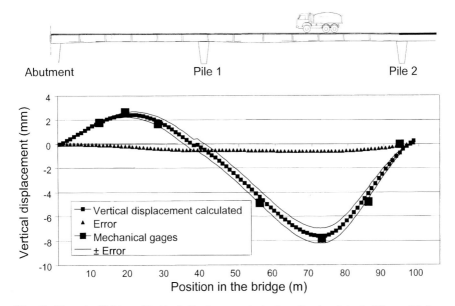

Fig. 8. Versoix Bridge: Vertical displacement during the load test. The solid line indicates the displacement calculated by the double-integration algorithm, while the points correspond to dial gages placed under the bridge (Courtesy of IMAC-EPFL, Lausanne, Switzerland.)

will induce bending in one-another. By measuring the intensity of the transmitted light it is therefore possible to reconstruct the deformation undergone by the structure on which the sensor is mounted.

A system based on this principle has been marketed for some years through Sicom and more recently by Deha-Com in France. This system was one of the earliest commercial applications of fiber-optic sensors for the monitoring of civil structures and was installed in different bridges, tunnels and high-rise structures. Typically obtainable resolutions are 30 µm for short periods (less than one day) and 100 µm for the long term [29]. Systems with measure the reflected light intensity with an optical time-domain reflectometer (OTDR) have also been proposed. These set-ups potentially allow for distributed deformation measurements.

Microbending sensors are conceptually simple. However, temperature, compensation, intensity drifts, system calibration and the inherently non-linear relationship between intensity and elongation still present some challenges. This type of sensor seems appropriate only for very short-term monitoring and for issuing alarms.

5 Brillouin-Scattering Sensors

Brillouin-scattering sensors show an interesting potential for distributed strain and temperature monitoring [30,31,32,33]. Systems able to measure the strain or temperature variations of fibers with lengths up to 50 km with spatial resolution in the meter range are now demonstrating their potential in the first field trials. For temperature measurements, the Brillouin sensor is a strong competitor to systems based on Raman scattering [34], while for strain measurements it has practically no rivals.

5.1 Principles

Brillouin scattering is the result of the interaction between optical and sound waves in optical fibers. Thermally excited acoustic waves (phonons) produce a periodic modulation of the refractive index. Brillouin scattering occurs when light propagating in the fiber is diffracted backwards by this moving grating, giving rise to frequency-shifted Stokes and anti-Stokes components. This process is called spontaneous Brillouin scattering.

Injecting into the fiber two counter-propagating waves with a frequency difference equal to the Brillouin shift can stimulate this process. Through electrostriction, these two waves will give rise to a travelling acoustic wave that reinforces the phonon population. This will in turn increase the scattering cross-section, effectively amplifying the lower frequency lightwave. Energy will therefore be transferred from the upper frequency lightwave (called the pump) to the lower one (called the probe). The amplification of the probe only depends on the pump intensity, which should be as intense as possible.

The probe wave should be much weaker to avoid excessive depletion of the pump photons and of the phonon populations. This process is called stimulated Brillouin amplification.

The Brillouin frequency shift can be seen as a Doppler shift of an optical wave reflected by a moving grating. The Brillouin shift V_B is given by

$$V_B = \frac{2nV_\alpha}{\lambda_0} \tag{5}$$

where V_α is the acoustic velocity, n the refractive index and λ_0 the vacuum wavelength of the incident light.

Typical Brillouin shifts are in the 12–13 GHz range at 1300 nm wavelenght and in the 10–11 GHz range at 1550 nm, mostly depending on the core doping concentration. The gain spectrum presents a Lorenzian profile centered at V_α and a full width at half-maximum in the 35 MHz range at 1300 nm wavelength and 25 MHz at 1550 nm wavelength [32].

If the probe signal consists of a short light pulse and its reflected intensity is plotted against its time of flight, it will be possible to obtain a profile of the Brillouin shift along the fiber length. When the frequency difference between probe and pump is scanned, we obtain the Brillouin gain as a function of the fiber position and of the frequency shift.

The most interesting aspect of Brillouin scattering for sensing applications lies in the temperature and strain dependence of the Brillouin shift. This is the result of the change of the acoustic velocity according to variation in the silica density. The temperature dependence of V_α is 1.36 MHz/°C at 1320 nm wavelength and decreases slightly for high GeO_2 core concentrations. The fiber coating and jacket is also known to have a minor influence on the value. The width of the Brillouin-gain spectrum is also temperature dependent. The strain dependence of V_α is close to 0.06 MHz/$\mu\varepsilon$ at 1320 nm wavelength.

It is interesting to note that the ratio of the strain to the temperature dependencies is 22 $\mu\varepsilon$/°C compared to 8.5 $\mu\varepsilon$/°C obtained for interferometric sensors. This means that Brillouin sensors are comparatively more temperature sensitive that interferometric ones. It also suggests to the possibility of obtaining reliable simultaneous measurement of strain and temperature by combining the two measurement methods.

5.2 Measurement Techniques

The measurement of the Brillouin shift can be approached in different ways. We will discuss here only the two main techniques that allow distributed measurements, i.e. measurements with some spatial resolution.

As discussed in the previous chapter, the Brillouin effect can be used in either the spontaneous or the stimulated mode. The main challenge in using spontaneous Brillouin scattering for sensing applications is in the extremely low level of the detected signal. This requires sophisticated signal processing and relatively long integration times. A commercial system based on spontaneous Brillouin scattering is available from ANDO (Japan).

Systems based the stimulated Brillouin amplification have the advantage of working with a relatively strong signal but face another challenge. To produce a meaningful signal the pump and the probe lightwaves must maintain an extremely stable frequency difference. This usually requires the synchronization of two laser sources that must inject the two signals at the opposite ends of the fiber under test [33]. To avoid the problem of synchronizing two remote lasers, the fiber is often looped-back, so that both lasers can reside at the same end of the fiber line. This approach has the disadvantage of practically halving the measurement range of the system. *Thévenaz* and *Robert* at EPFL proposed a more elegant approach [34]: generating both waves from a single laser source using an integrated-optics Mach-Zehnder modulator. The side-bands of the intensity modulated laser frequency generate the continuous probe signal. The modulation signal is provided by a stabilized microwave generator. This wave reaches the end of the fiber under test and is back-reflected by Fresnel reflection or by a mirror. At the same time, the same laser generates an intense and short light pulse (the pump signal) at its fundamental frequency. The pump signal will cross the probe on its back propagation. The Brillouin amplification will transfer energy from the pump to the probe signal only at the fiber locations where the Brillouin frequency shift corresponds to the frequency of the microwave generator. By scanning the microwave frequency it is therefore possible to obtain a map of the Brillouin frequency along the fiber. This arrangement offers the advantage of eliminating the need for two lasers and intrinsically insures that the frequency difference remains stable independently of the laser drift. A block diagram of this set-up is shown in Fig. 9.

AESA Cortaillod (Switzerland) markets a system based on this set-up (see Fig. 10). It has a measurement range of 10 km with a spatial resolution of 1 m or a range of 30 km with a resolution of 10 m. The strain resolution is 20 $\mu\varepsilon$ and the temperature resolution is 1 °C. The system is portable and can be used for field applications.

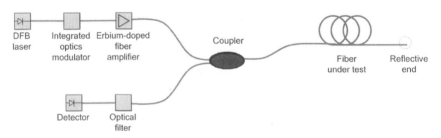

Fig. 9. Experimental set-up using stimulated Brillouin amplification. The pump and probe signals are generated from a single laser using an electro-optic modulator

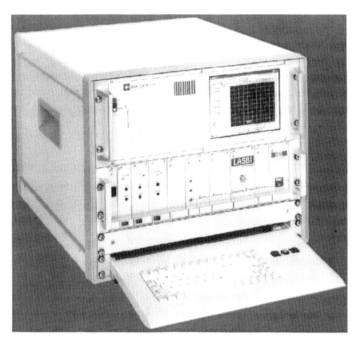

Fig. 10. Prototype system based on the stimulated Brillouin amplification (courtesy of AESA Cortaillod, Switzerland)

These values are close to the theoretical limits for a Brillouin system [35]. Obtaining a better spatial resolution would require shorter pump pulses. This will, however, increase the pulse bandwidth and reduce the interaction cross-section, leading to a decreased strain and temperature resolution.

5.3 Sensor Packaging

Since the Brillouin frequency shift depends on both the local strain and temperature of the fiber, the sensor set-up will determine the actual sensitivity of the system. For measuring temperatures it is sufficient to use a standard telecommunication cable. These cables are designed to shield the optical fibers from elongation of the cable (strain-relieved cables). The fiber will therefore remain in its unstrained state and the frequency shifts can be unambiguously assigned to temperature variations. If the frequency shift of the fiber is known at a reference temperature it will be possible to calculate the absolute temperature at any point along the fiber.

Measuring distributed strains requires a specially designed sensor. A mechanical coupling between the sensor and the host structure along the whole length of the fiber has to be guaranteed. The distributed coupling approach described in Sect. 3.4 seems the most appropriate for this task. To resolve the cross-sensitivity to temperature variations, it is also necessary to install a reference fiber along the strain sensor. As in the temperature measurement case, knowing the frequency shift of the unstrained fiber will allow an absolute strain measurement.

5.4 Application Example: Temperature Monitoring of the Luzzone Dam

Full-scale applications of Brillouin scattering systems to the measurement of distributed strain have yet to be made, the main obstacle being the realization of a suitable sensor packaging. Distributed temperature measurements are, however, highly interesting for structural monitoring of large structures. In the presented application, the system shown in Fig. 10 was used to monitor the temperature development of the concrete used to build a dam. The Luzzone dam, located in the southern Swiss alps, was recently raised by 17 m to increase the capacity of the reservoir. This was achieved by successively concreting 3 m thick blocks. The tests concentrated on the largest block to be poured, the one resting against the rock foundation at one end of the dam. The Brillouin sensor constited of an armored telecom cable installed in a serpentine (see Fig. 11) during concrete pouring.

The temperature measurements were started immediately after concrete pouring and extended over six months. The measurement system proved reliable even in the demanding environment present at the dam (dust, snow, and temperature excursions...). The temperature distributions 15 and 55 days after concrete pouring are shown in Fig. 12.

This example shows how it is possible to obtain a large number of measurement points with relatively simple sensors. The distributed nature of Brillouin sensing makes it particularly suitable for the monitoring of large structures where the use of more conventional sensors would require extensive cabling.

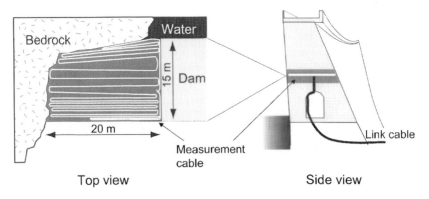

Fig. 11. Position of the temperature-sensing cable in the Luzzone dam (courtesy of L. Thévenaz and Ph. Robert, Metrology Lab. EPFL, Lausanne, Switzerland)

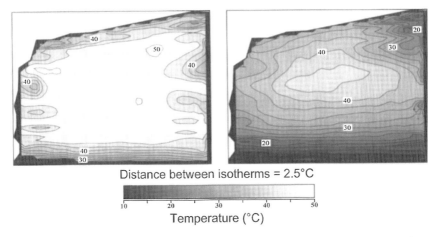

Fig. 12. Temperature distribution in the Luzzone dam after 15 and 55 days (courtesy of L. Thévenaz and Ph. Robert, Metrology Lab. EPFL, Lausanne, Switzerland)

6 Outlook

After an initial euphoric phase when fiber-optic sensors seemed on the verge of invading the whole world of sensing, it has appeared that these are only suitable in cases were the increased performance or functionality compensate for the often higher price tag. In the last few years the commercial applications of this technology have taken the first steps towards mainstream applications. The extensive research efforts in this area and the progress in the technology driven by the telecommunication industry will certainly make an increasing number of fiber-optic sensors available to the engineer.

7 Conclusions

Long-gage and distributed fiber-optic sensors take advantage of the characteristics that have made fibers attractive to the telecommunication industry. The possibility of guiding light signals over large distances and following sinuous paths allows the measurement of deformations and temperatures inside or at the surface of structures. The sensors can be arranged into a network that mimics the nervous system of the human body and monitors the health of the structure. Long-gage and distributed sensors are ideal for the general monitoring of large structures, where they allow a good coverage of the structure with a reduced number of sensors and little *a priori* knowledge of its degradation modes [36]. When more local information is required, the strain sensors described in the previous chapter constitute an ideal complement.

References

1. D. Inaudi: Fiber Optic Smart Sensing, in P. K. Rastogi (Ed.): *Optical Measurement Techniques and Applications*, Artech House, 255, Norwood, MA (1997)
2. R. Maaskant, T. Alavie, R.M. Measures, M. Ohn, S. Karr, D. Glennie, C. Wade, G. Tadros, S. Rizkalla: Fiber Optic Bragg Grating Sensor Network Installed in a Concrete Road Bridge, SPIE Proc. **2191**, 457 (1994)
3. S. T. Vohra, B. Althouse, G. Johnson, S. Vurpillot and D. Inaudi: Quasi-Static Strain Monitoring During the Push Phase of a Box-Girder Bridge Using Fiber Bragg Grating Sensors, in Europ. Workshop on Optical Fibre Sensors, Peebls Hydro, Scotland, July (1998)
4. R. D. Claus (Ed.): SPIE Proc. **1918** (1993)
5. J. S. Sirkis (Ed.): SPIE Proc. **2191** (1994)
6. W. B. Spillman (Ed.): SPIE Proc. **2444** (1995)
7. K. A. Murphy, D. R. Huston (Eds.): SPIE Proc. **2718** (1996)
8. N. Stubbs (Ed.): SPIE Proc. **3043** (1997)
9. V. V. Varadan (Ed.): SPIE Proc. **3323** (1998)
10. E. Udd: *Fiber-Optic Smart Structures*, Wiley, New York (1995)
11. J. Dakin, B. Culshaw: *Optical Fiber Sensors*, Artech House, Norwood, MA (1988)
12. Optical Fiber Sensor Conference Series OFS: 1) London (1983), 2) Stuttgart (1984), 3) San Diego (1985), 4) Tokyo (1986), 5) New Orleans (1988), 6) Paris (1989), 7) Sydney (1990), 8) Monterey (1991), 9) Florence (1993)
13. B. Culshaw, J. D. Jones (Eds.): SPIE Proc. **2360** (1994)
14. Eleventh International Conference on Optical Fiber Sensors OFS 11, Sapporo, Japan, May 21–24 (1996)
15. 12th International Conference on Optical Fiber Sensors OFS 12, Williamsbourgh, USA, October 28–31 1997, OSA Tech. Dig. Ser. **16** (1997)
16. A. J. Bruinsama, B. Culshaw (Eds.): Fiber Optic Sensors: Engineering and applications, SPIE Proc. **1511** (1991)
17. E. Udd: *Fiber Optic Sensors*, Wiley, New York (1991)
18. A. Dandrige et al.: Homodyne Demodulation Scheme for Fiber-Optic Sensors Using Phase Generated Carrier, IEEE, J. Quantum Electron. **18**, 1647 (1982)
19. A. Kersey: Optical Fiber Sensors, in P. K. Rastogi (Ed.): Optical Measurement Techniques and Applications, Artech House **217**, Norwood, MA (1997)
20. D. Inaudi, A. Elamari, L. Pflug, N. Gisin, J. Breguet, S. Vurpillot: Low-coherence Deformation Sensors for the Monitoring of Civil-engineering Structures, Sens. Actuators A **44**, 125 (1994)
21. D. Inaudi: Field Testing and Application of Fiber-Optic Displacement Sensors in Civil Structures, 12th International conference on OFS '97- Optical Fiber Sensors, Williamsbourg, OSA Tech. Dig. Ser. **16**, 596 (1997)
22. G. Kotrotsios, O. Parriaux: White Light Interferometry for Distributed Sensing on Dual Mode Fibers, Proc. Opt. Fiber Soc., Paris (1989) p. 568
23. G. Thursby, D. Walsh, W. C. Michie, B. Culshaw: An In-line Mode Splitter Applied to a Dual Polarimeter in Elliptical Core Fibers, SPIE Proc. **2360**, 339 (1994)
24. D. Inaudi, N. Casanova, P. Kronenberg, S. Marazzi, S. Vurpillot: Embedded and Surface Mounted Fiber-Optic Sensors for Civil Structural Monitoring, SPIE Proc. **3044**, 236 (1997)

25. S. Vurpillot, N. Casanova, D. Inaudi, P. Kronenberg: Bridge Spatial Deformation Monitoring with 100 Fiber-Optic Deformation Sensors, SPIE Proc. **3043**, 51 (1997)
26. D. Inaudi, S. Vurpillot, A. Scano: Mathematical Model for the Determination of the Vertical Displacement from Internal Horizontal Measurements of a Bridge, SPIE Proc. **2719**, 46 (1996)
27. D. Inaudi, S. Vurpillot, N. Casanova, P. Kronenberg: Structural Monitoring by Curvature Analysis Using Interferometric Fiber-Optic Sensors, SPIE Proc. **7**, 199 (1998)
28. L. Falco, O. Parriaux: Structural Metal Coatings for Distributed Fiber Sensors, Opt. Fiber Sens. Conf. Proc., 254 (1992)
29. Dehacom SA, Paris, France: Product Literature
30. T. Karashima, T. Horiguchi, M. Tateda: Distributed Temperature Sensing Using Stimulated Brillouin Scattering in Optical Silica Fibers, Opt. Lett. **15**, 1038 (1990)
31. J. P. Dakin et al.: Distributed Optical Fiber Raman Temperature Sensor Using a Semiconductor Light Source and Detector, Proc. IEE Colloq. Distributed Optical Fiber sensors (1986)
32. M. Niklès, L. Thévenaz, P. Robert: Brillouin Gain Spectrum Characterization in Single-Mode Optical Fibers, J. Lightwave Technol. **15**, 1842 (1997)
33. X. Bao, D. J. Webb, D. A. Jackson: 22-km Distributed Temperature Sensor Using Brillouin Gain in Optical Fiber, Opt. Lett. **18**, 552 (1993)
34. M. Nikles et al.: Simple Distributed Temperature Sensor Based on Brillouin Gain Spectrum Analysis, SPIE Proc. **2360**, 138 (1994)
35. A. Falley, L. Thévenaz, M. Facchini, M. Niklès, P. Robert: Distributed Sensing Using Stimulated Brillouin Scattering: Towards the Ultimate Resolution, OSA Tech. Dig. Ser. **16**, 324 (1997)
36. B. Culshaw: Structural Health Monitoring of Civil Engineering Structures, Prog. Struct. Engin. Mater. **1**, 308 (1998)

Techniques for Non-Birefringent Objects: Coherent Shearing Interferometry and Caustics

Sridhar Krishnaswamy

Department of Mechanical Engineering, Northwestern University
Evanston, IL 60208-3020, USA
s-krishnaswamy@nwu.edu

Abstract. Shearing interferometric techniques that can provide information about the stress or deformation states of non-birefringent transparent materials and opaque specularly reflective materials are described. Two types of coherent shearing interferometers that provide full-field information about the gradients of the stress state of an object are discussed. These techniques have recently found use in stress analysis, especially in experimental fracture mechanics studies. In addition, a simpler technique called the method of caustics which is also applicable to non-birefringent materials is described in brief. This technique does not provide full-field information but has been of historical interest in experimental fracture mechanics. Examples of application of all three techniques are given.

1 Introduction

The stress states of transparent objects that are optically birefringent can be studied using the classical techniques of photoelasticity [1]. Photoelasticity, however, cannot be used to monitor the stress state of objects that are essentially optically isotropic (i.e., materials which do not exhibit strong birefringence), and can only be used indirectly with birefringent coatings to monitor the deformation of opaque objects [1]. In this chapter, we shall look at some optical techniques that are available to the photomechanical technologist for use on non-birefringent materials. In this category, we include opaque specularly reflective specimens, such as those made of polished metals, and transparent materials which are essentially optically isotropic, such as glass and polymethylmethacrylate (PMMA). We exclude opaque diffusively reflective objects whose deformations can be studied using speckle interferometry [2] or geometric moiré techniques [3].

The extensive sets of tools that have been developed for optical-shop testing [4] to monitor aberrations in optical components such as transmissive phase objects (lenses), and specularly reflective objects (mirrors), can be readily adapted for stress/deformation analysis of non-birefringent materials. Some of the common interferometric methods that have been used in photomechanics are: Michelson, Mach-Zehnder, Twyman-Green, and holographic interferometry [4,5,6]. These methods are explained in the references cited as well as elsewhere in this book. In addition, moiré interferometry, which requires the deposition of a reflective grating on the test object and which has been widely used in experimental stress analysis, is also dealt with elsewhere [7].

In the Michelson, Mach-Zehnder and Twyman-Green interferometric methods listed above, an object wavefront containing information about the stress/deformation state of the object is made to interfere with a separate, typically planar, reference wavefront. As such, these techniques suffer from excessive sensitivity to rigid body motion, to air turbulence, and to any initial imperfections (non-planarity) of the test object. These factors, in conjunction with the fact that these interferometers are possibly *too* sensitive for many photomechanics applications, lead to the presence of an excessive number of fringes, most of which may be unrelated to the stress/deformation state of interest. This makes data interpretation extremely cumbersome. In contrast, interferometers that are self-referential, i.e. ones which derive the reference wavefront from the object wavefront itself, are actually ideal systems for stress analysis, especially for regions of stress concentration. Holographic and shearing interferometers fall into this category. Holographic interferometry typically requires the comparison of two different states of the test object, and is dealt with elsewhere. In this chapter, we shall consider coherent shearing interferometers, which are self-referential in that they interfere a wavefront with a *modified* (translated, rotated, or magnified) copy of the same wavefront [4]. (The coherent shearing interferometers discussed here should not be confused with *speckle correlation* shearing interferometry [2], also known as 'shearography' [8], which are applicable to diffusively scattering objects.)

Coherent shearing interferometers are sensitive to the *gradients* of the phase of the object wavefront (relatable to the object stress/deformation state). These interferometers are therefore insensitive to rigid-body motions and are also not very sensitive to initial imperfections of the test object. Some implementations of shearing interferometers are also common-path interferometers, providing enhanced protection against ambient vibration and air turbulence. These interferometers can have adjustable sensitivity, which enables user-selection of the sensitivity that is appropriate to a given situation. Coherent shearing interferometric methods have been used extensively in optical metrology for the measurement of aberrations in optical components such as lenses and mirrors [4,9]. Variations of the method have also been used in fluid mechanics applications [5], but it is only within the last decade or so that their potential has been exploited in experimental stress analysis of solids [10].

The bulk of this chapter is devoted to a description of two implementations of shearing interferometers for experimental stress analysis. This is followed by a brief description of a simple non-interferometric optical technique called the method of caustics, which like the shearing interferometers is also sensitive only to the gradients of the phase of the object wavefront, but which unfortunately does not provide full-field information about the object deformation. The method of caustics, which has been used extensively in the past especially in experimental fracture mechanics studies on non-birefringent materials [10], has since been supplanted by full-field shearing interferometers. However, an understanding of the method of caustics is useful in following the experimental fracture mechanics literature. First, prior to delving into the details of the optical set-ups, we shall consider how the object deformation/stress states affect an initially planar optical wavefront that interacts with a non-birefringent object.

2 Elasto-Optic Relations

When an optical beam transmits through a stressed transparent body or is reflected by a deformed reflective body, it experiences optical path length changes that are caused by the stressing/deformation of the body. These optical phase shifts can be related to the stress state in the body, and can be monitored interferometrically. In this section, we will obtain the optical phase shifts due to transmission through an optically isotropic thin solid plate, as well as due to reflection from an initially-flat highly-reflective thin plate. For simplicity, we shall assume that the plates are made of linear elastic materials that are loaded in a state of plane stress [11]. For transmission through thick plates, a plane-strain treatment may be appropriate [11]. For materials that do not respond linear elastically, or for transmission through fluids, it is possible to use the optical techniques described later in this chapter, but in these cases the stress-optic relations developed in this section need to be modified appropriately.

2.1 Optical Phase Shift due to Transmission through an Optically Isotropic Linear Elastic Medium

Consider a thin, transparent, optically isotropic, linear elastic solid plate of initially parallel sides of nominal thickness h and whose nominal refractive index is n_0. If the loading is in the plane $(x_1 - x_2)$ of a sufficiently thin plate, a plane state of stress will prevail where the stress variations through the thickness (x_3-direction) of the plate can be neglected [11]. Consider a parallel beam of monochromatic light (plane wave) that transmits through the plate along the x_3-direction. Within the paraxial approximation, the thin plate which acts as a phase object induces an optical path length change in the transmitted beam given by [12]

$$\delta S(x_1, x_2) = (n_0 - 1)\Delta h + h \Delta n , \tag{1}$$

where the first term is due to stress-induced dimensional changes (h is the nominal thickness, and Δh is the deformation-induced change in thickness) and the second term is due to stress-induced changes in the refractive index

(n_0 is the nominal refractive index and Δn is its deviation due to the stress-optic effect). The Maxwell-Neumann elasto-optic relations for an optically isotropic material under a state of plane stress relates the refractive index change to the stress-state through [12]

$$\Delta n = c_\sigma (\sigma_{11} + \sigma_{22}) , \tag{2}$$

where c_σ is a stress-optic constant for an optically isotropic material, and $\sigma_{11} + \sigma_{22}$ is the trace of the plane-stress stress tensor (the sum of the in-plane normal stresses) at the point. Furthermore, from elasticity theory the stress-induced thickness change under plane stress is given by [11]

$$\Delta h = -\frac{\nu h}{E}(\sigma_{11} + \sigma_{22}) , \tag{3}$$

where the Poisson's ratio ν and Young's modulus E are material elastic properties. Thus we find that the change in optical path length upon transmission through the specimen is

$$\delta S(x_1, x_2) = ch(\sigma_{11} + \sigma_{22}) , \tag{4}$$

where

$$c = c_\sigma - \frac{\nu}{E}(n_0 - 1) , \tag{5}$$

is a material-dependent stress-optic constant, which is tabulated in [12] for various materials and is typically of the order of $-1 \times 10^{-10} \, \text{m}^2/\text{N}$.

2.2 Optical Phase-Shift due to Reflection from a Polished Surface

Next, consider a parallel beam of monochromatic light (plane wave) that propagates in the x_3-direction, and which reflects off an optically reflective plate that is initially-flat in the (x_1, x_2) plane. If the object suffers a deformation, the reflector deviates from its initial planar shape due to the out-of-plane displacement $u_3(x_1, x_2)$ of the body. The normally incident planar beam will distort due to reflection at the surface of the object, and the emerging rays will no longer be parallel. For simplicity, we will restrict further attention to the paraxial approximation where the slope of the reflector (gradient of u_3) is assumed to be small so that the light rays are not too severely deflected from the horizontal x_3-direction. In this case, the change in optical path length is given by

$$\delta S(x_1, x_2) = 2u_3(x_1, x_2) = -\frac{\nu h}{E}(\sigma_{11} + \sigma_{22}) , \tag{6}$$

where the last relation follows from the plane-stress condition [11], assuming that the object undergoes only linear elastic deformation. Note also that any rigid-body motion can cause a spatially linearly varying path length change, but these are not included in the above as they are essentially unimportant as far as our discussion of shearing interferometers and caustics is concerned.

The deformation- or stress-induced change in optical path length for both the transmission and reflection cases, under linear elastic plane-stress conditions, can therefore be written as

$$\delta S(x_1, x_2) = \tilde{c} h \mathrm{Tr}(\tilde{\sigma}) \,, \tag{7}$$

where $\tilde{c} = c$ in transmission, and $\tilde{c} = -\nu/E$ in reflection; and $\mathrm{Tr}(\tilde{\sigma})$ is the trace of the plane-stress stress tensor. The optical phase shift incurred by the beams is, of course, related to the path length change through

$$\Phi_{\mathrm{sp}}(x_1, x_2) = \frac{2\pi}{\lambda} \delta S(x_1, x_2) \,, \tag{8}$$

where λ is the wavelength of the light used. Thus, by monitoring the phase change incurred in an initially planar optical wavefront, information about the stress/deformation state of an optically isotropic transmissive body or a specularly reflective body can be obtained.

3 Shearing Interferometry

The basic idea behind *lateral* shearing interferometry is to interfere an optical wavefront with a laterally-shifted (translated) copy of itself [13]. The result is an interference pattern that provides fringes related to the *gradients* (or more precisely, *spatial difference*) of the optical phase of the wavefront, which in photomechanics corresponds to the deformation or stress state of the object as we saw in the previous section. In experimental stress analysis, these techniques are most useful for the study of stress concentration such as near holes and cracks in a body.

Several methods of lateral shearing interferometry have been proposed utilizing different devices to create the shearing effect [4]. These include utilizing glass plates [13], ordinary prisms [14], birefringent prisms [15], and variations of Michelson and Mach-Zehnder set-ups [4]. The shearing interferometers that have found the greatest applications by far are those utilizing diffraction gratings to provide lateral shearing. In this category fall the single-grating shearing interferometers of *Ronchi* [16], the double-frequency grating interferometer of *Wyant* [17], and the dual-grating interferometers of *Patorski* [18], *Hariharan* et al. [19], and *Tippur* et al. [20].

While most of the schemes mentioned above have been used predominantly in optical-component testing, it is possible to configure many of them for use in photomechanics. In this section, we first describe the set-up of *Tippur* et al. [20], called the Coherent Gradient Sensor (CGS), since this has been used quite extensively over the last decade in the study of fracture phenomena. This is then followed by a description of a different implementation of a lateral shearing interferometer that utilizes a calcite beam-displacement crystal [21]. This scheme, which is a combination Polariscope / Shearing Interferometer (PSI) in a single device, can be used to test both birefringent and non-birefringent test objects. The descriptions that follow of the CGS set-up, which uses a grating-based shearing device, and the PSI set-up, which uses a polarization-based shearing device, are typical of other similar optical arrangements that can be configured for photomechanics applications.

3.1 A Dual-Grating Shearing Interferometer – The Coherent Gradient Sensor (CGS)

CGS, which is a lateral shearing interferometer with on-line spatial filtering, can be used in transmission and reflection modes, both statically and dynamically [20]. In transmission mode, the CGS fringe patterns can be related to the planar stress gradients in the specimen. For specularly reflective specimens, the fringe patterns can be related to the gradients of the specimen surface displacement.

The experimental set-up used for CGS in transmission mode is shown in Fig. 1. A collimated laser beam is incident on an optically isotropic transparent specimen of interest. Before transmission through the specimen, the beam is a planar wavefront; after transmission the wavefront is no longer planar but is perturbed from its planarity due to the stress-optic effect arising from any mechanical deformation of the specimen. This perturbed wavefront is then diffracted upon transmission through a pair of high-density sinusoidal diffraction gratings G_1 and G_2. All of the diffracted wavefronts arising from the gratings are then collected by a filtering lens L_1. At the back focal-plane of the filtering lens a two-dimensional aperture is placed such that only two of these diffracted wavefronts are selected for subsequent imaging through the imaging lens L_2 that is used to put the specimen in focus at the image plane.

Figure 2 shows the modification of the above set-up for measuring surface slopes of specularly reflective opaque specimens when CGS is used in reflection mode. In this case, the specularly reflecting object surface is illuminated

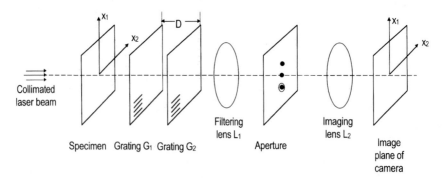

Fig. 1. CGS set-up: transmission mode

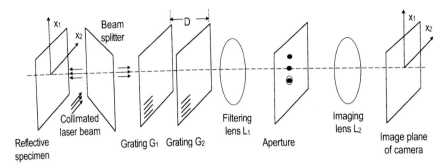

Fig. 2. CGS set-up: reflection mode

normally by a collimated beam of laser light using a beam splitter. The reflected beam then gets processed, as in the transmission mode, through an optical arrangement that is identical to the one shown in Fig. 2 from this point onwards.

Figure 3 explains the working principle of the CGS method. Let the gratings G_1 and G_2 have their rulings parallel to, say, the x_2-axis. A plane wave transmitted through or reflected from an undeformed specimen and propagating along the optical axis is diffracted into three plane wavefronts E_0, E_1 and E_{-1} by the first grating G_1. (If a non-sinusoidal grating is used, higher diffraction orders will be present, but the substance of the analysis below is still valid). The angle between the propagation directions of E_0 and $E_{\pm 1}$ is given by the diffraction condition $\theta = \sin^{-1}(\lambda/p)$, where λ is the optical wavelength and p is the pitch of the diffraction gratings. Upon incidence on the second grating G_2, these three wavefronts are further diffracted to provide a total of nine wavefronts: $E_{(0,0)}, E_{(0,1)}, E_{(1,-1)}, E_{(1,0)}, E_{(1,1)}$ etc. These wavefronts which propagate in different directions, are then brought to focus at spatially separated (in the x_1-direction) diffraction spots on the back

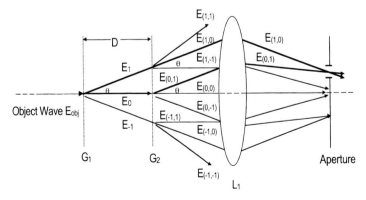

Fig. 3. CGS working principle

focal-plane of the filtering lens. The spacing between these diffraction spots is directly proportional to $\sin\theta$ or inversely proportional to the grating pitch.

Now, consider a plane wave normally incident on a *loaded* specimen. The resulting transmitted or reflected wavefront will no longer be planar but will be distorted either due to changes of refractive index of the material (in transmission) and/or due to non-uniform surface deformations (in reflection). The object wavefront that is incident on G_1 now carries information regarding the specimen deformation. Assuming that the deformation-induced optical phase change is not too severe, the deformed wavefront consists of light rays traveling with perturbations to their initial direction parallel to the optical axis. Within this paraxial approximation, which is valid for most practical situations, we can now express the object wavefront as

$$E_{\text{obj}} = E_0 e^{i\Phi_{\text{sp}}(x_1, x_2)}, \tag{9}$$

where E_0 is the amplitude and the phase $\Phi_{\text{sp}}(x_1, x_2)$ includes the stress- or deformation-induced phase shift in the initially planar optical wavefront. Since the object wavefront is no longer planar, each of the diffraction spots on the focal plane of L_1 will now be locally surrounded by a halo of dispersed light field due to the deflected rays containing the information of interest. The extent of this halo depends on the severity of the object deformation. By using a two-dimensional aperture at the filtering plane, information located around one of the spots can be further imaged. For the two-dimensional aperture located as shown in Fig. 3, the wavefronts that are selected for imaging are $E_{(1,0)}$ and $E_{(0,1)}$. These wavefronts are identical copies of the object wavefront except that one is laterally translated ("sheared") with respect to the other along the x_1-direction by an amount (see Fig. 3)

$$\Delta x_1 = D\theta = D\lambda/p, \tag{10}$$

where D is the spacing between the gratings, and the angle θ is assumed to be small.

The selected wavefronts that interfere in the camera can therefore be expressed as

$$E_{(0,1)} = E_{01} e^{i\Phi_{\text{sp}}(x_1, x_2)}, \tag{11}$$

$$E_{(1,0)} = E_{10} e^{i\Phi_{\text{sp}}(x_1 + \Delta x_1, x_2)}. \tag{12}$$

The interference fringe pattern is thus given by

$$\begin{aligned} I &= (E_{(0,1)} + E_{(1,0)})(E_{(0,1)} + E_{(1,0)})^*, \\ &= E_{(01)}^2 + E_{(10)}^2 + 2 E_{(01)} E_{(10)} \cos\left[\Phi_{\text{sp}}(x_1 + \Delta x_1, x_2) - \Phi_{\text{sp}}(x_1, x_2)\right], \\ &= E_{(01)}^2 + E_{(10)}^2 \\ &\quad + 2 E_{(01)} E_{(10)} \cos\left(\Delta x_1 \frac{\Phi_{\text{sp}}(x_1 + \Delta x_1, x_2) - \Phi_{\text{sp}}(x_1, x_2)}{\Delta x_1}\right), \\ &\approx E_{(01)}^2 + E_{(10)}^2 + 2 E_{(01)} E_{(10)} \cos\left(\frac{D\lambda}{p} \frac{\partial \Phi_{\text{sp}}}{\partial x_1}\right), \end{aligned} \tag{13}$$

where the last equality arises if the phase difference is treated as a derivative of the phase, provided the shearing amount is sufficiently small. The CGS bright fringes are therefore spatial locations where

$$\frac{D\lambda}{p}\frac{\partial \Phi_{\text{sp}}}{\partial x_1} = 2m\pi \qquad m = 0, \pm 1, \pm 2, \ldots , \tag{14}$$

where m is the fringe order number. CGS fringes are therefore related to the gradients of the deformation- or stress-induced optical phase shift, which can in turn be related more directly to the object stress or deformation states using (7) and (8). CGS bright fringes are thus given by

$$\tilde{c}h\frac{\partial}{\partial x_1}(\sigma_{11} + \sigma_{22}) = m\xi \qquad m = 0, \pm 1, \pm 2, \ldots , \tag{15}$$

where $\xi = p/D$ is a user-selectable shearing-parameter that governs the sensitivity of the CGS interferometer. Figure 4 shows a typical CGS transmission-mode fringe pattern for the case of a compressive line load on the edge of a thin plate.

Some salient points about CGS are:

(1) The shearing direction and the amount of shearing can be adjusted by appropriate orientation and spacing of the diffraction gratings. Thus the sensitivity and the direction of stress- or deformation-gradients may be adjusted to suit the needs of the experiment.

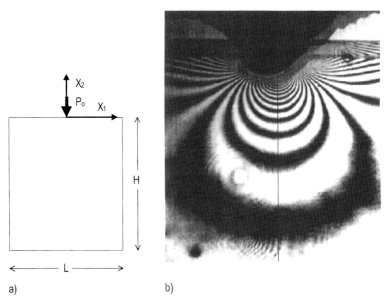

Fig. 4. Compressive line load at the edge of a plate: (**a**) schematic (**b**) shearing fringes for shearing along the direction of the applied force

(2) CGS fringes have a contrast that is close to unity. This is because in most situations, the amplitudes of the two interfering beams are essentially the same, both having been derived from the same beam and having been once diffracted and once transmitted through identical gratings; i.e. $E_{10} \sim E_{(01)}$ in (13).

(3) The above simplified analysis of the CGS technique captures the basic principles of operation of the method. Its validity is of course limited to the paraxial, small shear cases. If the shear Δx_1 is not small, the approximation in (13) of the phase difference by a derivative no longer holds, but the technique can still be used with fringe interpretation following along the lines of *Bruck* and *Rosakis* et al. [22]. More in-depth analyses of the CGS technique and its applications can be found in *Tippur* et al. [20,23].

3.2 A Polarization-Based Shearing Interferometer: Compact Polariscope / Shearing Interferometer (PSI)

We now look at another class of lateral shearing interferometers that use birefringent beam-displacement crystals [15,21] to provide copies of an incident object wavefront. Specifically, we describe the set-up of *Lee* and *Krishnaswamy* [21], which is a single optical device that can operate either as a shearing interferometer (for non-birefringent materials) or as a polariscope (for birefringent materials). Furthermore, unlike CGS and other grating-based shearing interferometers where more than two-thirds of the incident object light intensity may be wasted in the diffraction orders not used, PSI loses only about half the incident light intensity, thereby making it more attractive for dynamic applications in conjunction with high-speed photography.

Figure 5 shows the optical component layout for the PSI device in transmission mode. An unexpanded laser beam with vertical polarization passes through a first quarter-wave plate prior to being expanded by a beam expander and spatial filter. The expanding beam is collimated by a 200 mm f/4 lens to provide an $H = 50$ mm diameter planar beam which illuminates the transparent test object. The output beam is then reduced in size to a diameter d by a beam-reducer system consisting of a 300 mm f/4.5 lens in combination with a 70–210 mm zoom-lens in a reverse telescopic combination which provides a reduction ratio $R = d/H$ in the range 0.25-0.75 by varying the zoom-lens focal length. A second quarter-wave plate and a half-wave plate are placed between the two lenses. The beam then traverses through a custom-made calcite beam-displacement prism (40 mm diameter, 8 mm thick) which provides two parallel but laterally-displaced orthogonally-polarized copies of the input beam. The output beams pass through an analyzer prior to being imaged by a 50 mm lens (with additional spacers to obtain the desired magnification) onto a camera. The camera lens is adjusted such that the specimen plane is brought into focus at the camera image plane.

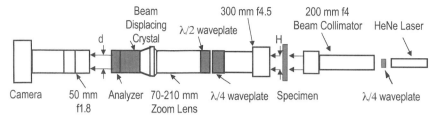

Fig. 5. Optical layout of the PSI device

The waveplates, the polarizers and the beam-displacing prism are all mounted such that they can be rotated about the optical axis of the PSI. The direction of shear for the shearing-mode of PSI can be controlled by rotating the calcite beam-displacement crystal. The amount of shear on the image plane can be controlled by adjusting the zoom lens of the beam-reducer system. The single beam of diameter d that is incident on the crystal produces two output copies that are displaced by a fixed amount Δ_c determined only by the crystal properties which are not adjustable. This, however, represents a *variable* shear amount of $\Delta = \Delta_c H/d = \Delta_c/R$ on the specimen object plane, where R is adjustable by changing the magnification of the system via the beam-reducer arrangement.

The operation of this device – as is the case for any other optical device involving polarization-dependent components – is most generally described by use of the Jones matrix formalism [24]. Using this, one can easily obtain the phase retardations induced along different polarizations of an input optical beam and the consequent alteration of the electric-field vector due to an optic component. The output beam from an optic component is given by its Jones matrix operating on the input electric-field vector. Choosing the $(\omega t - 2\pi x_3/\lambda)$ form for the phase of the incoming planar light beam propagating along the x_3-direction with frequency ω and wavelength λ, it can be shown that the Jones matrix for a wave-plate whose fast axis makes an angle α to the x_1-direction and which introduces a relative phase retardation of amount Φ_Δ between its fast and slow axes is given by [24]

$$\boldsymbol{J}_{\text{wave}}\{\Phi_\Delta, \alpha\} = \begin{bmatrix} \cos^2\alpha + e^{-i\Phi_\Delta}\sin^2\alpha & \frac{1}{2}\sin 2\alpha(1 - e^{-i\Phi_\Delta}) \\ \frac{1}{2}\sin 2\alpha(1 - e^{-i\Phi_\Delta}) & \sin^2\alpha + e^{-i\Phi_\Delta}\cos^2\alpha \end{bmatrix}. \quad (16)$$

With appropriate parameters, the above Jones matrix can be used for the half- and quarter-wave plates in the set-up, and also for birefringent test specimens. The Jones matrix for a linear polarizer oriented at an angle α to the x_1-direction is similarly [24]

$$\boldsymbol{J}_{\text{linear}}\{\alpha\} = \begin{bmatrix} \cos^2\alpha & \frac{1}{2}\sin 2\alpha \\ \frac{1}{2}\sin 2\alpha & \sin^2\alpha \end{bmatrix}. \quad (17)$$

The appropriate Jones matrix for thin plates of optically isotropic materials can be expressed as [24]

$$\boldsymbol{J}_{\text{iso}}\{\Phi_{\text{sp}}\} = e^{-i\Phi_{\text{sp}}} \begin{bmatrix} 1 & 0 \\ 0 & 1 \end{bmatrix}, \tag{18}$$

where Φ_{sp} is now the *total* phase retardation at any point due to transmission through the specimen and it is related to the principal stresses at that point through (7 and 8).

The PSI device can be used in a variety of different configurations. For optically isotropic materials the shearing-mode of the PSI device should be used, one possible arrangement of which is shown in Fig. 6. In this case, the electric-field vector just before the beam-displacement prism is given by [21]

$$\begin{aligned}\boldsymbol{E}_{\text{wol}-} &= \boldsymbol{J}_{\text{wave}}\{\pi, \pi/8\}\boldsymbol{J}_{\text{wave}}\{\pi/2, 3\pi/4\} \\ &\quad \boldsymbol{J}_{\text{iso}}\{\Phi_{\text{sp}}\}\boldsymbol{J}_{\text{wave}}\{\pi/2, \pi/4\} \begin{bmatrix} 0 \\ E_0 e^{i\omega t} \end{bmatrix},\end{aligned} \tag{19}$$

where E_0 is the amplitude and ω the frequency of the input optical beam. The above simplifies to an electric field with linear polarization at 45° to the horizontal. Upon transmission through the beam-displacement prism, the vertical component of the above gets spatially translated in the x_1-direction by an amount Δx_1, and the output electric vector field is therefore given by:

$$\boldsymbol{E}_{\text{wol}+} = \frac{i}{\sqrt{2}} E_0 e^{i\omega t} \begin{bmatrix} e^{-i\Phi_{\text{sp}}(x_1, x_2)} \\ e^{-i\Phi_{\text{sp}}(x_1+\Delta x_1, x_2)} \end{bmatrix}. \tag{20}$$

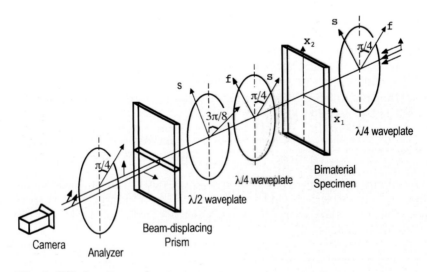

Fig. 6. PSI shearing-mode set-up

The output electric vector field after passage through the analyzer oriented at 45° to the horizontal is therefore

$$\boldsymbol{E}_{\text{out}} = \boldsymbol{J}_{\text{linear}}\{\pi/4\}\boldsymbol{E}_{\text{wol}+}. \tag{21}$$

The resulting intensity imaged in the camera then becomes

$$I = E_0^2 \cos^2\left\{\frac{1}{2}[\Phi_{\text{sp}}(x_1 + \Delta x_1, x_2) - \Phi_{\text{sp}}(x_1, x_2)]\right\},$$
$$\approx E_0^2 \cos^2\left(\frac{\Delta x_1}{2}\frac{\partial \Phi_{\text{sp}}}{\partial x_1}\right), \tag{22}$$

where the derivative is an approximation to the difference if the shearing distance Δx_1 is suitably small. The shearing-mode fringe relation yielding bright fringes is therefore again given by (15), but with the shearing-parameter for PSI being $\xi = 2\lambda/\Delta x_1$. A typical sequence of fringe patterns using the PSI device in its shearing mode is shown in Fig. 7, for the case of a quasi-statically propagating subinterfacial crack in a three-point bend PMMA/Al bimaterial fracture specimen. The PSI device can easily be operated in reflection mode in order to measure gradients of the surface displacement of a specimen. This is done by appropriate reconfiguration of the optical set-up similar to the front-end of the CGS reflection-mode set-up.

For the sake of completeness, we note that the PSI device can also be used as a conventional polariscope, the details of which may be found in *Lee* and *Krishnaswamy* [21]. This requires rotating some of the optical components about the optic axis. It may be worth noting that in its shearing mode of operation described above, the wave-plates in the PSI set-up are essentially functionally ineffective, leaving only the beam-displacement crystal to provide the sheared images. Conversely, in its polariscope mode of operation, the beam-displacement crystal is made to be functionally ineffective, and the wave-plates are the only functional components [21]. If a test specimen is made of only optically isotropic materials, the PSI device can be used as a shearing interferometer similar to CGS. If the specimen has only optically birefringent materials, the PSI can be used as a conventional circular polariscope in either dark or light field arrangement [21]. The real advantage of the PSI device is in applications where the test specimen is a composite made of two different materials, one of which exhibits optical birefringence (say, material 1) and the other is optically isotropic (material 2). The PSI device can then be used in the photoelastic mode to obtain isochromatic fringes from material 1 of the specimen, and in this mode no information will be obtained from material 2 (since photoelasticity requires birefringence). The shearing mode of PSI can be used to obtain shearing fringes from material 2. A typical composite fringe pattern from the vicinity of a crack on the interface of a PMMA/epoxy three-point bend fracture specimen is shown in Fig. 8. The PMMA side fringes are shearing fringes, whereas the epoxy side fringes are photoelastic isochromatics.

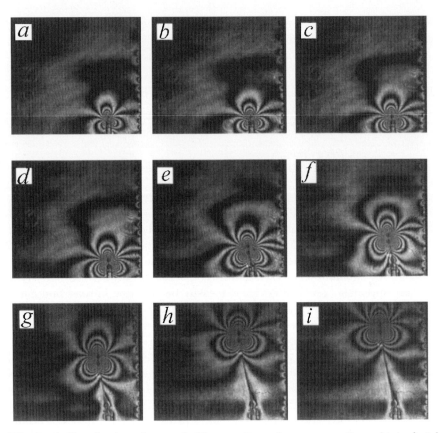

Fig. 7a–i. Shearing interferometric fringe patterns for a propagating subinterfacial crack in a PMMA/Al plate. The shearing direction is vertical

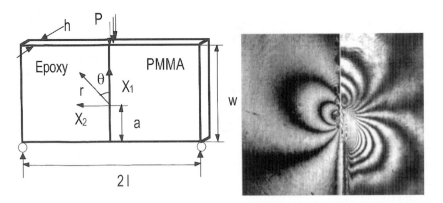

Fig. 8. Interfacial fracture in bimaterial composite specimens: (**a**) schematic (**b**) composite fringe pattern

4 Optical Caustics

The last optical technique that we will describe here is the method of caustics. While this technique is very simple to configure and is therefore readily accessible to virtually any laboratory, it does suffer one major drawback in that it is not a full-field technique like the shearing interferometers that are supplanting the method of caustics in photomechanics. Historically, however, this technique has played a major role in experimental stress analysis, especially of fracture phenomena in non-birefringent materials, and a basic familiarity with the technique is helpful in understanding some of the literature.

Optical caustics are formed due to the reflections and/or refractions of light that lead to focusing of a light beam. Some familiar examples of optical caustics include the bright lines of light formed on a swimming-pool floor due to the rippling of the water surface, rainbows, and the bright streaks of light seen from the scattering of a street light by water droplets on a window pane [25]. The earliest studies on optical caustics were reported in the early to mid-1800s [26,27]. The idea of using optical caustics as a metrological tool is more recent. In experimental solid mechanics, *Manogg* [28] was the first to report the use of optical caustics to study dynamic crack propagation in a transparent medium. Caustics in both reflection and transmission have since been extensively used by *Theocaris* [29], *Kalthoff* [30], and *Rosakis* et al. [31] among others to study a wide range of problems in elastic and elasto-plastic, static and dynamic fracture mechanics. *Theocaris* [32] has used optical caustics to study torsion and bending problems in mechanics. *Rossmanith* [33] has developed a technique based on optical caustics to investigate Rayleigh wave propagation in solids. In experimental fluid mechanics, the related technique of optical shadowgraphs has found application, particularly in the early part of this century [34]. Optical caustics have since then also been used to study heat transfer [35], and to visualize flows in liquids [36] and in liquid crystals [37].

The phenomenum of optical caustics can be analyzed using geometric optics to provide the shape of the envelope of the rays that form a bright caustic [30]. For photomechanical applications, this geometrical optics approach is usually sufficient when the primary interest is to relate a characteristic dimension of the caustic to the stress/deformation quantity that is to be measured. Even though caustics can be obtained using white light [38], it is usual in photomechanics to use planar monochromatic light, and we shall restrict attention to this case.

Consider a monochromatic plane wave propagating in the x_3-direction (Fig. 9). Upon transmission through a thin phase object (in the $x_1 - x_2$ "specimen" plane), the plane wavefront distorts due to refraction inside the phase object, and the emerging rays are no longer parallel. Under certain conditions shadow regions are formed when light that transmits through the phase object is deflected away from such regions, and bright caustic surfaces are formed by the focusing of several of the deflected rays (Fig. 9). As before,

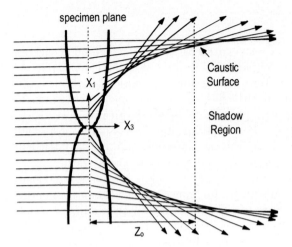

Fig. 9. Caustics set-up in transmission mode

if the phase object is thin enough, the optical phase delay and resulting beam deflection incurred in the transmission of the plane beam can be thought of as being lumped at the specimen mid-plane. (For a thick phase object, it will be necessary to trace the rays through the phase-object up to the point where it emerges from the object). The distorted wavefront immediately after transmission through the specimen plane can then be expressed as

$$S(x_1, x_2, x_3) = x_3 + \delta S(x_1, x_2) , \qquad (23)$$

where $\delta S(x_1, x_2)$ is the spatially non-uniform change in optical path length given in (7). The rays emerging from the specimen plane will propagate normally to the wavefront $S(x_1, x_2, x_3)$. As shown in Fig. 10, the rays emerging from position $\boldsymbol{x} = (x_1, x_2)$ at the specimen plane will propagate along a straight line [in the direction of the gradient of the wavefront surface $S(x_1, x_2, x_3)$] to get mapped to position $\boldsymbol{X} = (X_1, X_2)$ in a "reference" plane located a distance z_0 from the specimen plane. The reference plane can be either real or virtual. Traditionally, we choose: $z_0 < 0$ if the reference plane is between the observation position and the specimen, and $z_0 > 0$ if the specimen is between the reference plane and the observation position. If the rays are not too severely deflected from their original propagation direction along the x_3-direction, it is easy to see that the mapping from the specimen to the reference plane is given by [10]

$$X_i = x_i - z_0 \frac{\partial}{\partial x_i} [\delta S(x_1, x_2)], \qquad i = 1, 2. \qquad (24)$$

Several rays from the specimen plane will focus onto the same points (forming a caustic) at the reference plane if the mapping in (24) is not one-to-one. This occurs when the Jacobian of the mapping in (24) vanishes within

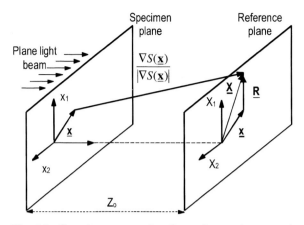

Fig. 10. Caustic-ray mapping from the specimen to the reference plane

the region of interest. That is, a caustic will be formed if

$$J = \frac{\partial X_1}{\partial x_1}\frac{\partial X_2}{\partial x_2} - \frac{\partial X_1}{\partial x_2}\frac{\partial X_2}{\partial x_1} = 0 \ . \tag{25}$$

The above defines an "initial curve" on the specimen plane. Rays emerging from the initial curve at the specimen plane will map onto the caustic curve at the reference plane, where the caustic curve is the intersection of the three-dimensional caustic surface with the reference plane (see Fig. 9). The details of the initial curve and the shape of the optical caustics that are formed (if any) can be obtained knowing the optical path change $\delta S(x_1, x_2)$.

Figure 11 shows a typical caustic pattern obtained from the stress-concentration region in the vicinity of a statically-loaded crack in a PMMA specimen. Note that the central "shadow-spot" is surrounded by a bright caustic, and in turn these are ringed by fringes, an explanation of which requires a more detailed wave-optic analysis of the method of caustics and is beyond the scope of this review [25].

Optical caustics can also be obtained in reflection from a mirror-polished reflective surface [10]. Here the mapping of (24) is again applicable, but now the non-uniform optical length change $\delta S(x_1, x_2)$ arises from the non-planar shape of the reflector and is given by (6). The rest of the analysis is identical to the above.

5 Applications

We shall now look at some common applications of the three optical methods of stress analysis of non-birefringent materials described above to obtain quantitative information of interest in experimental solid mechanics. As may be expected from the fact that these techniques are sensitive to the gradients

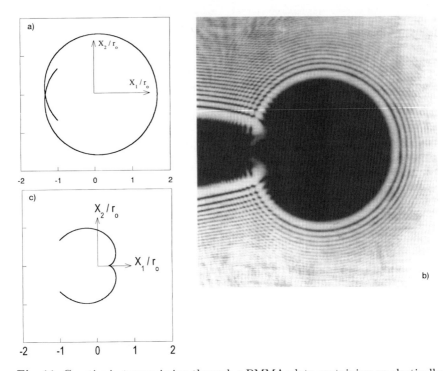

Fig. 11. Caustics in transmission through a PMMA plate containing an elastically-loaded crack: (**a**) theoretical real caustic (**b**) experimental real caustic (**c**) theoretical virtual caustic

of the trace of the stress tensor, it is not readily possible to obtain the individual stress components directly from the transmission fringe patterns (for CGS and PSI) or from the caustics. Therefore, the analysis approach typically used is indirect in that the functional form of the underlying stress/deformation field from which these patterns are obtained should be known except for some undetermined coefficients. These undetermined parameters are then determined by analyzing the fringe/caustic patterns.

5.1 Compressive Line Load on the Edge of a Plate

Consider the case of a compressive line load of magnitude P_0 (load per unit length) acting along the edge of a thin plate, as shown in Fig. 4a. Assuming that the other boundaries of the plate are sufficiently far away and that the plane-stress approximation is appropriate to this situation, the stress-field under the load is given by the theory of linear elasticity to be [11]

$$\sigma_{rr} = \frac{2P_0 \sin\theta}{\pi \, r} \,,$$
$$\sigma_{r\theta} = \sigma_{\theta\theta} = 0 \,, \qquad\qquad -\pi \leq \theta \leq 0 \,. \qquad (26)$$

A plane light beam that transmits through or reflects from the specimen will undergo distortion due to the stress-induced change in optical refractive index and/or change in thickness of the plate. The resulting change in path length is given by (7), which in this case simplifies to

$$\delta S(r,\theta) = \tilde{c}h \frac{2P_0}{\pi} \frac{\sin\theta}{r}, \qquad -\pi \leq \theta \leq 0. \qquad (27)$$

5.1.1 Caustics

The mapping of optical rays from the specimen plane (r,θ) to the reference plane (X_1, X_2) is then given from the above in (24) as

$$X_1 = r\cos\theta + \tilde{c}hz_0 \frac{2P_0}{\pi} \frac{\sin(2\theta)}{r^2},$$
$$X_2 = r\sin\theta - \tilde{c}hz_0 \frac{2P_0}{\pi} \frac{\cos(2\theta)}{r^2}. \qquad (28)$$

The initial curve on the specimen plane that maps onto the caustic in the reference plane is found, using (25), to be a circle of radius given by [30]

$$r_0 = \left(\frac{4}{\pi}|\tilde{c}z_0|hP_0\right)^{\frac{1}{3}}. \qquad (29)$$

The shape of the caustic at the reference plane is therefore obtained by mapping the above points on the initial curve to the reference plane using (24). The resulting caustic is a generalized epicycloid of the form

$$X_1 = r_0\left[\cos\theta + \frac{1}{2}\text{sign}(\tilde{c}z_0)\sin(2\theta)\right],$$
$$X_2 = r_0\left[\sin\theta - \frac{1}{2}\text{sign}(\tilde{c}z_0)\cos(2\theta)\right], \qquad (30)$$

The theoretically expected transmission caustic for a real reference plane is shown in Fig. 12a and the corresponding experimental results for a polymethylmethacrylate (PMMA) specimen are shown in Fig. 12b.

5.1.2 Shearing Interferometry

The shearing interferometric fringes for a compressive edge-line load can be obtained using (27) directly in (15)

$$-\tilde{c}h\frac{2P_0}{\pi}\frac{\sin(2\theta)}{r^2} = m\xi, \qquad m = 0, \pm 1, \pm 2, \ldots. \qquad (31)$$

Typical PSI shearing fringes for this case are shown in Fig. 4b. By obtaining the fringe orders m and locations (r,θ), it is possible to relate the experimental data to theory. In this case, the zeroth order fringe is easily located

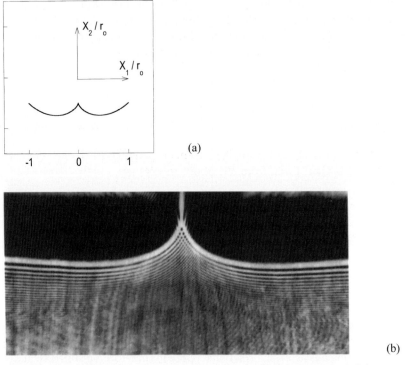

Fig. 12. Caustics due to compressive edge line load: (**a**) theoretical (**b**) experimental in transmission through a PMMA plate

because the gradients of the stresses far away from the load point should vanish. Subsequent fringes can be numbered in increasing integral numbers (half integral fringe orders should be used if both the white and black fringes are considered) towards the load point. The fringe locations are the center points of each white or black band. More sophisticated processing can be done using the entire gray level information, but this is typically unnecessary since it is usually possible to obtain many more data points than there are unknown parameters to be obtained from the fringe patterns. In the case under discussion, there is actually only one undetermined parameter, namely the load P_0 in (31). This can be solved by knowing the fringe order and location at any one spatial location. More often, one uses an over-deterministic approach by calculating the undetermined parameters from a set of experimentally measured fringe locations and orders, which provides an array of data triplets (r, θ, m), which are then used to minimize the error in (31) in a least-squares sense [1].

Experiments such as these are done not so much as to obtain the load P_0 which can of course be far more easily obtained with a load-cell, but rather to get the material-dependent elasto-optic constant 'c' in transmission [39].

5.2 Fracture Mechanics

Consider a bimaterial thin plate specimen containing a crack on the interface between the two materials (Fig. 8a). Let the plate be in a state of plane stress. The asymptotic stress field from linear elastostatic theory can be written as [40]

$$\sigma_{ij}(r,\theta) = \frac{1}{\sqrt{2\pi}} \sum_{n=0}^{\infty} r^{\frac{n-1}{2}} \left\{ \text{Re}\left[k_n \left(\frac{r}{\hat{r}}\right)^{i\varepsilon_n}\right] \hat{\sigma}^I_{ijn}(\theta) \right. $$
$$\left. + \text{Im}\left[k_n \left(\frac{r}{\hat{r}}\right)^{i\varepsilon_n}\right] \hat{\sigma}^{II}_{ijn}(\theta) \right\} , \tag{32}$$

where the k_n are undetermined complex amplitudes, $\hat{\sigma}^I_{ijn}(\theta)$ and $\hat{\sigma}^{II}_{ijn}(\theta)$ are dimensionless functions that are completely known [39], $\hat{r}$ is a characteristic length, and

$$\varepsilon_n = \begin{cases} 0 & \text{for} \quad n = 1, 3, 5, \ldots \\ \varepsilon = \frac{1}{2\pi} \ln\left(\frac{\kappa_1/\mu_1 + 1/\mu_2}{\kappa_2/\mu_2 + 1/\mu_1}\right) & \text{for} \quad n = 0, 2, 4, \ldots \end{cases} \tag{33}$$

where ε is called the materials mismatch-parameter involving the elastic properties of the two materials; $\kappa_i = \frac{3-\nu_i}{1+\nu_i}$, where ν_i is the Poisson's ratio and μ_i is the shear modulus of the two materials. The only undetermined parameters are the values of k_n and the purpose of the optical experiments is to obtain these. Of particular interest is the amplitude $k_0 = K_1 + iK_2 \equiv K$ of the singular term (when $n = 0$), and this is called the stress-intensity factor. The above asymptotic relation is also valid for a crack in a homogeneous medium (in which case the mismatch parameter ε becomes zero, and the equations simplify to that of a mixed-mode Williams' expansion [41]).

The optical path length change for a planar optical beam either transmitting through or reflecting from such a specimen is then given by [21]

$$\delta S(r,\theta) = \frac{\check{c}h}{\sqrt{2\pi}} \sum_{n=0,2,4,\ldots}^{\infty} \frac{2 e^{-\varepsilon(\pi-\theta)}}{\cosh(\pi\varepsilon)} r^{(n-1)/2}$$
$$\times \left[\text{Re}(k_n \hat{r}^{i\varepsilon}) \cos\left(\frac{n-1}{2}\theta - \varepsilon \ln\frac{r}{\hat{r}}\right) \right.$$
$$\left. + \text{Im}(k_n \hat{r}^{i\varepsilon}) \sin\left(\frac{n-1}{2}\theta - \varepsilon \ln\frac{r}{\hat{r}}\right) \right]$$
$$+ \frac{4}{1+\omega} \frac{\check{c}h}{\sqrt{2\pi}} \sum_{n=1,3,5,\ldots}^{\infty} r^{\frac{(n-1)}{2}}$$
$$\times \left[\text{Re}(k_n \hat{r}^{i\varepsilon}) \cos\left(\frac{n-1}{2}\theta - \varepsilon \ln\frac{r}{\hat{r}}\right) \right.$$
$$\left. + \text{Im}(k_n \hat{r}^{i\varepsilon}) \sin\left(\frac{n-1}{2}\theta - \varepsilon \ln\frac{r}{\hat{r}}\right) \right] , \tag{34}$$

where $\omega = \frac{\kappa_1+1}{\kappa_2+1}\frac{\mu_2}{\mu_1}$ is a combination of material elastic properties.

5.2.1 Caustics

By far the greatest use of optical caustics in experimental stress analysis has been in linear elastic-fracture mechanics. The approach typically used is to assume that the functional form of the change in optical path length $\delta S(x_1, x_2)$ caused by structural deformation in the vicinity of a crack is predominantly given by the singular term alone (with the terms $n > 0$ not being important in (34). Under this assumption of "$\boldsymbol{K}$-dominance" of the crack-tip fields, the objective then is to determine the stress intensity factor $\boldsymbol{K}$ from the dimensions of the optical caustics.

As an illustrative example, consider the case of a tensile-loaded crack in a thin transparent homogeneous solid plate. This is known as mode-I loading in fracture mechanics terminology [41], and the stress-intensity factor is just $\boldsymbol{K} = K_1$. The change in path length near the crack tip for this case is given by

$$\delta S(r,\theta) = \tilde{c}h \frac{2K_1}{\sqrt{2\pi}} r^{-\frac{1}{2}} \cos \frac{3\theta}{2} + O(1) \quad \text{as} \quad r \to 0, \quad -\pi \le \theta \le \pi, \quad (35)$$

where (r, θ) is the polar coordinate system located at the crack-tip. The initial curve for the caustics formed in this case is a circle of radius [30]

$$r_0 = \left(\frac{3}{2\sqrt{2\pi}} |\tilde{c}hz_0|K_1 \right)^{2/5}. \quad (36)$$

The mapping of optical rays from the initial curve on the specimen leads to a caustic on the reference plane given by

$$X_1 = r_0 \left[\cos\theta + \frac{2}{3}\text{sign}(\tilde{c}z_0) \cos \frac{3\theta}{2} \right],$$
$$X_2 = r_0 \left[\sin\theta + \frac{2}{3}\text{sign}(\tilde{c}z_0) \sin \frac{3\theta}{2} \right]. \quad (37)$$

This is an epicyloid, as shown in Fig. 11a,c for both a real and a virtual reference plane. The experimental caustic obtained in transmission using a real image plane for a crack in a polymethylmethacrylate specimen is shown in Fig. 11b. Since the size of the caustic is proportional to the initial curve radius r_0, which in turn is related to the stress intensity factor K_1 through (36), it is possible to calculate the latter from the measured maximum dimension (H) along the x_2-direction of the caustic [10]

$$K_1 = \frac{2\sqrt{2\pi}}{3(3.17)^{5/2}} \frac{H^{5/2}}{|cz_0h|}. \quad (38)$$

The method of caustics can also be used under mixed-mode loading [12], in the study of interfacial fracture phenomena [42], and in the study of dynamically propagating cracks [30,43,44]. Since caustics do not provide full-field fringe data such as those obtained with shearing interferometry or photoelasticity, it is very important that the limitations of the method of caustics discussed in [45,46] be kept in mind when using this technique. In particular, it is important to use the method of caustics to extract stress-intensity factors *only when the underlying region from which the caustics are obtained is known to be K-dominant*. Recently, attempts have been made to extend the method of caustics to situations where K-dominance does not hold [10], but shearing interferometric methods provide a more direct alternative as described below.

5.2.2 Shearing Interferometry

The shearing fringes for an interfacial crack can be analyzed by substituting (34) in (15)

$$m\xi = \frac{\tilde{c}h}{\sqrt{2\pi}} \sum_{n=0,2,4,\ldots}^{\infty} \frac{2\,e^{-\varepsilon(\pi-\theta)}}{\cosh(\pi\varepsilon)} r^{(n-3)/2}$$

$$\times \left\{ \operatorname{Re}(k_n r^{i\varepsilon}) \left[(n-1)\cos\left(\frac{n-3}{2}\theta - \varepsilon \ln \frac{r}{\hat{r}}\right) \right.\right.$$

$$\left. +2\varepsilon \sin\left(\frac{n-3}{2}\theta - \varepsilon \ln \frac{r}{\hat{r}}\right) \right]$$

$$+\operatorname{Im}(k_n r^{i\varepsilon}) \left[(n-1)\sin\left(\frac{n-3}{2}\theta - \varepsilon \ln \frac{r}{\hat{r}}\right) \right.$$

$$\left.\left. -2\varepsilon \cos\left(\frac{n-3}{2}\theta - \varepsilon \ln \frac{r}{\hat{r}}\right) \right] \right\}$$

$$+\frac{2}{1+\omega} \frac{\tilde{c}h}{\sqrt{2\pi}} \sum_{n=1,3,5,\ldots}^{\infty} r^{(n-3)/2}$$

$$\left[\operatorname{Re}(k_n)(n-1)\cos\left(\frac{n-3}{2}\theta\right) + \operatorname{Im}(k_n)(n-1)\sin\left(\frac{n-3}{2}\theta\right) \right] \quad (39)$$

Once again, the above equations are applicable to the case of a crack in a homogeneous solid by setting the material mismatch parameter ε to zero.

To give an idea of the accuracy with which the stress-intensity factor can be obtained using shearing interferometry, results are shown in Fig. 13a for a three-point bend specimen made of PMMA. The experimentally-obtained fringe order m and fringe location (r,θ) can again be used in an over-deterministic least-squares procedure [1,21] to obtain the value of k_n to the required number of terms in (39). The zeroth-order fringe is located from knowledge of the far-field stress state. In this example it is the white fringe that comes into the crack-tip (which should be located at the mid-point of the two sheared crack-tips visible in Fig. 13a at about a 60° angle to the horizontal. Furthermore, it is necessary to exclude a near-tip region of about one half-plate thickness where strong three-dimensional effects prevail, and therefore the plane-stress assumption does not hold here [20]. This is also the region where determining the location of the fringe centers is prone to greatest error, and exclusion of such regions is actually beneficial. The accuracy with which the stress-intensity factor K and the other higher-order parameters in the stress-field can be obtained depends on the number of terms needed to adequately describe the stress-state in the regions from which the fringe data are collected. This generally varies with the specimen geometry, and it is therefore advisable to analyze the same set of fringes using increasing numbers of terms in the expansion till convergence is obtained [20].

For the example shown in Fig. 13a, a three-term expansion in (39) was necessary for convergence, and this yielded a stress-intensity factor value of $K_I = 0.54\,\mathrm{MPa}\sqrt{\mathrm{m}}$. A finite element analysis of the same geometry and loading yielded a value of $K_I = 0.55\,\mathrm{MPa}\sqrt{\mathrm{m}}$, which indicates a discrepancy of less than 2%. Shown in Fig. 13b are the reconstructed fringes using the experimentally-determined values of k_n back in the fringe relation (39). The good all-round agreement between the experimental data and the reconstructed fringes indicates that the three-term expansion is indeed sufficient in this region to adequately describe the crack-tip stress-fields.

Shearing interferometry has been used successfully not only in its initial application to static fracture mechanics [10,20], but has also since been used in

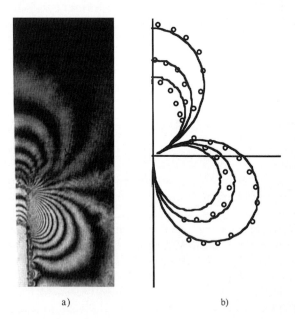

Fig. 13. Shearing mode fringe pattern from the PMMA side of an interfacial crack in a bimaterial composite epoxy/PMMA specimen: (**a**) fringe pattern (**b**) comparison of data (circles) and reconstructed fringes (lines) using (39)

conjunction with a high-speed camera to monitor dynamic crack propagation in homogenous [47] and heterogeneous [48,49] media.

6 Conclusion

In this review we have described two implementations of shearing interferometers and also briefly the method of caustics. These optical techniques have found application in experimental stress analysis of non-birefringent transparent and reflective opaque specimens. It has been shown that all three of these optical techniques provide information about the gradients of the stress/deformation of the test object. As such, they are less sensitive to initial imperfections (such as non-planarity) in the specimens. The shearing interferometers described here are common-path systems that are quite robust and provide adjustable sensitivity. They provide an ideal complement to the full-field technique of photoelasticity that has been extensively used on birefringent materials.

References

1. J. W. Dally, W. F. Riley: *Experimental Stress Analysis*, McGraw-Hill, New York (1991)
2. R. Jones, C. Wykes: *Holographic and Speckle Interferometry*, Cambridge Univ. Press., London (1983)
3. A. J. Durelli, V. J. Parks: *Moiré Analysis of Strain*, Prentice-Hall, Englewood Cliffs, NJ (1970)
4. D. Malacara: *Optical Shop Testing*, Wiley, New York (1992)
5. L. H. Tanner: Some Laser Interferometers for Use in Fluid Mechanics, J. Sci. Instrum. **42**, 834 (1965)
6. P. K. Rastogi: *Holographic Interferometry-Principles and Methods*, Springer, Berlin, Heidelberg (1984)
7. D. Post, B. Han, P. Ifju: *High-Sensitivity Moiré: Experimental Analysis for Mechanics and Materials*, Springer, Berlin, Heidelberg (1994)
8. Y. Y. Hung: Measurement of Slopes of Structural Deflections by Speckle-Shearing Interferometry, Exp. Mech. **17**, 281 (1974)
9. W. J. Bates: A Wavefront Shearing Interferometer, Proc. Phys. Soc. **59**, 196 (1947)
10. A. J. Rosakis: Two Optical Techniques Sensitive to Gradients of Optical Path Difference, in J. S. Epstein (Ed.): *Experimental Techniques in Fracture*, VCH, Weinheim (1993) p. 327
11. S. P. Timoshenko, J. N. Goodier: *Theory of Elasticity*, McGraw-Hill, New York (1970)
12. J. Beinert, J. F. Kalthoff: Experimental Determination of Dynamic Stress-Intensity Factors by Shadow Patterns, in G. Sih (Ed.): *Mechanics of Fracture* **7**, Nijhoff, Hingham, MA (1981) p. 281
13. M. V. R. K. Murty: The Use of a Single Plane Parallel Plate as a Lateral Shearing Interferometer with a Visible Gas Laser Source, Appl. Opt. **3**, 531 (1964)

14. J. D. Briers: Prism Shearing Interferometry, Opt. Technol. **1**, 196 (1969)
15. P. H. Lee, G. A. Woolsey: Versatile Polarization Interferometers, Appl. Opt. **20**, 3514 (1981)
16. V. Ronchi: Forty Years of History of a Grating Interferometer, Appl. Opt. **3**, 437 (1964)
17. J. C. Wyant: Double Frequency Grating Lateral Shear Interferometer, Appl. Opt. **12**, 2057 (1973)
18. K. Patorski: Talbot Interferometry with Increased Shear, Appl. Opt. **24**, 4448 (1985)
19. P. Hariharan, W. H. Steel, J. C. Wyant: Double Grating Interferometer with Variable Lateral Shear, Opt. Commun. **11**, 317 (1974)
20. H. V. Tippur, S. Krishnaswamy, A. J. Rosakis: Optical Mapping of Crack Tip Deformations Using the Methods of Transmission and Reflection Coherent Gradient Sensing: a Study of Crack Tip K-dominance, Int. J. Fract. **52**, 91 (1991)
21. H. S. Lee, S. Krishnaswamy: A Compact Polariscope and Shearing Interferometer for Mapping Stress Fields in Bimaterial Systems, Exp. Mech. **36**, 404 (1996)
22. H. A. Bruck, A. J. Rosakis: On the Sensitivity of Coherent Gradient Sensing. Part 1. A Theoretical Investigation of Accuracy in Fracture-Mechanics Applications, Opt. Lasers Eng. **17**, 83–101, 1992; Part.2. An Experimental Investigation Of Accuracy In Fracture-Mechanics Applications, Opt. Lasers Eng. **18**, 25 (1993)
23. H. V. Tippur: Coherent Gradient Sensing - A Fourier Optics Analysis and Applications to Fracture, Appl. Opt. **31**, 4428 (1992)
24. E. Hecht, A. Zajac: *Optics*, Addison-Wesley, Reading, MA (1974)
25. J. A. Lock, J. H. Andrews: Optical Caustics in Natural Phenomena, Am. J. Phys. **60**, 397–407 (1992)
26. G. B. Airy: On the Intensity of Light in the Neighbourhood of a Caustic, Trans. Cambridge Philos. Soc. **6**, 379, (1838)
27. A. A. Cayley: A Memoir Upon Caustics, Philos. Trans. R. Soc. **147**, 273, (1857)
28. P. Manogg: Anwendung der Schattenoptik zur Untersuchung des Zerreissvorganges von Platten, Dissertation, University of Freiburg, Germany (1964)
29. P. S. Theocaris: Elastic Stress Intensity Factors Evaluated by Caustics, in G. C. Sih (Ed.): *Mechanics of Fracture* **7**, 189, Nijhoff, Hingham, MA (1981)
30. J. F. Kalthoff: Shadow Optical Method of Caustics, in A. S. Kobayashi (Ed.): *Handbook on Experimental Mechanics*, McGraw-Hill, New York (1987), Chap. 9, p. 430
31. A. J. Rosakis, C. C. Ma, L. B. Freund: Analysis of the Optical Shadow Spot Method for a Tensile Crack in a Power-Law Hardening Material, J. Appl. Mech. **50**, 777 (1983)
32. P. S. Theocaris: Multicusp Caustics Formed from Reflections of Warped Surfaces, Appl. Opt. **27**, 780 (1988)
33. H. P. Rossmanith: The Caustic Approach to Rayleigh-Waves and their Interaction with Surface Irregularities, Opt. Lasers Eng. **14**, 115 (1990)
34. H. W. Liepmann, A. Roshko: *Elements of Gas Dynamics*, Wiley, New York (1957)
35. G. Dacosta, R. Escalona: Time Evolution of the Caustics of a Laser Heated Liquid Film, Appl. Opt. **29**, 1023 (1990)
36. S. Mahaut, D. K. Lewis, G. Quentin, M. Debilly: Visualization of Internal Caustics in Cylindrical Liquid Cavities with a Pulsed Schlieren System, J. Phys. IV, Paris, **4**, 741 (1994)

37. A. Joets, R. Ribotta: Caustics and Symmetries in Optical Imaging: The Example of Convective Flow Visualization, J. Phys. I, Paris, **4**, 1013 (1994)
38. M. V. Berry, A. N. Wilson: Black-and-White Fringes and the Colors of Caustics, Appl. Opt. **33**, 4714 (1994)
39. H. S. Lee: Quasi-static Subinterfacial Crack Propagation, Ph.D. thesis, Northwestern University, Chicago, IL (1998)
40. J. R. Rice: Elastic Fracture Mechanics Concepts for Interfacial Cracks, J. Appl. Mech. **55**, 418 (1988)
41. K. Hellan: *Introduction to Fracture Mechanics*, McGraw-Hill, New York (1985)
42. P. S. Theocaris, C. Stassinakis: Experimental Solution of the Problem of a Curvilinear Crack in Bonded-Dissimilar Materials, Int. J. Fract. **13**, 13 (1977)
43. K. Ravi-Chandar, W. G. Knauss: On the Characterization of Transient Stress Fields Near the Tip of a Crack, J. Appl. Mech. **54**, 72 (1987)
44. A. J. Rosakis, J. Duffy, L. B. Freund: The Determination of the Dynamic Fracture Toughness of AISI 4340 Steel by the Shadow Spot Method, J. Mech. Phys. Solids **34**, 443 (1984)
45. S. Krishnaswamy, A. J. Rosakis: On the Extent of Dominance of the Asymptotic Elastodynamic Crack-Tip Fields; Part 1: an Experimental Study Using Bifocal Caustics, J. Appl. Mech. **58**, 87 (1990)
46. H. Nigam, A. Shukla: Comparison of the Techniques of Transmitted Caustics and Photoelasticity as Applied to Fracture, Exp. Mech. **28**, 123 (1988)
47. S. Krishnaswamy, H. V. Tippur, A. J. Rosakis: Measurement of Transient Crack-Tip Deformation Field Using the Method of Coherent Gradient Sensing, J. Mech. Phys. Solids **40**, 339 (1992)
48. J. Lambros, A. J. Rosakis: Dynamic Decohesion of Bimaterials - Experimental-Observations and Failure Criteria, Int. J. Solids Struct. **32**, 2677 (1997)
49. R. P. Singh, J. Lambros, A. Shukla, A. J. Rosakis: Investigation of the Mechanics of Intersonic Crack Propagation along a Bimaterial Interface Using Coherent Gradient Sensing and Photoelasticity, Proc. R. Soc. London A **453**, 2649 (1997)

Advances in Two-Dimensional and Three-Dimensional Computer Vision

Michael A. Sutton, Stephen R. McNeill, Jeffrey D. Helm, and Yuh J. Chao

Department of Mechanical Engineering, University of South Carolina
Columbia, SC 29208, USA
sutton@sc.edu

Abstract. The foundations of two- and three-dimensional image correlation, as well as recent developments, are described in detail. The versatility and robustness of these methods are illustrated through application examples from diverse areas including fracture mechanics, biomechanics, constitutive property measurement in complex materials, model verification for large, flawed structures and nondestructive evaluation. A detailed description of experimental and data-reduction procedures is presented for the application of the two-dimensional image-correlation method to thin-sheet mixed-mode I/II fracture studies, local crack-closure measurements using optical microscopy and the measurement of constitutive properties. Application examples using three-dimensional image correlation include profiling of components for reverse engineering and manufacturing and the measurement of full-field surface deformation during wide cracked panel tensile tests for verification of buckling and crack-growth models. Results from nearly sixteen years of use have demonstrated that both two-dimensional and three-dimensional image-correlation methods are robust and accurate tools for deformation measurements in a variety of applications. The range of uses for the two- and three-dimensional image-correlation methods is growing rapidly as scientists and engineers begin to understand their true capabilities.

1 Introduction

Computer vision is a broad term, referring to a wide range of engineering applications from robotic vision to object recognition. In this review, the term computer vision will refer to computer-based, non-contacting, surface deformation measurement methods used for the study of solid structures. Even with this restriction, a wide variety of methods have been advanced in recent years, and the pace of development continues to increase rapidly. This introduction will present a brief survey of current computer-vision methods. Emphasis will be placed on recent work in the area of in-plane deformation measurement of planar surfaces using digital image correlation (DIC-2D) and the use of digital image correlation for the measurement of general deformation of planar and curved surfaces in three-dimensional space (DIC-3D).

One of the earliest papers to propose the use of computer-based analysis for deformation measurements was written by *Peters* and *Ranson* [1] in 1981. Interestingly, the original application envisioned was for analyzing images of internal structure obtained by using ultrasonic waves. Their work described how digital ultrasound images of a solid, subjected to two-dimensional loading, could be analyzed to determine the average, through-thickness planar displacements and displacement gradients for an object. They proposed to obtain full-field measurements by comparing the locations of small regions, referred to as subsets, in a digital image of an object before loading to the subset's location in a later image of the object while under load. In their paper [1], they stated that fundamental continuum-mechanics concepts governing the deformation of small areas would be ideally suited to the development of numerical algorithms. Over the next ten years, the concepts proposed in their work have been modified, for use with optical ilumination and image acquisition numerical algorithms developed [2,3,4,5,6,7,8] and the method applied successfully to the measurement of the in-plane components of surface deformations [9,10,11,12,13,14,15,16,17,18,19,20,21,22,23,24,25,26,27,28].

Among the early papers, three are of particular significance. First, in the work of *Chu* et al. [8], a description of the basic theory was outlined and experimental results were presented which conclusively demonstrated that the method could be used to accurately measure simple deformations of a solid body. Secondly, a series of papers by *Sutton* et al. [4,5] presented a new approach for determining surface deformations employing Newton-Raphson methods for optimization. The resulting method, which remains the foundation of DIC-2D to this day, resulted in a twenty-fold increase in speed with no loss in accuracy. Finally, modeling work performed by *Sutton* et al. [3] demonstrated that the primary parameters affecting the accuracy of the surface deformations obtained by the Newton-Raphson based DIC-2D method are (a) the number of quantization levels in the digitization process (i.e. the number of bits in the A/D converter for converting light intensity into a digital value), (b) the ratio of the sampling frequency to the frequency of the intensity signal (i.e. the number of sensors in the camera used to record a given pattern) and (c) use of interpolation functions to reconstruct the intensity pattern at sub-pixel locations.

Researchers continued to improve and gain a deeper understanding of the DIC-2D method, and others began using the method to address a wide range of problems [9,10,11,12,13,14,15,16,17,18,19,20,21,22,23,24,25,26,27,28]. DIC-2D was applied to the measurement of velocity fields both in two-dimensional seeded flows [9] and in rigid-body mechanics [11]. DIC-2D principles were used successfully to measure strain fields in retinal tissue overcoming the difficulties inherent in measuring strains in flexible materials and to determine the tissue's mechanical properties under monotonic and cyclic loading [13,15]. The non-contacting nature of DIC-2D also allowed the measurement of strains in thin paper specimens [16] by researchers at the University of South Carolina. Most significantly, the DIC-2D method continues to be used extensively in fracture-mechanics studies [10,14,19,20,21,22,23,24,25,26], including measurement of (a) strain fields near stationary and growing crack-tips [22,23,24],

(b) crack-tip opening displacement during crack growth [25,26,27,28] and (c) strain measurements near crack-tips at high temperatures [31,32]. Recently, the method has been successfully extended to the study of deformations from scanning tunneling electron microscopy images [33,34,35] as well as deformation in concrete during compressive loading [36].

It is worth noting that the paper by *Chao* et al. [16] has led to the adoption of the DIC-2D method by many researchers in the wood, paper and forest products area. The method has been used to (a) study the deformation of single wood cells [37], (b) determine the mechanical properties of small specimens made of wood products [38,39,40] and ASTM standard-sized wood/wood composite specimens [41,42,43,44], (c) study the drying process in wood specimens [45,46,47] and (d) determine the mechanical properties of wooden connections [48,49].

Though all of the applications described above employ DIC-2D concepts, including the use of direct image-correlation principles for determination of displacements and displacement gradients, Fast Fourier Transforms [29,30] (FFT's) were shown to be a viable alternative for those applications where in-plane strains and rigid body rotations are relatively small. The FFT method uses discrete Fourier transforms of the intensity pattern in both deformed and undeformed sub-regions to determine the cross-correlation function. The displacement of the deformed sub-region is then estimated by locating the peak of the cross-correlation function. The FFT approach was found to be both fast and accurate for a large number of applications. However, the method does not allow subset deformation and hence experiences a loss of accuracy, even in the presence of only moderate strains or in-plane rotations. Applications have included determination of deformations in shock mounts and the mechanical properties of both fibers and the fiber-matrix interface region in a composite using images from an scanning electron microscope (SEM).

One fundamental limitation of both the DIC-2D and FFT methods is their basis in two-dimensional concepts. The change in magnification due to a relatively small out-of-plane motion, can introduce significant errors in the measured in-plane displacement. To minimize these effects, *Sutton* et al. [22,23,24] utilized an optical system that is relatively insensitive to small magnitudes of out-of-plane motion. This insensitivity was achieved by increasing the distance from the camera to the surface of the object, requiring the use of long-focal-length lenses, and thereby minimizing any change in magnification due to small out-of-plane movements of the object. Long-focal-length lens systems continue to be used successfully, including the use of a far-field microscope to measure crack-tip opening displacements within a few micrometers of the crack-tip [50,51] during cyclic loading to quantify crack-closure effects. Attempts to correct for large out-of-plane movements have proven extremely difficult if not impossible. For example, early attempts by *Peters* et al. [19] to correct the in-plane deformations using approximate estimates of the out-of-plane motion for a pressurized cylinder were not successful.

In addition, early work in the paper industry [53] attempted to infer out-of-plane buckling from the "measured" in-plane deformations using the DIC-2D system with minimal success. More recently, *Lu* et al. [52] have shown that correcting the "measured" displacements for nominally cylindrical specimens will result in strain errors of the order of 2000 μstrain. These limitations provided the motivation to develop a method to measure full three-dimensional displacements.

The first three-dimensional displacement measurements were performed in 1988 by McNeill and resulted in an unpublished internal report [54]. Using a known horizontal translation of one camera to obtain two views of an object, McNeill calibrated a simple stereo vision system and demonstrated that the shape of an inclined planar object could be accurately measured. In 1991, *Luo* et al. [55] successfully developed a two-camera stereo vision system for deformation measurements and applied it to fracture problems [56,57]. To overcome some of the key limitations of the method (square subsets that remained square in both cameras, mismatch in the triangulation of corresponding points, and a calibration process that was laborious and time-consuming), *Helm* et al. [58,59] successfully developed a two-camera stereo vision system that (a) included the effects of perspective on subset shape (b) constrained the analysis to epipolar lines and (c) simplified the process of calibrating the system. The measurement system, 3D Digital Image Correlation (DIC-3D), is being used for a wide range of applications to both large and small structures.

In Sect. 2, the theory for DIC-2D and DIC-3D will be described. In Sect. 3, recent applications of both DIC-2D and DIC-3D will be presented. In Sect. 4, a discussion of the advantages and disadvantages of the DIC-2D and DIC-3D methods is provided. In Sect. 5, a summary of recent advances will be given, as well as a discussion of future applications that are envisioned.

2 Theory and Numerical Implementation

2.1 Two-Dimensional Video Image Correlation

2.1.1 Basic Concepts

The basis of two-dimensional video image correlation for the measurement of surface displacements is the matching of one point from an image of an object's surface before loading (the undeformed image) to a point in an image of the object's surface taken at a later time/loading (the deformed image). Assuming a one-to-one correspondence between the deformations in the image recorded by the camera and the deformations of the surface of the object, an accurate, point-to-point mapping from the undeformed image to the deformed image will allow the displacement of the object's surface to be measured. Two main requirements must be met for the successful use of DIC-2D. First, in order to provide features for the matching process, the surface of the object must have a pattern that produces varying intensities of diffusely reflected

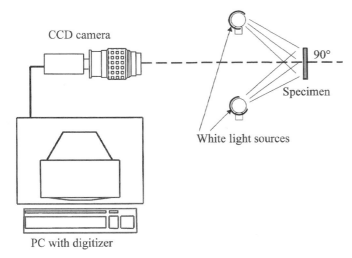

Fig. 1. Schematic of a data-acquisition system for two-dimensional video image correlation

light from its surface. This pattern may be applied to the object or it may occur naturally. Secondly, the imaging camera must be positioned so that its sensor plane is parallel to the surface of the planar object, as shown in Fig. 1.

The imaging process of the camera converts the continuous intensity field reflected from the surface $O(X,Y)$ into a discrete field $I(X,Y)$ of integer intensity levels. In a CCD camera, this transformation occurs when the light incident on a sensor (commonly known as a pixel) is integrated over a fixed time period. The rectangular array of sensors in a charge-coupled device (CCD) array converts the continuous intensity pattern into a discrete array of integer intensity values, $I(i,j)$, where i denotes the row number and j denotes the column number in the sensor plane. The displacement field for an object is obtained at a discrete number of locations by choosing subsets from the initial image and searching throughout the second image to obtain the optimal match. Details of this process are outlined in the following paragraphs.

The process of deformation in two dimensions is shown schematically in Fig. 2. The functions are defined as follows; (a) $O(X,Y)$ denotes the continuous intensity pattern for the undeformed object, (b) $O'(X,Y)$ is the continuous intensity pattern for the deformed object, (c) $I(X,Y)$ is the discretely sampled intensity pattern for the undeformed object and (d) $I'(X,Y)$ is the discretely sampled intensity pattern for the deformed object. It is important to note that a basic tenet of the DIC-2D method is that points in $I(X,Y)$ and $I'(X,Y)$ are assumed to be in one-to-one correspondence with points in $O(X,Y)$ and $O'(X,Y)$, respectively. Thus, one can use $I(X,Y)$ and $I'(X,Y)$ to determine the displacement field for the object $O(X,Y)$.

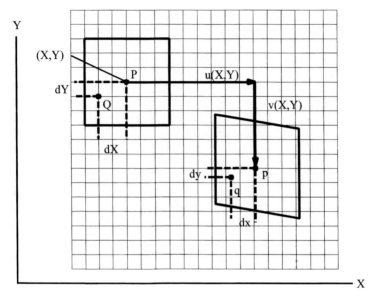

Fig. 2. Schematic of the deformation process in two dimensions

Recalling basic continuum-mechanics concepts for the deformation of a region, one would expect that the object point imaged onto the center of an integer pixel location in the undeformed image will be imaged to locations between pixels of the stationary CCD camera after deformation. Thus, obtaining accurate estimates for surface deformations using $I(X,Y)$ and $I'(X,Y)$ requires an interpolation scheme to reconstruct a continuous intensity function. The intensity pattern for a typical 10×10 pixel subset is shown in Fig. 3, where black and white represent low and high intensity, respectively A wide variety of interpolation methods has been used successfully and Fig. 4 shows a plot of the raw intensity data, bilinear interpolation fit, bicubic interpolation fit and a bicubic-spline fit for the 10×10-subset shown in Fig. 3.

It is noted that accurate measurement of surface deformations requires that the mapping from the undeformed image to the deformed image be as accurate as possible. Thus, it is important to "oversample" the intensity pattern, using several sensors to sample each feature of the intensity pattern. Through a combination of oversampling, interpolation and quantization with at least 8 bits, the original intensity pattern can be reconstructed with reasonable accuracy and the displacement field estimated with accuracy of ± 0.02 pixels or better.

Fig. 3. Intensity values for a typical 10 × 10 pixel subset

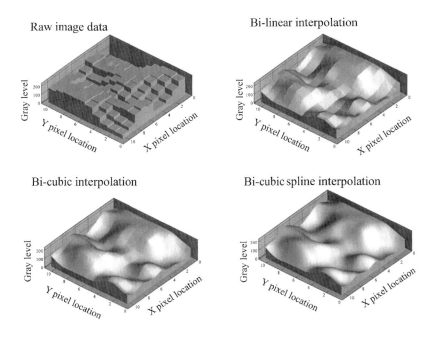

Fig. 4. Four representations of the intensity values over the 10 × 10 pixel region

2.1.2 Two-Dimensional Image Correlation: Mathematical Formulation

For the small subset centered at (X, Y) on the undeformed object in Fig. 2, the discretely sampled and continuously interpolated intensity pattern at points P and Q, located at positions (X, Y) and $(X+\mathrm{d}X, Y+\mathrm{d}Y)$ respectively, can be written as

$$I(P) = I(X, Y),$$
$$I(Q) = I(X + \mathrm{d}X, Y + \mathrm{d}Y), \qquad (1)$$

where $(\mathrm{d}X, \mathrm{d}Y)$ represent small distances in the (X, Y) coordinate system. Note that, if the values for $\mathrm{d}X$ and $\mathrm{d}Y$ are integer pixel values, then no interpolation is required for the undeformed image. As shown in Fig. 2, after deformation of an object, points P and Q are deformed into positions p and q, respectively. Assuming that the intensity pattern recorded after deformation is related to the undeformed pattern by the object deformations, and defining $\{u(X, Y), v(X, Y)\}$ as the displacement vector field, we can write $x = X + u(X, Y), y = Y + v(X, Y)$ so that

$$I'(x, y) = I(X + u(X, Y), Y + v(X, Y)),$$
$$I'(x + \mathrm{d}x, y + \mathrm{d}y) = I[X + \mathrm{d}X + u(X + \mathrm{d}X, Y + \mathrm{d}Y), Y + \mathrm{d}Y + v(X + \mathrm{d}X, Y + \mathrm{d}Y)],$$
$$= I\left[X + u(X, Y) + \left(1 + \frac{\partial u}{\partial X}\right)\mathrm{d}X + \frac{\partial u}{\partial Y}\mathrm{d}Y,\right.$$
$$\left. Y + v(X, Y) + \frac{\partial v}{\partial X}\mathrm{d}X + \left(1 + \frac{\partial v}{\partial Y}\right)\mathrm{d}Y\right]. \qquad (2)$$

Assuming that the subset is sufficiently small so that the displacement gradients are nearly constant throughout the region of interest, each subset undergoes uniform strain resulting in the parallelogram shape for the deformed subset shown in Fig. 2. Conceptually, determining $u, v, \partial u/\partial X, \partial u/\partial Y, \partial v/\partial X, \partial v/\partial Y$ for each subset is simply a matter of determining all six parameters so that the intensity values at each point in the undeformed and deformed regions match. To obtain these values, there are several measures of intensity pattern correspondence that could be used, including the following summations for a series of selected points, Q_i

(a) $\Sigma_i |I'(q_i) - I(Q_i)|$ (magnitude of intensity value difference)

(b) $\Sigma_i (I'(q_i) - I(Q_i))^2$ (sum of squares of intensity value differences)

(c) $\dfrac{\Sigma_i(I'(q_i)I(Q_i))}{(\Sigma_i I'(q_i)^2)^{\frac{1}{2}}(\Sigma_i I(Q_i)^2)^{\frac{1}{2}}}$ (normalized cross-correlation)

(d) $\Sigma_i(I'(q_i)I(Q_i))$ (cross-correlation)

Minimization of (a,b) and maximization of (c,d) for each subset will provide the optimal estimates for all six parameters [2,4,5,8,11]. It is noted that, when a range of values are chosen for the displacements of the centerpoint, $(u_l(P), v_l(P))$, and the displacement gradients in the subset, $\{u/\partial X, \partial u/\partial Y, \partial v/\partial X, \partial v/\partial Y\}$, the positions of the points will fall between integer values in the deformed subset. Thus, the intensity values at $M \times M$ points in the deformed subset must be obtained from the interpolation function, since non-integer locations correspond to positions that are between pixel locations. As noted previously, a wide range of interpolation methods have been used successfully, including bilinear, bicubic and bicubic-spline. In most of the results to be discussed in the following sections, a normalized cross-correlation function is used to obtain best estimates for the displacement field and it is defined as

$$1.0 - C_l(u, v, \partial u/\partial jX, \partial u \partial Y, \partial v/\partial X, \partial v/\partial Y) =$$

$$1.0 - \frac{\sum_{i=1}^{i=N} \sum_{j=1}^{j=M} I(X_i, Y_j) I'[X_i + u(X_i, Y_j), Y_j + v(X_i, Y_j)]}{\sum_{i=1}^{i=N} \sum_{j=1}^{j=M} \{I^2(X_i, Y_j) I'^2[X_i + u(X_i, Y_j), Y_j + v(X_i, Y_j)]\}^{\frac{1}{2}}} \cdot (3)$$

The values for $u(P), v(P), \partial u/\partial X, \partial u/\partial Y, \partial v/\partial X, \partial v/\partial Y)$ which minimizes the quantity $(1.00 - C_l)$ are assumed to represent the best estimates of the subset's displacement and strain components. It is noted that the quantity $(1.00 - C_l)$ is zero for a perfect match and one for orthogonal (completely mismatched) patterns, providing a quantitative measure of the accuracy of the match between deformed and undeformed subsets. A wide range of optimization methods has been used successfully to obtain the optimal value for $(1.00 - C_l)$, including Newton-Raphson [2,3,4,5,7,8], coarse-fine [2,3,4,5] and recently the Levenburg-Marquardt method. Currently, most of the analyses are performed with the Newton-Raphson method, since it is at least twenty times faster than the coarse-fine search. Recent analyses performed using the Levenburg-Marquardt method indicate that this method is as fast as the Newton-Raphson approach and has better convergence characteristics.

To obtain optimal estimates for all six parameters in (3), the iteration process is as follows:

1. Obtain an initial estimate of centerpoint translation for the subset of interest from visual inspection of the images, denoting it as $[u_l(P), v_l(P)]$,
2. Allow the gradients to be non-zero and perform a full, six-parameter search process to minimize (3). The values of u, v and the displacement gradients which minimize $(1 - C_l)$ are the optimal estimates for the displacements and displacement gradients,
3. Using results from the previous subset as the initial guess, repeat (2) for the next subset,
4. Repeat steps (2) and (3) until data is obtained throughout the region of interest.

With regard to accuracy of the parameters obtained by this method, several points should be noted:

(a) Local gradient terms for each subset vary by ±0.005 in a uniform strain region due to a combination of factors including interpolation errors, quantization levels and noise in the intensity pattern. Due to this, gradients for each subset are only used to improve estimates of the centerpoint displacement $[u_l(P), v_l(P)]$.
(b) By obtaining the displacements of many subset centers, P_I, a displacement field is generated. Typical errors are ±0.02 pixels for each displacement component. The data field is smoothed using one of many techniques available [7,60,61] to reduce noise in the measured displacement fields. Using 8-bit digitizers and bilinear interpolation, typical point-to-point accuracy in the displacement gradients obtained from smoothed displacement data is $\pm 200 \times 10^{-6}$ in a non-uniform strain field.
(c) With regard to surface fitting, bilinear interpolation is the simplest approach for estimating intensity values between pixel locations. It has been used effectively in a wide range of applications. However, previous experiments have shown that bicubic and bicubic-spline interpolation methods provide improved accuracy, especially when the displacement gradients are small (e.g. displacement gradients < 0.002 throughout the image area).
(d) The use of zero gradients for all subsets to obtain a displacement field is computationally efficient. For small-strain cases, the choice of zero displacement gradients has no discernible bias and gives reasonable accuracy (typical accuracy in strains is ±200 strain). However, for strain or rotation of the order of 0.03, errors in the measured displacements in using this approach will increase rapidly and the use of local displacement gradients is recommended.

Equations (1,2,3) define a procedure for obtaining the optimal affine transformation between an undeformed and deformed image pair under the restrictions imposed on the transformation by continuum-mechanics principles. A similar procedure is commonly used in computer graphics as part of an image "warping" process [62], where the transformations are generally not restricted by continuum requirements.

With regard to the various correlation formulae described above, the following remarks are to be noted. First, all forms have been used successfully by the authors to determine the optimal affine transformation for a subset. However, the sum of squares of differences, the sum of magnitude of differences and the cross-correlation have two disadvantages. First, all are somewhat sensitive to changes in lighting, with errors in the measured displacements increasing as both contrast and background intensity levels change. Secondly, since the correlation coefficient increases substantially as lighting changes occur, it is difficult to determine whether accurate pattern matches have been obtained. For these reasons, we have used the normalized cross-correlation (3) in all of our analyses. Our studies have shown that (3) can be used to obtain optimal affine transformation parameters with background lighting changes up to 30%. Furthermore, the normalized correlation coefficient obtained during the analysis can be used to determine when "matching" has degraded.

With regard to the accuracy of DIC-2D relative to other approaches, it is noted that *James* [63] has developed a digital image-processing methodology for estimating in-plane displacements from SEM and optical photomicrographs. Typically, the error in each displacement for his work is ±0.50 pixels. However, due to the large magnification factors associated with the photomicrograph (5000×), the dimensional error in displacement can be as low as 0.10 µm. For comparison, using a 2400 line/mm grating in moiré interferometry, *Post* [64] has shown that the displacement sensitivity of the method is of the order of 0.42 µm with accuracy in strain of approximately 50 µstrain. However, *McKelvie* [65] noted that lens aberrations and the loss of frequency information due to the finite lens size might limit the actual displacement sensitivity to about five grating pitches or a displacement error of ±2 µm.

2.1.3 Calibration Process for DIC-2D System

In general, calibration of a DIC-2D system would require a process that is somewhat similar to the procedure outlined in Sect. 2.2.2 and 2.2.3 for three-dimensional vision. One major difference is that, for two-dimensional calibration, *in-plane translations and/or rotations* are performed for the planar object to determine (a) the aspect ratio λ (ratio of the row and column dimensions for each sensor in the camera), which is constant for a specific camera-digitizing board interface, (b) the magnification factor for one direction in pixels/mm, (c) the location of the lens center, and (d) the distortion coefficient for the lens. Fortunately, if high quality lenses are used in the imaging process, the effects of image distortion can be neglected so that two-dimensional calibration requires the determination of only one magnification factor and the aspect ratio.

One method for determining the aspect ratio for a camera-digitizing board combination is to perform a series of in-plane rotation tests for a planar object having a high contrast speckle pattern. Using a least squares procedure, the location of the center of rotation, the unknown rotation angles and λ are determined. Accuracy of $\pm 5 \times 10^{-4}$ for λ has been obtained using this method. Another method, typically employed for DIC-3D studies where a precision-ground grid is used as part of the calibration process, is described in Sect. 2.2.3.

The magnification factors, relating the size of the region being imaged to the pixel dimensions of the sensor plane, are generally determined by one of two methods. A highly accurate method employs a series of translations for a planar surface having a high contrast speckle pattern. By performing image correlation to determine the translation in pixels, the magnification factors for both the row and column directions are obtained. Another method uses a measurement standard attached to the surface of a planar object. By performing edge detection to locate the pixel position of two marks having a known spacing, the magnification factor can be determined with an accuracy of 0.2% in most cases.

2.2 Three-Dimensional Video Image Correlation

2.2.1 Basic Concepts

Single camera DIC systems are limited to planar specimens that experience little or no out-of-plane motion. This limitation can be overcome by the use of a second camera observing the surface from a different direction. Three-dimensional Digital Image Correlation (DIC-3D) is based on a simple binocular vision model. In principle, the binocular vision model is similar to human depth perception. By comparing the locations of corresponding subsets in images of an object's surface taken by the two cameras, information about the shape of the object can be obtained. In addition, by comparing the changes between an initial set of images and a set taken after load is applied, full-field, three-dimensional displacement can be measured. Both the initial shape measurement and the displacement measurement require accurate information about the placement and operating characteristics of the cameras being used. To obtain this information, a camera calibration process must be developed and used to accurately determine the model parameters. In the following paragraphs, key aspects of the DIC-3D method are outlined.

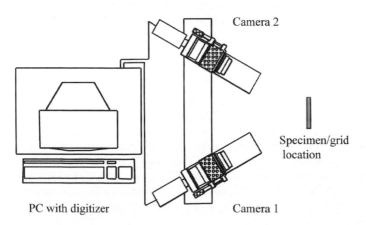

Fig. 5. Schematic of a complete three-dimensional measurement system

A schematic of a typical DIC-3D system is shown in Fig. 5. A photograph indicating typical orientations for the cameras and specimen in a DIC-3D system is shown in Fig. 6. In this work, the camera and lens system are modeled as a pin-hole device. To increase the accuracy of the pin-hole model, a correction term for Seidel lens distortion [55,56,57,58] is included. The imaging characteristics of a camera modeled in this manner can be described by

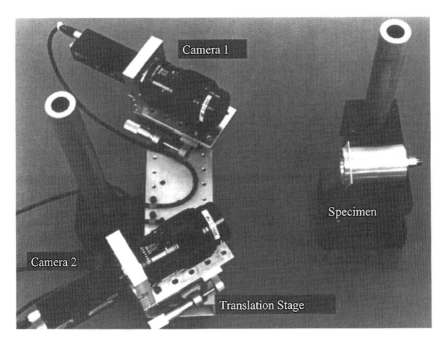

Fig. 6. Photograph of typical camera components and specimen in a three-dimensional measurement system

five parameters. The parameters are (a) the pinhole distance (phd), (b) the two-dimensional location of the intersection of the optical axis and the sensor plane, denoted as the center of the image (C_x, C_y), (c) a lens distortion factor, κ, and (d) the aspect ratio, λ. The parameters $C_x, C_y, \text{phd}, \kappa$ and λ describe the internal properties of a camera and are *intrinsic* parameters. In addition, the parameters $\alpha, \beta, \gamma, X_0, Y_0, Z_0$ describe the overall orientation and position of a camera and are *extrinsic* parameters. Thus, a total of eleven parameters are required to describe the imaging process and global orientation of each camera.

Figure 7 shows the seven coordinate systems that are used to describe the location of points both in space and on images. As shown in Fig. 7, the system-wide coordinate system $(X_{\text{sys}}, Y_{\text{sys}}, Z_{\text{sys}})$ serves as a bridge between the two cameras. Coordinates $(X_{\text{cam1}}, Y_{\text{cam1}}, Z_{\text{cam1}})$ and $(X_{\text{cam2}}, Y_{\text{cam2}}, Z_{\text{cam2}})$ describe spatial locations in the camera coordinate system for Camera 1 (cam1 system) and Camera 2 (cam2 system), respectively, with the origin of each camera system located at each camera's pinhole. The four remaining coordinate systems are used to transform positions from pixel coordinates in each camera into physical dimensions in the cam1 and cam2 systems. It should be noted that the calibration process (a) determines the position and orientation of the system-wide coordinate system $(X_{\text{sys}}, Y_{\text{sys}}, Z_{\text{sys}})$ on a calibrated grid, which also serves as the global coordinate system and (b) determines

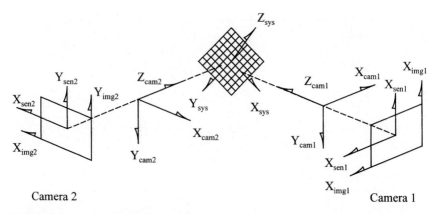

Fig. 7. Schematic of coordinate systems employed in three-dimensional measurements

the location and orientation of each camera relative to the system coordinate system.

The relationship between the camera coordinates and the system coordinates can be obtained by initially aligning the camera and system coordinate systems. Then, the rotational matrix relating these systems is obtained by consecutively rotating the system coordinate system by (a) α about Z_{cam}, (b) γ about Y_{cam} and (c) β about X_{cam}. This yields the following rotational transformation that will convert system coordinates into camera coordinates

$$[R] = \begin{bmatrix} \cos\alpha\cos\gamma & -\sin\alpha\cos\gamma & \sin\gamma \\ \sin\alpha\cos\beta + \cos\alpha\sin\gamma\sin\beta & \cos\alpha\cos\beta - \sin\alpha\sin\gamma\sin\beta & -\cos\gamma\sin\beta \\ \sin\alpha\sin\beta - \cos\alpha\sin\gamma\cos\beta & \cos\alpha\sin\beta + \sin\alpha\sin\gamma\cos\beta & -\cos\gamma\cos\beta \end{bmatrix} \quad (4)$$

Defining the vector $[T]$ to be the translation (X_0, Y_0, Z_0) of the origin of the system coordinates with respect to the rotated camera coordinates, one has

$$[T] = \begin{bmatrix} X_0 \\ Y_0 \\ Z_0 \end{bmatrix}. \quad (5)$$

The transformation from system to camera coordinates can be written as

$$\begin{bmatrix} X_{cam} \\ Y_{cam} \\ Z_{cam} \end{bmatrix} = [R] \begin{bmatrix} X_{sys} \\ Y_{sys} \\ Z_{sys} \end{bmatrix} + [T]. \quad (6)$$

The projection from camera coordinates to sensor coordinates can be derived from similar triangles as

$$\frac{X_{cam}}{X_{sen}} = \frac{Y_{cam}}{Y_{sen}} = \frac{Z_{cam}}{phd}, \quad (7)$$

which can be rearranged to give the following projection equations:

$$X_{sen} = \frac{phd X_{cam}}{Z_{cam}},$$

$$Y_{sen} = \frac{phd Y_{cam}}{Z_{cam}}. \tag{8}$$

Since (8) demonstrates that the projection of a known sensor coordinate into a three-dimensional position results in a one-to-many transformation, the mapping of sensor-plane positions to a unique three-dimensional location requires additional information. Thus, two cameras are most often employed to resolve the ambiguity in the mapping process.

To account for lens distortion, an approximate form for the relationship between distorted and undistorted (X_{sen}, Y_{sen}) locations are

$$\overline{X_{sen}} = \frac{X_{sen}}{1 + \kappa \sqrt{X_{sen}^2 + Y_{sen}^2}},$$

$$\overline{Y_{sen}} = \frac{Y_{sen}}{1 + \kappa \sqrt{X_{sen}^2 + Y_{sen}^2}}, \tag{9}$$

and an approximate inverse of (9) can be written

$$X_{sen} = \frac{2\overline{X_{sen}}}{1 + \sqrt{1 - 4\kappa(\overline{X_{sen}}^2 + \overline{Y_{sen}}^2)}},$$

$$Y_{sen} = \frac{2\overline{Y_{sen}}}{1 + \sqrt{1 - 4\kappa(\overline{X_{sen}}^2 + \overline{Y_{sen}}^2)}}, \tag{10}$$

to convert from distorted to undistorted sensor plane locations.

To access the image data stored in the computer, the sensor coordinates must be related to image coordinates. This is accomplished with a simple transformation of the sensor coordinates. This transformation requires the location of the image center (C_x, C_y) and the aspect ratio (λ) of the camera/digitizer system. The transformation from the sensor plane to the image coordinates is written

$$X_{img} = \overline{X_{sen}} + C_x,$$

$$Y_{img} = \lambda \overline{Y_{sen}} + C_y. \tag{11}$$

Equations (4 - 11) relates coordinates for an object point $(X_{sys}, Y_{sys}, Z_{sys})$ to its two-dimensional position in the camera sensor plane. The DIC-3D method employs two cameras to determine three-dimensional positions of a point through use of the two-dimensional images of the same point from both cameras. It must be emphasized that accurate three-dimensional measurements imply that (4 – 11) accurately model the imaging process. Thus, the calibration process used to determine the eleven parameters ($\alpha, \beta, \gamma, X_0, Y_0, Z_0, C_x, C_y$, phd, κ and λ) in (4 – 11) for each camera in the DIC-3D system must be accurate and reliable so that three-dimensional surface measurements can be obtained with optimal accuracy.

2.2.2 Calibration Process for DIC-3D System

As noted previously, λ is solely a function of the camera-digitizing board interface and its value can be obtained independently. Furthermore, if the camera-digitizing board combination is the same for both cameras, the value obtained for λ from one camera can be used for both cameras.

After determining λ, each camera is calibrated separately to obtain the remaining ten parameters using a precision grid common to both cameras. Since both cameras are calibrated using simultaneously recorded images of the same grid (i.e. both cameras have the same system coordinate system), the positions of the cameras relative to each other are known after the calibration. As part of the calibration process for each camera, the relative position of the grid with respect to each camera must be altered at least one time. One method to achieve this goal is to translate each camera perpendicular to the sensor plane by a known distance. It is noted that several translations can be used in the calibration process, but only one translation along the direction perpendicular to the sensor plane is required.

2.2.3 Determination of Aspect Ratio, λ

To obtain an accurate value for λ, one can use the procedure outlined briefly in Sect. 2.1.3. However, in most of our DIC-3D tests, the following procedure was used. First, a precision ground grid was mounted (a) in the camera's field of view and (b) approximately perpendicular to the sensor plane. Secondly, at least twenty images are acquired. Thirdly, for each image, the sensor-plane locations for all of the grid's intersection points are determined using image thresholding and least-squares curve fitting. Fourthly, using several 5×5 square sub-group of the intersection points within each image, several average values for the ratio ($\Delta Y_{\text{img}}/\Delta X_{\text{img}}$) of alternate diagonal subset vertices in the 5×5 sub-group are obtained for each image. By averaging the N ratio values obtained from all twenty grid images, one obtains $\lambda \cong [1/N \Sigma_{i=1,N}(\Delta Y_{\text{img}}/\Delta X_{\text{img}_i})]$. As noted earlier, this experiment must be performed only once for a given computer-camera-digitizer combination. For a Sony XC-77 camera-Data Translation digitizing board combination, the value for λ obtained by performing these experiments is $\lambda \approx 0.9620$.

Though this value for λ can be used to obtain accurate three-dimensional displacement data, in many applications it is important to achieve optimal accuracy. One method used previously [58] to improve the accuracy in λ is as follows. First, after determining λ and calibrating the cameras (see next section), a DIC-3D system is constructed. Secondly, a precision-ground flat plane with random pattern is placed in the field of view and translated by a known amount perpendicular to the plane. Thirdly, images are acquired and analyzed to obtain data related to the object motion. An improved value for λ is obtained that minimizes the difference between the measured planar position and the known translation over the field of view. Using this method, an improved estimate for $\lambda \approx 0.9617$ was obtained.

2.2.4 Determination of Camera Parameters

After estimating λ, the remaining ten parameters for each camera must be determined to complete the calibration process. After calibration cameras must remain in the same position relative to each other. As shown in Fig. 7, the calibration grid is positioned near the specimen location. The calibration grid establishes the position and orientation of the system coordinate system, with the initial plane of the grid defined to have $Z_{sys} = 0$. The camera system is positioned so that images of the grid can be acquired. Lighting on the grid is adjusted to produce roughly even illumination of the background, with the grid lines clearly visible. Each camera is focused on the grid using a large F-number to maximize the depth of field.

As with the determination of λ, the camera calibration process is based on a series of images of the calibration grid. The camera calibration process is as follows. First, images are acquired simultaneously by both cameras in the initial position. Secondly, both cameras are translated perpendicular to the sensor plane by ΔZ_1 and a new set of images of the grid is acquired. Thirdly, translations are repeated for several ΔZ_J and images recorded by each camera. Fourthly, intersection points for the grid are obtained using the procedure described in Sect. 2.2.3. Since the intersection points provide a series of known positions $(X_{sys}, Y_{sys}, 0)_J$, these locations form the foundation for the calibration process. Fifthly, the sensor position for all of the known grid positions is determined using a combination of (4,5,6,7)

$$Y_{sen} = \text{phd} \times \frac{(\sin\alpha\cos\beta + \cos\alpha\sin\gamma\sin\beta)X_{sys} + (\cos\alpha\cos\beta - \sin\alpha\sin\gamma\sin\beta)Y_{sys} + Y_0}{(\sin\alpha\sin\beta - \cos\alpha\sin\gamma\cos\beta)X_{sys} + (\cos\alpha\sin\beta + \sin\alpha\sin\gamma\cos\beta)Y_{sys} + Z_0 + \Delta Z_{cam}} \quad (12)$$

$$X_{sen} = \text{phd} \times \frac{\cos\alpha\cos\gamma X_{sys} - \sin\alpha\cos\gamma Y_{sys} + X_0}{(\sin\alpha\sin\beta - \cos\alpha\sin\gamma\cos\beta)X_{sys} + (\cos\alpha\sin\beta + \sin\alpha\sin\gamma\cos\beta)Y_{sys} + Z_0 + \Delta Z_{cam}} \quad (13)$$

Sixth, the $(X_{sen}, Y_{sen})_J$ positions are corrected for lens distortion and converted to image coordinates using (8,9,10,11) and the locations of the intersections in the image $(X_{int}, Y_{int})_J$ are acquired. This allows a direct comparison of the projected intersection points to the imaged positions for each point. Seventh, the error function is written:

$$\text{Error} = \sum_{J=1}^{n}[(x_{\text{img}_J} - x_{\text{int}_J})^2 + (y_{\text{img}_J} - y_{\text{int}_J})^2] \qquad (14)$$

and (14) is minimized for each camera in the DIC-3D system using a Levenberg-Marquardt non-linear optimization method [58,59] to establish the ten parameters for each camera.

2.2.5 Profile and 3-D Displacement Measurements

Profile measurements for an object typically use a random pattern, either projected onto or bonded to the surface. In either case, after calibration is complete, the object to be profiled is placed in the field of view and images are acquired simultaneously by both of the calibrated cameras. The process of determining the three-dimensional profile for an object is shown in Fig. 8. Since curved surfaces can be approximated by a finite number of planar patches, the image intensity values obtained from Camera 1 are projected onto a virtual plane. The plane is described in Camera 1 coordinates by two direction angles θ, ϕ and the variable distance Z_p denoting the location of the intersection of the plane and the optic axis for Camera 1. An equation for the virtual plane can be written in the form:

$$Z_p\sqrt{1 - \cos^2\phi - \cos^2\theta} = X_{\text{cam1}}\cos\theta + Y_{\text{cam1}}\cos\phi$$
$$+ Z_{\text{cam1}}\sqrt{1 - \cos^2\phi - \cos^2\theta} \qquad (15)$$

where (ξ, ϕ, θ) = angles orienting the normal to a candidate plane, $\xi = \sqrt{1 - \cos^2\phi - \cos^2\theta}$, Z_p = location on candidate plane of the point $(0, 0, Z_p)$ where the optic axis intersects the candidate plane.

To obtain the location of the virtual plane which corresponds to the location of the planar patch on the true object, the image coordinates $(X_{\text{img2}}, Y_{\text{img2}})$ for each pixel in the selected subset are converted to sensor coordinates and corrected for lens distortion using (6,7,8,9,10,11). After calculating $(X_{\text{sen1}}, Y_{\text{sen1}}, \text{phd}_1)$, the coordinates $(Y_{\text{cam1}}, Z_{\text{cam1}})$ in the Camera 1 system can be obtained using (7,8) to give

$$Y_{\text{cam1}} = X_{\text{cam1}}\frac{Y_{\text{sen1}}}{X_{\text{sen1}}},$$

$$Z_{\text{cam1}} = \frac{X_{\text{cam1}}\text{phd}_1}{X_{\text{sen1}}}. \qquad (16)$$

Combining (14) and (15), expressions for $X_{\text{cam1}}, Y_{\text{cam1}}$ and Z_{cam1} are obtained in terms of known parameters $X_{\text{sen1}}, Y_{\text{sen1}}, \text{phd}_1$ and the candidate plane's parameters, θ, ϕ and Z_p:

$$X_{\text{cam1}} = \frac{X_{\text{sen1}}Z_p\sqrt{1 - \cos^2\phi - \cos^2\theta}}{\text{phd}_1\sqrt{1 - \cos^2\phi - \cos^2\theta} + Y_{\text{sen1}}\cos\phi + X_{\text{sen1}}\cos\theta},$$

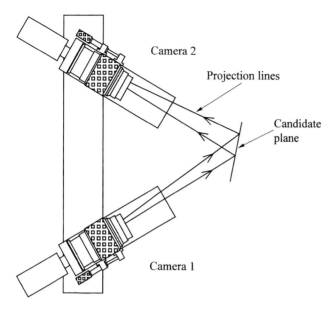

Fig. 8. Schematic representing the projection and back-projection processes used in VIC-3D to determine the orientation and three-dimensional position for subsets

$$Y_{\text{cam1}} = \frac{Y_{\text{sen1}} Z_p \sqrt{1 - \cos^2 \phi - \cos^2 \theta}}{\text{phd}_1 \sqrt{1 - \cos^2 \phi - \cos^2 \theta} + Y_{\text{sen1}} \cos \phi + X_{\text{sen1}} \cos \theta},$$

$$Z_{\text{cam1}} = \frac{\text{phd}_1 Z_p \sqrt{1 - \cos^2 \phi - \cos^2 \theta}}{\text{phd}_1 \sqrt{1 - \cos^2 \phi - \cos^2 \theta} + Y_{\text{sen1}} \cos \phi + X_{\text{sen1}} \cos \theta}. \quad (17)$$

The three-dimensional projected position in the cam1 system of the pixel that has position $(X_{\text{sen1}}, Y_{\text{sen1}})$ is given by (17). Fourthly, the position $(X_{\text{cam1}}, Y_{\text{cam1}}, Z_{\text{cam1}})$ is converted into image coordinates in the cam2 system, $(X_{\text{img2}}, Y_{\text{img2}})$, using (4 – 11) and the known relationship between the cam1 and cam2 coordinate systems. This process, denoted as back-projection, includes a correction for Camera 2 lens distortion. It is worth noting that the process of back-projecting the points to the Camera 2 sensor plane from the candidate plane requires only the position $(X_{\text{cam1}}, Y_{\text{cam1}}, Z_{\text{cam1}})$ and parameters determined by calibration of the two cameras. Finally, the optimal position of the candidate plane is obtained by varying Z_p, θ and ϕ and minimizing a cross-correlation error function, similar to the one used for 2D-correlation, for the subsets of intensity values in both Camera 1 and Camera 2. Initial estimates for Z_p, θ and ϕ, the candidate plane's parameters, are obtained through a graphical user interface. The optimization process uses sub-pixel interpolation to improve the accuracy of the estimated gray levels in Camera 2. A measure of cross-correlation error [58] is employed for the optimization process to reduce the effects of lighting variations between cameras. To obtain

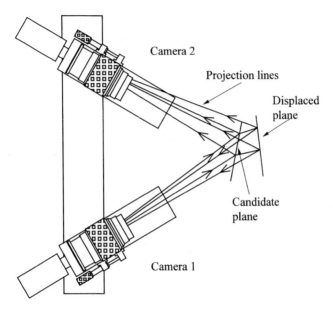

Fig. 9. Projection and back-projection used to determine three-dimensional displacements for subsets

full-field data, initial estimates for the position and orientation of adjoining subsets are the optimal locations obtained for the previous subset.

The process used to obtain three-dimensional surface displacement measurements is shown graphically in Fig. 9. In Fig. 9, the candidate plane is allowed to undergo translations and rotations. Therefore, the points established by the projection of any subset from Camera 1 onto the candidate plane will translate and rotate with the candidate plane. These displaced points are then back-projected into Camera 1 and Camera 2. The gray levels for the projections are compared with the recorded gray levels from images taken by both Camera 1 and Camera 2 after loading the specimen. A cross-correlation error function similar to the one used for profiling is employed to determine the position of the surface point before deformation, as well as translations and rotations at the point.

The error function is similar to those described in Sect. 2.1.2 for two-dimensional correlation. However, for three dimensions, the error function is modified to include intensity data from both cammeras. The data includes: (a) a subset from the initial image in Camera 1, (b) a subset from the initial image in Camera 2 obtained by projection of the subset from Camera 1, (c) a subset from Camera 1 at some later time, t_2, and (d) a subset from Camera 2 at the same time t_2 obtained by projection of the subset from Camera 1. By minimizing the error function for all four images simultaneously, the best-fit candidate plane and its displacements are established. This process includes perspective distortion directly through the pin-hole camera model used for projection. This is a substantial improvement over previous approaches [55,56,57], which did not include subset distortion due to perspective distortion. It should be noted that the DIC-3D system can handle relatively large in-plane and out-of-plane rotations.

General error estimates for DIC-3D systems are difficult to determine since there are several model parameters that have an effect on the accuracy of the measurements. However, estimates can be provided for a specific range of parameter values. Specifically, the error estimates given below are based on the following parameter ranges; (a) pan angle between cameras in the range $30°$ to $60°$, (b) twist and tilt angles in the range $-5°$ to $5°$, (c) distance from camera to object, Z, in the range 0.3 m to 3 m, (d) size of region imaged, D, in the range 10 mm to 0.5 m and (e) $Z/D > 5$ for all experiments.

Results from several experiments performed using model parameters in this range indicate that the error in each component of displacement can be approximated as follows:

(a) Error in in-plane displacement is of the order of $D/10\,000$, where D is the in-plane dimension of the object that is imaged by the camera,
(b) Error in out-of-plane displacement is of the order of $Z/50\,000$, where Z is the approximate distance between the sensor plane and the object.

3 Applications

3.1 Two-Dimensional Video Image Correlation

3.1.1 Measurement of Crack-Tip Opening Displacement During Mixed Mode I/II Crack Growths Using DIC-2D

Background. Aging of the commercial aircraft fleet resulting in the increased potential for multi-site damage in the fuselage skin, has heightened the need for methods to predict the residual strength of aircraft components [66,67]. Since stable crack extension occurs in most aircraft fuselage structures prior to collapse, primary emphasis has been placed on deformation-based fracture criteria. Since the fuselage of an aircraft is subjected to various combinations of loading, flaws in the fuselage experience local conditions that include various levels of Mode I, Mode II and/or Mode III behavior; such a state is referred to as mixed mode. Recent studies have shown that a critical Crack-Tip Opening Angle (CTOA) (or equivalently, Crack-Tip Opening Displacement (CTOD) at a specific distance behind the crack-tip) is a viable parameter for predicting crack growth in thin sheet 2024-T3 aluminum under both Mode I [25,26] and Mode II [28,68] loading conditions.

DIC-2D has been used successfully in measuring the in-plane deformations of thin sheet aerospace materials under mixed-mode I/II loading. Specifically,

CTOD, surface displacement and strain field measurements have all been acquired through the use of DIC-2D in laboratory tests [25,28,68]. The subsequent paragraphs will summarize the work done in this area with a detailed explanation of the application of DIC-2D in obtaining CTOD measurements.

Experimental Approach. The test fixture used in this work was a modified Arcan test fixture [28]. The fixture with specimen bolted into place is shown in Fig. 10. The initial value of a/w was 0.17 throughout the tests, where a is the initial crack length and w is the specimen width. Tests were performed for a variety of loading angles, φ, to obtain various levels of K_{II}/K_I in the crack-tip region, where K_I and K_{II} are the Mode I and Mode II stress intensity factors, respectively. The range of K_I and K_{II} values, as well as the second term in the elastic crack-tip solution, are given in [69]. To assess the effect of grain orientation on critical CTOD, specimens were machined in both LT (initial crack perpendicular to rolling direction) and TL (crack parallel to rolling direction) orientations, as shown in Fig. 11.

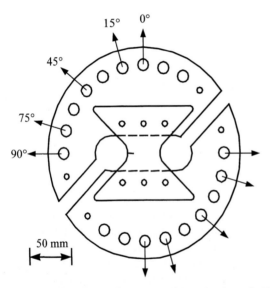

Fig. 10. Arcan test fixture used in mixed-mode I/II tests

During the tests, which involved predominantly in-plane deformations, a single camera system similar to that shown in Fig. 1 was used to record images of the crack-tip region. During the test, in-plane, mixed mode, crack-tip-opening displacement was measured accurately both at initiation and during crack growth using the DIC-2D system, as outlined in [28,68].

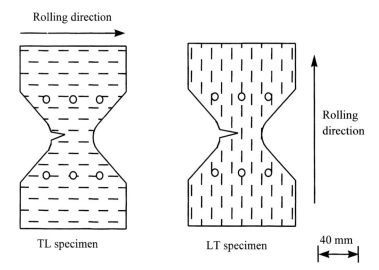

Fig. 11. Arcan test specimens and orientation of crack relative to rolling direction for 2024-T3 aluminum sheet material

Pattern Application. To obtain CTOD at a fixed distance d behind the crack-tip using DIC-2D, the specimen is most-often fatigue pre-cracked to the initial a/w value required for each specimen geometry. Once the initial flaw size is obtained, the specimen is removed from the load-frame and a high contrast, black and white, random pattern is applied to one side of the specimen. Since oversampling is essential for optimum accuracy, each small black region should occupy approximately a 5×5 pixel area. If an area 5 mm $\times$ 5 mm square is imaged by a 512×512 pixel array, each small region should be approximately 49 μm in size.

To achieve a usable random pattern of the required size, a thin coating of white enamel spray paint was applied to the surface (white epoxy spray paint was used if the plastic deformation was large and de-bonding of the enamel paint occurred). To apply a high quality black pattern, two methods have been used successfully: (a) a light spray of black paint, and (b) a light coating of Xerox toner powder. When black paint was used as an over-spray, the white paint was allowed to dry and then a fine mist of black paint was applied. Both fine-tipped attachments for airbrushes and spray cans have been used successfully and the method of choice was refined through trial applications. To apply a Xerox toner powder pattern, the surface was lightly spray painted in white. Then, a mist of toner powder was applied to the wet surface until a pattern of sufficient density was observed. Regardless of the method used, the random pattern was evaluated by using the DIC-2D set-up to record an image of the pattern and compute a histogram of the image; optimal patterns have a relatively broad, flat histogram. If the pattern was not satisfactory, it was removed and the surface repainted.

Experimental Image Acquisition. To obtain CTOD during crack growth all experiments were performed in displacement control. The camera was mounted on a two-dimensional translation stage to track the moving crack-tip region and images of the crack-tip region were acquired during crack growth. To minimize the effects of out-of-plane translation during the test, the distance between the camera system and the specimen was maximized. For a distance of 0.60 m to 1 m from the camera to the specimen, a 200 mm lens with a 2× adapter was used to obtain the required magnification.

Prior to initiating the crack-growth test, images of a ruler were acquired to determine the image magnification factor. Typical magnification factors range from 80 to 140 pixels per mm. An image of the crack-tip region, along with two subsets located on opposite sides of the crack approximately 1 mm behind the current crack-tip, is shown in Fig. 12. Images of the crack-tip region were acquired (a) with a small pre-load on the specimen, (b) for load values approximating the value at which crack growth occurs and (c) after a small amount of stable crack growth occurred, where crack growth was less than approximately 1 mm.

Assuming crack growth occurred from left to right, upon completion of crack growth the camera was translated until the current crack tip was located approximately 150 pixels from the left side of the screen. Loading was then increased and images acquired until stable crack growth occurred; the process described above was repeated until sufficient data was obtained or the crack grew outside the imaged area.

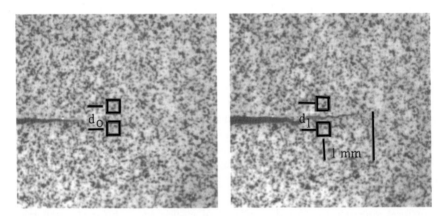

Fig. 12. Location of subsets along crack line for mixed-mode I/II crack-tip opening displacement measurement

Image analysis procedure to obtain CTOD. The measurement of CTOD is shown graphically in Fig. 12, where CTOD = $d_1 - d_0$ at 1mm behind the current crack-tip. To compute the displacements of points located on either side of the crack line, subsets centered about the points of interest were chosen from the current configuration of the fracture specimen. For most CTOD measurements, subsets ranging in size from 10×10 to 30×30 pixels on both sides of the crack line were used.

Results and Conclusions. As shown in Fig. 13a–c, CTOD for thin rolled 2024-T3 aluminum alloy materials under tension-shear loading varies whith grain orientation. Data indicates that critical the CTOD is approximately 20% larger when the crack grows nominally perpendicular to the grain orientation (LT). This observation is consistent with previous experimental results [28,68]. It is worth noting that for shear-type crack growth ($\phi \geq 75°$), the crack surfaces always interfered during crack propagation. Thus, even though the critical CTOD is approximately constant during crack growth, the crack-tip driving force will be reduced for shear-type crack growth, due to frictional losses along the crack line.

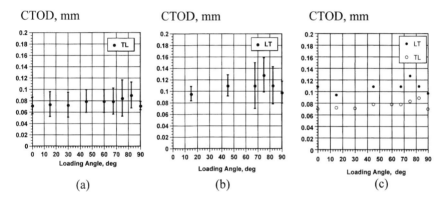

Fig. 13a–c. CTOD results for both TL and LT specimens and all loading angles

Results from mixed-mode I/II loading indicate that (a) the value of critical the CTOD during crack growth approaches a constant after approximately 5 mm of crack extension, (b) the critical CTOD for 2024-T3 is a strong function of the direction of crack growth relative to the rolling direction, with the critical CTOD for cracks growing in nominally the LT direction being 20% larger than for cracks growing nominally in the TL direction.

3.1.2 Local Crack-Tip Closure Measurements Using DIC-2D

Background. The measurement of key quantities in the crack-tip region has been an area of active research for many years. In particular, a simple and effective method for obtaining high accuracy, local measurements of the crack-opening displacements (COD) in regions 0 to 250 µm behind the crack-tip has been lacking.

A wide range of methods has been developed for measuring COD [70] including clip gages, strain gages and interferometric methods. Clip gages are easy to use, but they only provide information at positions that are remote from the crack-tip location. Thus, these measurements provide little information about the current closure conditions at the crack-tip. Strain gages applied to various locations near the crack-tip, including back-face and side-face positions, have been used to characterize closure loads. However, these methods provide data at a fixed position on the specimen, which means that the data cannot be obtained at pre-specified distances behind the growing crack-tip. Interferometric methods include moiré interferometry [71] and the indentation method [70]. Moiré methods are quite accurate and can be used to quantify COD along the crack line. However, the application of a grid to the polished surface remains a drawback of the method. The indentation method has been used successfully to measure COD in a wide range of applications, including elevated temperature crack growth. As with other methods, the position of the indentations is fixed on the specimen, providing data at positions that vary in distance behind the crack-tip.

New Approach to COD Measurement. To simplify the process of obtaining COD at arbitrary distances behind the moving crack-tip, a vision-based methodology has been employed. The set-up consists of a far-field microscope, three-dimensional translation stage, a camera-digitizing board-PC combination for recording and storing images of the region and the DIC2-D software for obtaining the relative displacements of points on either side of the crack line.

Using a far-field microscope lens, the system is capable of imaging areas 0.8 mm ×0.8 mm or smaller. The specimen surface was polished to obtain maximum contrast. By applying a dark, random pattern to the polished surface and illuminating the surface using an on-axis light source, the polished region appears much lighter than the dark pattern, providing a high contrast, random intensity variation for later analysis using the DIC2-D software.

Two methods were developed to obtain a high-contrast intensity pattern of the appropriate density on the polished surface. First, a modification of the method used for applying Xerox toner powder in Mode I/II testing was used. In this case the toner powder was filtered, using 11 µm filter paper and a compressed air source, to obtain only the smaller particles for application to the surface. By applying the filtered powder onto the polished surface and heating it for a few minutes to 100 °C, the toner powder bonds to the surface. A typical pattern obtained this way is shown in Fig. 14. Average size of the random pattern was ≈ 20 µm after heating.

Since the pattern developed this way is suitable for tests in air, it was used to obtain all the results for this work. However, to obtain a smaller,

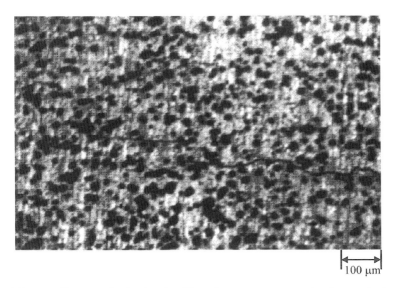

Fig. 14. Pattern obtained using filtered xerox toner powder for crack closure measurements at high magnification

more durable pattern for use in vacuum or corrosive environments, a mask was made using electron-beam lithography. The mask is manufactured only once and then used each time a new experiment is to be performed. A typical pattern obtained using a lithographic mask is shown in Fig. 15. The pattern shown in Fig. 15 was produced using the following process. Firstly, the polished specimen's surface was spin-coated with photoresist. Secondly, a lithographic mask with 8 µm features was placed on the coated surface and an ultraviolet light source was used for illumination. Thirdly, the exposed portions of the photoresist pattern were removed through a development process. Fourthly, a thin layer of tantalum was evaporated onto the specimen surface. Fifthly, the remaining photoresist was removed through a final development process, leaving a tantalum pattern on the polished surface. Since any type of pattern can be obtained by electron-beam lithography, much smaller patterns can be placed on the specimen for use at higher magnification.

System Accuracy. A simple tension test was performed using a dog-bone specimen made of 2024-T3 aluminum. By measuring the relative displacements along the direction of loading for three 91×91 pixel subsets in a 0.5 mm by 0.30 mm area that were separated by 0.16 mm, the value for strain along the line was determined to have a standard deviation of 200 µstrain. Thus, an estimate for the displacement error in the method is approximately 30 nm or 0.05 pixels at the magnification used in this work.

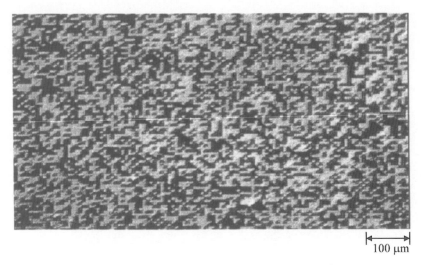

Fig. 15. Typical pattern obtained using a lithographic mask for crack closure measurements at high magnification

Crack Closure Measurements. Figure 16 presents the measurement of crack-opening displacement as a function of cyclic load at distances of 0.075 mm and 0.224 mm behind the current crack-tip. The specimen was machined into an extended compact tension geometry from aluminum alloy 8009, with width of 38.1 mm, thickness of 2.3 mm and $a/w = 0.54$. During the test, $\Delta K = 4.4$ MPa m$^{\frac{1}{2}}$ and the R-ratio was held constant at 0.10.

As shown in the measured COD curves for two locations behind the crack-tip, a clearly defined change in slope was observed, corresponding to the elimination of crack surface contact at each position behind the crack-tip [72,73]. Furthermore, the data shows that the load at which crack surface contact was eliminated at the free surface of the specimen is a function of position behind the crack-tip.

3.1.3 Large Specimen Tests

Background. In recent years, new materials have been developed for use in a wide range of applications: functionally gradient materials for advanced applications in the transportation and defense industries, toughened ceramics for defense applications, and advanced composite materials for aerospace uses. In many cases, the development of new materials is driven by the conflicting demands of simplicity in manufacturability, adequate strength and reduced weight.

Recently, after receiving congressional approval to build the Space Station, NASA has embarked upon a program to reduce the weight of the Space

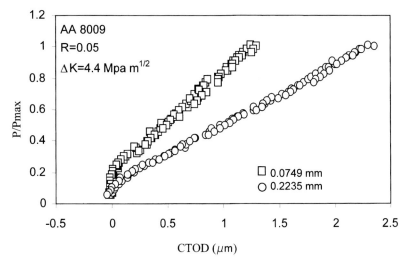

Fig. 16. Measured crack-opening displacement as a function of cyclic load at two locations behind the current crack-tip

Shuttle so that additional Space Station components can be transported into space. In particular, NASA proposed to manufacture the external tanks of the Space Shuttle using AL-2195, an aluminum-lithium alloy, that would reduce weight by 33.4 kN.

To demonstrate that the proposed alloy had the required structural properties, NASA initiated an extensive test program. Results from early biaxial tests indicated that there was a soft spot in the center of the manufactured cruciform specimen. After several metallurgical studies were completed, it was conjectured that the soft spot was due to a combination of residual stresses in the material and the machining techniques used to make the specimen. To understand better how the machining processes affected the local material response, the full-field deformations throughout the center portion of the test section were measured during loading and unloading of the specimen. In the following sections, the results from using a DIC-2D system to measure the full-field deformations in a critical region of an AL-2195 test specimen are presented.

Experimental Set-up. The test specimen is shown in Fig. 17. The cruciform test specimen, which has four grip regions, was machined for use in a biaxial test frame. For this work, the specimen was uniaxially loaded in a 0.448 mN test frame, using one set of grips. The center region of the cruciform specimen was milled to a depth of 3 mm on both sides to obtain a central, square, recessed test section that was 6.4 mm thick.

During the tests, a DIC-2D system similar to the one shown in Fig. 1 was used to record images. Even though the specimen was expected to undergo

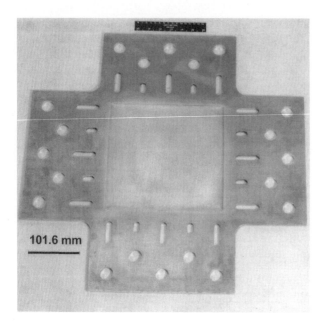

Fig. 17. Cruciform specimen machined from Al-2195

predominantly in-plane deformation, the DIC-2D system was placed 2 m from the test specimen to minimize the effects of any out-of-plane motion on the measured displacements. To increase mobility and adaptability in the test facility, the digital camera was mounted on a heavy, professional Bogen tripod during the test. To stabilize the lens system, a lens support mount was machined and used during the test.

Due to concerns about the center region of the test specimen, the region imaged was reduced to 46 mm by 43 mm; based on strain gage data from previous tests, this region appeared to have large gradients in strain. A Canon 200 mm lens was used for imaging the central region.

Prior to loading the specimen, and after the optical set-up was complete, images of a ruler mounted to the surface of the specimen were acquired. Using these images, the magnification factor was approximately 11.5 pixels per mm. After acquiring ruler images, the central region was lightly spray painted with white and black paint to obtain a high-contrast random pattern.

All experiments were performed under displacement control. The loading process was as follows: (a) the load was increased incrementally to a maximum of 320.3 kN, (b) the load was decreased incrementally until fully unloaded, and (c) the load was increased incrementally to the maximum load. Images were acquired every 4.448 kN throughout the loading-unloading-reloading process.

To obtain the strain field, $\varepsilon_{yy}(x, y)$, along the loading direction, the in-plane displacement field was determined using the DIC-2D algorithm. Data

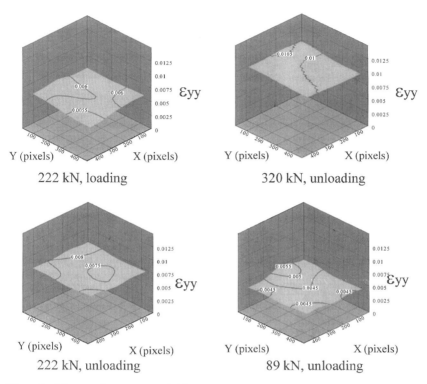

Fig. 18. Measured ε_{yy} strain field in center region of cruciform specimen during uniaxial tensile loading and unloading

was acquired for the central 400×400 pixel region of the image. The displacement field was obtained at 1681 locations using 21×21 pixel subsets (overlapped by 10 pixels) on a 41×41 grid in each image. After computing the 2D displacement field at each point, the vertical displacement field was fitted with a smoothed surface using the procedures described previously [5,7]. The resulting ε_{yy} field at four different loads (two ε_{yy} fields during initial loading and two ε_{yy} fields during unloading) is shown in Fig. 18. Table 2 presents both the average strain value over the field, $\varepsilon_{yy}^{\text{avg}}$, and the standard deviation in the strain for all loading.

Discussion of Results. The data in Fig. 18 for the strain field $\varepsilon_{yy}(x,y)$ indicates that the strain gradients were negligible over the center region. Furthermore, the data clearly showed that the deviations in strain throughout the loading-unloading-reloading processing were small. Thus, the results did not indicate the presence of a soft spot in the material, demonstrating that the material did not exhibit unwanted, localized deformations due to either residual stresses or machining.

After discussions with NASA personnel, it was concluded that the strain gage data from this test, and previous tests, was suspect due to a combination of wire mislabeling and gage malfunctioning. Thus, the DIC-2D data provided a high degree of confidence that AL-2195 could be used for Space Shuttle applications.

3.2 Three-Dimensional Video Image Correlation

3.2.1 Profiling of Structures for Reverse Engineering

Background. Reverse engineering of components has become increasingly important to many government agencies, particularly since the downsizing of federal government has led to the loss of established contractors and associated component documentation. To address this concern, a wide range of methods has been developed for accurately measuring the shape of a part. The techniques include contacting methods (e.g. precision coordinate-measurement machines) as well as non-contacting methods (e.g. optical systems, computer-aided tomographic systems, and sonar-based systems).

The goal of each measurement system is to accurately measure the shape of an object as rapidly as possible. In the following sections, the procedure described in Sect. 2.2.5 is used to accurately measure the shape of an object. To increase the accuracy of each measurement, several partially overlapping regions of the object are measured. To obtain a single representation for the object, advanced data processing algorithms are used to combine the measurements.

Experimental Set-up. The three-dimensional measurement system used in this work is similar to the one shown in Fig. 5 and Fig. 6. The cameras were Pulnix digital cameras with Bitflow Raptor digital data acquisition boards. Canon 100 mm F4 macro lenses were used for imaging the object. A custom camera mount was developed [58] that served as an adapter from the C-mount to the Canon FD mount, as well as the main support for the lens and camera. Both cameras were mounted on Newport translation stages having a least count accuracy of 0.10 μm. The camera-translation-stage combinations were firmly mounted on an extruded aluminum bar, which provided maximum rigidity and reduced relative motion between the two cameras. Calibration was performed using a 10 mm ×10 mm grid divided into 100 squares, with point-to-point accuracy of the grid being ±4 μm. The grid was oriented at ≈ 45° from the row direction during the calibration process, which gives the best accuracy in locating the grid intersection points with the current algorithm. To enhance contrast during calibration the grid was mounted on a translucent plate and illuminated from the rear. The accuracy of the set-up (as indicated by the calibration process) was ±4 μm, which is consistent with the reported accuracy of the calibration grid.

To obtain a random pattern of appropriate size for use in imaging portions of a 12 mm area, Xerox toner powder and white paint were used. The procedure used was the same as outlined in Sect. 3.1.1.

For this work, a twenty-five cent coin was used as the object; feature heights were less than 0.200 mm across the coin. To obtain maximum accuracy in the measurements, several overlapping images of the object were obtained at high magnification. To achieve this goal, the object was moved three times, resulting in four overlapping data sets.

Registration of surface patches. Multiple patches are often required when profiling objects due to (a) partial occlusion of the object surface, (b) limited field of view for the experimental set-up, or (c) requirements for increased resolution for measurements. Any combination of these factors will make it impossible to profile the entire surface using one pair of images. For the object used in this experiment, a twenty-five cent coin, multiple patches were used to improve the accuracy of the final surface profile, which consisted of a set of (X, Y, Z) coordinates. Figure 19 shows the four patches of data obtained from the profiling system used in this experiment. Each patch consists of 5040 data points obtained by correlating overlapping 21×21 pixel subsets from the images. To obtain adjacent patches, the object was translated and rotated so that the new set of images had approximately 20% overlap with the previous set of images. Based on the calibration error estimates, the accuracy of the profile measurements for this set-up is estimated to be $\pm 4\,\mu m$ for each patch.

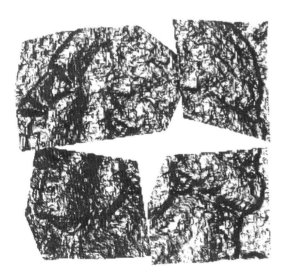

Fig. 19. Measured three-dimensional positions for four overlapping regions of an object

Ideally, the object or camera movement before acquiring a new set of images would be accurately quantified to simplify the registration process. Unfortunately, extremely high accuracy in the object or camera motion would require expensive equipment and an elaborate experimental set-up. In this work, an automatic surface registration algorithm was developed which does not require high accuracy motion data for the object or camera. In fact, the algorithm only requires approximate overlapped positioning of surface patches as initial guesses. Estimated relative position (initial guess) among surface patches can be obtained either from the measurement process (e.g., data from a robot-arm control-system used to move the object or camera system) or by operator interaction. In the latter case, the operator identifies geometric features on the surface patches and drags them across the computer graphic screen to obtain approximate alignment of the patches. Figure 20 shows the positions of the four patches in Fig. 19 after using crude initial guesses for the object motion to begin the registration process.

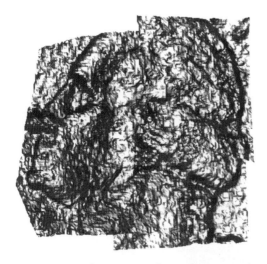

Fig. 20. Initial estimate of translations and rotations for the recombination of regions

After applying the initial guesses, each of the surface patches is represented by a triangularized mesh generated from the measured data sets. The triangularized meshes for the various patches must overlap so that the algorithm can compare geometric shapes in the overlapped area and find the best match. The algorithm is responsible for accurately determining the true rigid body transformation between surface patches so that they can be registered to form a complete model of the object surface.

The registration algorithm developed for this work detects overlapped areas among surface patches and creates pseudo-corresponding point pairs along the surface normal in the overlapped area. Then, an optimization process is performed to minimize the distance between pseudo-corresponding point pairs by varying the rigid body transformation parameters for three of the surface patches. Final positions are achieved by iteratively updating the position of the surface patches and corresponding point pairs. It should be noted that the algorithm also attempts to reduce registration errors that may be produced by the order in which the patches were registered Eventually, the algorithm allows all patches to adjust their final position to minimize gaps due to accumulated registration errors between patches.

After completion of the registration, overlapped surface patches undergo mesh topology operations so that they are merged together to generate a single, consistent polygon representation of the object surface. Figure 21 shows the final object profile, obtained by reconstruction of the surface patches.

Fig. 21. Representation of the final positions of data points after optimal recombination of four overlapping regions

Summary. Three-dimensional image correlation algorithms can be used to obtain surface profile data with high accuracy. By combining the power of advanced data processing techniques with DIC-3D measurement methods, it is now possible to obtain full, 360° surface representations for complex objects. The geometric models obtained from this approach can be used to create CNC code for precision manufacturing of mechanical parts, thereby facilitating the process of reverse engineering.

3.2.2 Measurement of Three-Dimensional Surface Deformations in Wide Panels Using DIC-3D

Background. As the current fleet of commercial, military, and private aircraft continues to age, and with many aircraft exceeding their original design life, techniques are being developed to aid in the evaluation of aircraft structures and materials. As noted in earlier sections, two-dimensional digital image correlation has been used for many years to accurately measure the in-plane displacements of planar structures under a variety of loads. Because aircraft components and typical aerospace laboratory test specimens deform in a complex manner when loaded, surface measurement systems must be able to make accurate measurements in three-dimensional space. For example, for wide, thin panels fabricated from 2024-T3 aluminum and subjected to tensile loading, the presence of a flaw in the structure will cause the panel to buckle out-of-plane during crack growth. For this case, the DIC-3D system described in Sect. 2.2 is required for accurate determination of the full-field, three-dimensional displacements during deformation. In the following sections, results obtained using the DIC-3D system for measurement of 3-D displacements during tensile loading of center-cracked fiber-reinforced composite and 2024-T3 aluminum panels will be presented.

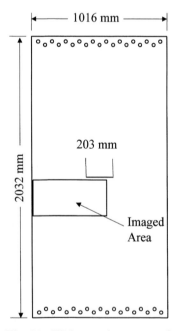

Fig. 22. Wide panel geometry for the 2024-T3 aluminum panel

Test Specimen and Testing Procedure. All tests were performed at NASA Langley Research Center in Hampton, Virginia; the aluminum panels were tension tested in a 1.34 MN tension test frame within the Mechanics of Materials Branch and the fiber-reinforced composite panels were compression tested in a similar test frame in the Structural Mechanics Branch.

Fig. 23. Photograph of wide panel specimen and imaging set-up in 1.34 MN test frame

The wide panel geometry for the 2024-T3 aluminum panel is shown in Fig. 22 and a photograph of a wide panel specimen with high contrast pattern already applied is shown in Fig. 23. The area imaged on the aluminum sheet is approximately 500 mm wide by 260 mm tall and is located just below the center crack, extending from one edge of the sheet past the centerline of the specimen. All tests were performed under displacement control. Images were acquired every 8.9 kN until crack growth occurred. After initial crack growth, images were acquired at intervals of approximately 4 mm of crack growth until final fracture.

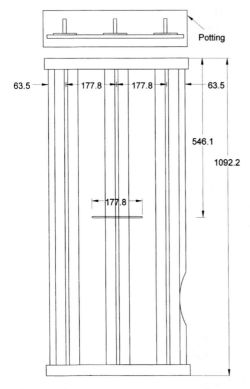

Fig. 24. Geometry for integrally-stiffened composite panel with end potting for distribution of compression loading

Compression testing was performed on a composite reconstruction of a stub box upper wing cover panel. The geometry for the integrally-stiffened composite panel is shown in Fig. 24. The composite panel was 1.09 m long, 0.482 m wide, nominally 12 mm thick and made of Kevlar-stitched, resin-film-infused, graphite-epoxy. Stiffening was provided by integral T-sections on the back of the composite panel. The panel was potted in 38.1 mm thick concrete on each end to improve the uniformity of compressive load distribution. A flaw was introduced into the panel by cutting a 2.38 mm radius notch through the panel and stiffener in the center region. Since strain gages covered most of the skin side of the panel, the pattern was applied around the gages and lead wires, as shown in Fig. 25.

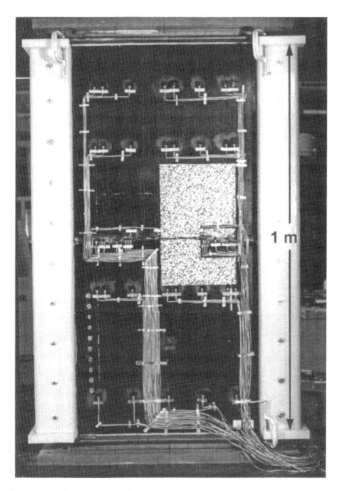

Fig. 25. Photograph of composite panel with applied 3M vinyl pattern, strain gages and knife edges to prevent wrinkling on edges

Image Acquisition System for Large Area Interrogation. Due to the physical size of the region to be imaged for both panels, several modifications were required so that a DIC-3D system could be adapted for this work. First, to improve the accuracy of the measured three-dimensional displacement components, the camera-translation-rotation stage combinations were mounted near opposite ends of a 2 m long, extruded aluminum bar. The bar was then mounted vertically to a 2 m by 0.7 m by 0.7 m modular frame structure. Secondly, smaller focal length 28 mm Nikon lenses were used for imaging of the large areas. Thirdly, due to the inaccessibility of the camera-translation-rotation stage units, computer-controlled translation stages accurate to ±0.10 μm were used to move the cameras for the calibration process. Fourthly, to simplify the process of obtaining a high-contrast random pattern for large areas, a pattern was electrostatically printed onto a 3M adhesive-backed vinyl material. The vinyl material was then temporarily placed in the region of interest for alignment and then firmly pressed onto the surface to release micro-encapsulated epoxy for a more permanent application. Fifthly, a range of lighting methods employed successfully, including halogen flood lights, fluorescent lamps, fiber-optic bundles and incandescent light sources. In each case, the primary goal was to obtain an even distribution of lighting throughout the area of interest.

Calibration of the modified DIC-3D system followed the procedure outlined in Sect. 2.2.2-2.2.4. Due to the size of the experimental set-up, all experimental equipment was placed in its final position and calibrated using a 0.4 m square calibration grid placed against the surface of the test specimen. After calibration, the grid was removed and testing was begun.

All images were analyzed using DIC-3D software which implements the procedure described in Sect. 2.2.5. The software computes the three-dimensional positions and displacements for a field of points for each load level of interest.

Wide Panel Results: Tension of Center-cracked Aluminum Panel. The load-crack extension curve for the 2024-T3 aluminum sheet is shown in Fig. 26. As can be seen in the Fig. 26, crack growth began at 150 kN, continued in a stable manner up through the maximum load of 269 kN until unstable crack growth fractured the panel at a load of 241 kN.

Figure 27 presents contour plots for all three components of displacement for a load of 240.2 kN just before final fracture. Figure 28 presents the out-of-plane displacement for a horizontal line $Y = -25$ mm, below the crack line in Fig. 22. As shown in Fig. 28, the buckling of the panel continued to increase as the crack grew, reaching a maximum of 25 mm just before failure. Furthermore, inspection of Fig. 27 and 28 indicates that the edge of the panel appears to be flat during loading, with the centerline region showing increased buckling. This observation suggests that load transfer into the remaining ligament for a center-cracked sheet will tend to flatten these areas.

Wide Panel Results: Compression of Center-notched Composite Panel. For the composite panel shown in Fig. 25, full-field contours for all three displacement components at the maximum load of 1.34 mN are shown

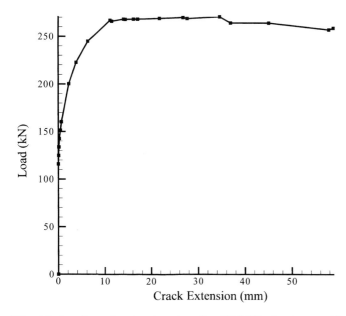

Fig. 26. Load-crack extension data for 2024-T3 aluminum wide panel test

in Fig. 29, 30, and 31, respectively. The maximum measured displacement in the U, V and W fields were $U_{max} = 0.701$ mm, $V_{max} = -3.56$ mm and $W_{max} = 2.77$ mm, respectively.

The DIC-3D data for the out-of-plane displacement was compared to DCDT measurements from a single point on the panel. The DCDT was located on the web of the central stiffener on the back of the panel, approximately 46 mm above the notch centerline. The DIC-3D measurements were obtained at the same vertical and horizontal position on the opposite side of the panel. Figure 32 presents a comparison of the results. It can be seen from this comparison that the data agree quite well. The initial movement of the specimen can be seen in the data, as well as the linear variation in the out-of-plane displacement after initial straightening occurred. The difference between the image correlation data and the DCDT data near the maximum displacement position can be explained by the fact that the positions monitored by the two methods are not exactly the same. The DCDT tracks a point in space, while the image correlation technique tracks a point on the specimen surface. Thus, as the panel is compressed to a greater degree, these two points become farther apart, resulting in a small difference in the measurements.

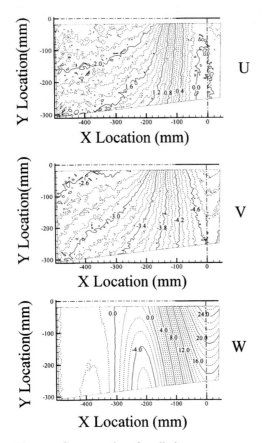

Fig. 27. Contour plots for all three components of displacement for tensile loading of 2024-T3 wide panel just prior to final fracture

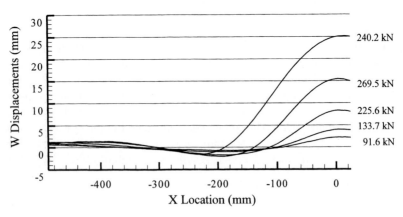

Fig. 28. Out-of-plane displacement as a function of loading along a horizontal line $Y = -25$ mm below the crack line in 2024-T3 wide panel test

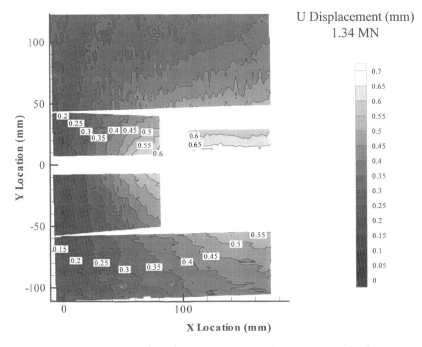

Fig. 29. U-Displacement field for composite panel at maximum load

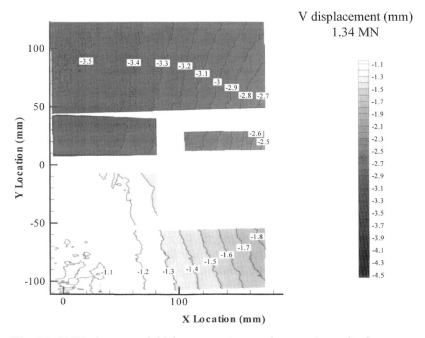

Fig. 30. V-Displacement field for composite panel at maximum load

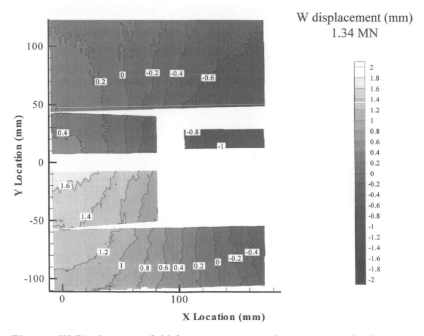

Fig. 31. W-Displacement field for composite panel at maximum load

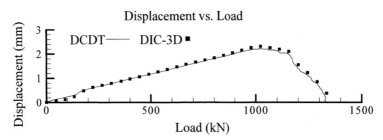

Fig. 32. Comparison of DCDT and W-Displacement measurements at one position for all loads

4 Discussion

With regard to the DIC-2D and DIC-3D methods, a synopsis of the advantages and disadvantages are provided in the following paragraphs. First, the advantages of the DIC-2D method are (a) simplicity of system calibration, (b) ease in applying random pattern (vinyl pre-manufactured pattern, light spray paint) for displacement and strain measurements, (c) user-friendly software and imaging system for real-world applications, (d) well-defined displacement error estimates for current DIC-2D systems (± 0.02 pixels), and (e) ability to use the system for both large and small areas. The primary disadvantages of any two-dimensional imaging and measurement system are (a) the out-of-plane displacements relative to the camera must be minimized to ensure that the measured displacements are accurate and (b) the object should "approximate" a planar surface both before and after deformation.

With regard to the DIC-3D method, the advantages are (a) ability to make measurements on both curved and planar surfaces, (b) ability to measure both surface profiles and full-field, three-dimensional surface displacements, (c) availability of user-friendly software to simplify the process of acquiring and analyzing images, (d) potential of method in field applications for both large and small structural measurements, and (e) fully digital recording platform, which provides potential for real-time measurements as computer technology advances. Disadvantages of the current method are (a) expense of the equipment required for a fully-integrated system and (b) need for precision calibration grids (or precision, long-travel, three-dimensional translation stages) to calibrate the DIC-3D system over a range that spans the size of the object to be studied.

With regard to commercial level digital image correlation software, several programs have appeared on the market to meet the need of those who wish to use image correlation without having to develop their own software. The availability of such software eliminates the time-consuming task of developing and validating user-friendly algorithms that can acquire and analyze images to obtain displacements and strains for components under load.

5 Summary

Computer vision methods have been successfully developed and applied to the solution of a wide range of scientific and engineering problems, including diverse areas such as fracture mechanics, bio-mechanics, constitutive property measurements in complex materials, model verification for large flawed structures and non-destructive evaluation. Two-dimensional computer vision has been applied successfully to study the deformations of planar regions ranging in size from 0.5 mm to 1 m, in environments that include water, high-temperature air and standard laboratory conditions. Three-dimensional computer vision was used to accurately measure the deformations of curved and planar structures ranging in size from 4 mm to 1.3 m in both field and laboratory conditions.

Applications currently being pursued include (a) use of specially designed cameras for simultaneous measurement of surface deformations and temperature, (b) development of new methods for calibration of multiple camera systems to simplify the process of measuring accurate surface deformations on very large structures, and (c) adaptation of the two-dimensional method for use in measuring surface profiles from multiple scanning electron microscope images.

From our perspective, the last sixteen years have shown conclusively that both the DIC2-D and DIC3-D systems are robust, accurate and relatively simple to use in most applications. Thus, the range of applications for which the methods can be applied successfully seems to be growing rapidly as scientists and engineers use the method and begin to understand the true capabilities of the DIC2-D and DIC3-D systems.

Acknowledgements

The authors wish to thank Mr Michael Boone, Mr Hubert Schreier, Mr Glen Hanna, Dr David S. Dawicke, Dr Wei Zhao and Dr Audrey Zink for their assistance in completing this manuscript. In addition, the support of Dr Charles E. Harris and Dr James C. Newman, Jr in the Mechanics and Materials Branch at NASA Langley Research Center is deeply appreciated. Through their unwavering technical and financial assistance, the potential of the DIC-3D method is now being realized. Finally, the support of Dr Pramod Rastogi throughout the development process is gratefully acknowledged.

References

1. W. H. Peters, W. F. Ranson: Digital Imaging Techniques in Experimental Stress Analysis, Opt. Eng. **21**, 427 (1981)
2. M. A. Sutton, W. J. Wolters, W. H. Peters, W. F. Ranson, S. R. McNeill: Determination of Displacements Using an Improved Digital Correlation Method Image Vis. Comput. **1**, 133 (1983)
3. M. A. Sutton, S. R. McNeill, J. Jang, M. Babai: The Effects of Subpixel Image Restoration on Digital Correlation Error Estimates, Opt. Eng. **27**, 173 (1988)
4. M. A. Sutton, M. Cheng, S. R. McNeill, Y. J. Chao, W. H. Peters: Application of an Optimized Digital Correlation Method to Planar Deformation Analysis, Image Vis. Comput. **4**, 143 (1988)
5. M. A. Sutton, H. A. Bruck, S. R. McNeill: Determination of Deformations Using Digital Correlation with the Newton-Raphson Method for Partial Differential Corrections, Exp. Mech. **29**, 261 (1989)
6. M. A. Sutton, H. A. Bruck, T. L. Chae, J. L. Turner: Development of a Computer Vision Methodology for the Analysis of Surface Deformations in Magnified Images, ASTM STP-1094 on MICON-90, Advances in Video Technology for Microstructural Evaluation of Materials, 109 (1990)
7. M. A. Sutton, J. L. Turner, T. L. Chae, H. A. Bruck: Full Field Representation of Discretely Sampled Surface Deformation for Displacement and Strain Analysis, Exp. Mech. **31**, 168 (1991)
8. T. C. Chu, W. F. Ranson, M. A. Sutton, W. H. Peters: Applications of Digital Image Correlation Techniques to Experimental Mechanics, Exp. Mech. **25**, 232 (1895)
9. W. H. Peters, He Zheng-Hui, M. A. Sutton, W. F. Ranson: Two-Dimensional Fluid Velocity Measurements by Use of Digital Speckle Correlation Techniques, Exp. Mech. **24**, 117 (1984)

10. S. R. McNeill, W. H. Peters, M. A. Sutton, W. F. Ranson: A Least Square Estimation of Stress Intensity Factor from Video-Computer Displacement Data, Proc. 12th Southeastern Conference of Theoretical and Applied Mechanics, (1984) p. 188
11. J. Anderson, W. H. Peters, M. A. Sutton, W. F. Ranson, T. C. Chu: Application of Digital Correlation Methods to Rigid Body Mechanics, Opt. Eng. **22**, 238 (1984)
12. W. H. Peters, W. F. Ranson, J. F. Kalthoff, S. R. Winkler: A Study of Dynamic Near-Crack-Tip Fracture Parameters by Digital Image Analysis, J. Phys. IV Paris, Colloq **5**, suppl. n°8, Tome 46, 631 (1985)
13. W. Wu, W. H. Peters, M. Hammer: Basic Mechanical Properties of Retina in Simple Tension, Trans. ASME, J. Biomech. Eng. **109**, 1 (1987)
14. S. R. McNeill, W. H. Peters, M. A. Sutton: Estimation of Stress Intensity Factor by Digital Image Correlation, Eng. Fract. Mech. **28**, 101 (1987)
15. B. R. Durig, W. H. Peters, M. A. Hammer: Digital Image Correlation Measurements of Strain in Bovine Retina, in: *Optical Testing and Metrology*, Proc. SPIE **954**, 438 (1988)
16. Y. J. Chao, M. A. Sutton: Measurement of Strains in a Paper Tensile Specimen Using Computer Vision and Digital Image Correlation, Part 1: Data Acquisition and Image Analysis System and Part 2: Tensile Specimen Test, Tappi J. **70** No.3, 173 and **70** No.4 153 (1988)
17. S. R. Russell, M. A. Sutton: Image Correlation Quantitative NDE of Impact and Fabrication Damage in a Glass Fiber Reinforced Composite System, J. Mater. Eval. **47**, 550 (1989)
18. S. R. Russell, M. A. Sutton, W. H. Peters: Strain Field Analysis Acquired through Correlation of X-ray Radiographs of Fiber-Reinforced Composite Laminates, Exp. Mech. **29**, 237 (1989)
19. W. H. Peters, W. M. Poplin, D. M. Walker, M. A. Sutton, W. F. Ranson: Whole Field Experimental Displacement Analysis of Composite Cylinders, Exp. Mech. **29**, 58 (1989)
20. B. Durig, H. Zhang, S. R. McNeill, Y. J. Chao, W. H. Peters: A Study of Mixed Mode Fracture by Photoelasticity and Digital Image Analysis, J. Opt. Lasers Engin. **14**, 203 (1991)
21. C. Hurschler, J. T. Turner, Y. J. Chao, W. H. Peters: Thermal Shock of an Edge Cracked Plate: Experimental Determination of the Stress Intensity Factor by Digital Image Correlation, in: Proc. SEM Spring Conference on Experimental Mechanics, (1990) p. 374
22. M. A. Sutton, H. A. Bruck, T. L. Chae, J. L. Turner: Experimental Investigations of Three-Dimensional Effects Near a Crack-Tip Using Computer Vision, Int. J. Fract. **53**, 201 (1991)
23. G. Han, M. A. Sutton, Y. J. Chao: A Study of Stable Crack Growth in Thin SEC Specimens of 304 Stainless Steel, Eng. Fract. Mech. **52**, 525 (1995)
24. G. Han, M. A. Sutton, Y. J. Chao: A Study of Stationary Crack-Tip Deformation Fields in Thin Sheets by Computer Vision, Exp. Mech. **34**, 357 (1994)
25. D. S. Dawicke, M. A. Sutton: CTOA and Crack Tunneling Measurements in Thin Sheet 2024-T3 Aluminum Alloy, Exp. Mech. **34**, 357 (1994)
26. D. S. Dawicke, M. A. Sutton, J. C. Newman, C. A. Bigelow: Measurement and Analysis of Critical CTOA for an Aluminum Alloy Sheet, ASTM STP 1220 on Fracture Mechanics **25**, 358 (1995)

27. S. R. McNeill, M. A. Sutton, Z. Miao, J. Ma: Measurement of Surface Profile Using Digital Image Correlation, Exp. Mech. **37**, 13 (1997)
28. B. E. Amstutz, M. A. Sutton, D. S. Dawicke: Experimental Study of Mixed Mode I/II Stable Crack Growth in Thin 2024-T3 Aluminum, ASTM STP 1256 on Fatigue and Fracture **26**, 256 (1995)
29. D. J. Chen, F. P. Chiang, Y. S. Tan, H. S. Don: Digital Speckle-Displacement Measurement Using a Complex Spectrum Method, Appl. Opt. **32**, 1839 (1993)
30. F. P. Chiang, Q. Wang, F. Lehman: New Developments in Full-Field Strain Measurements Using Speckles, ASTM STP 1318 on Non-Traditional Methods of Sensing Stress, Strain and Damage in Materials and Structures, 156 (1997)
31. J. Liu, M. A. Sutton, J. S. Lyons: Experimental Characterization of Crack-Tip Deformations in Alloy 718 at High Temperatures, J. Engr. Mat. Technol. **120** (1998)
32. J. S. Lyons, J. Liu, M. A. Sutton: Deformation Measurements at 650 °C with Computer Vision, Exp. Mech. **36**, 64 (1996)
33. G. Vendroux, W. G. Knauss: Submicron Deformation Field Measurements: Part I, Developing a Digital Scanning Tunneling Microscope, Exp. Mech. **38**, 18 (1998)
34. G. Vendroux, W. G. Knauss: Submicron Deformation Field Measurements: Part II, Improved Digital Image Correlation, Exp. Mech. **38**, 86 (1998)
35. G. Vendroux, N. Schmidt, W. G. Knauss: Submicron Deformation Field Measurements: Part III, Demonstration of Deformation Determination, Exp. Mech. **38**, 154 (1998)
36. S. Choi, S. P. Shah: Measurement of Deformations on Concrete Subjected to Compression Using Image Correlation, Exp. Mech. **37**, 304 (1997)
37. L. Mott, S. M. Shaler, L. H. Groom: A Technique to Measure Strain Distributions in Single Wood Pulp Fibers. J. Wood Fiber Sci. **28**, 429 (1995)
38. D. Choi: Failure Initiation and Propagation in Wood in Relation to its Structure, PhD thesis SUNY, Syracuse, NY (1991)
39. D. Choi, J. L. Thorpe, R. B. Hanna: Image Analysis to Measure Strain in Wood and Paper, Wood Sci. Technol. **25**, 251 (1991)
40. D. Choi, J. L. Thorpe, W. A. Cote, R. B. Hanna: Quantification of Compression Failure Propagation in Wood Using Digital Image Pattern Recognition, J. Forest Prod. **46**, 87 (1996)
41. C. P. Agrawal: Full-field Deformation Measurements in Wood Using Digital Image Processing, MS thesis, Virginia Tech., Blacksburg, VA (1989)
42. A. G. Zink, R. W. Davidson, R. B. Hanna: Strain Measurement in Wood Using a Digital Image Correlation Technique, Wood Fiber Sci. **27**, 346 (1995)
43. A. G. Zink, R. B. Hanna, J. W. Stelmokas: Measurement of Poisson's Ratios for Yellow-Poplar, J. Forest Prod. **47**, 78 (1997)
44. A. G. Zink, R. W. Davidson, R. B. Hanna: Effects of Composite Structure on Strain and Failure of Laminar and Wafer Composites, Compos. Mater. Struct. **4**, 345 (1997)
45. G. Kifetew: Application of the Deformation Field Measurement Method on Wood During Drying, Wood Sci. Technol. **20**, 455 (1997)
46. G. Kifetew, H. Lindberg, M. Wiklund M: Tangential and Radial Deformation Field Measurements on Wood During Drying, Wood Sci. Technol. **31**, 34 (1997)
47. A. G. Zink, L. C. Pita: Strain Development in Wood During Kiln Drying, in: 52nd Annual Meeting of Forest Products Society, Mexico (1998)

48. A. G. Zink, R. W. Davidson, R. B. Hanna: Experimental Measurement of the Strain Distribution in Double Overlap Wood Adhesive Joints, J. Adhes. **56**, 27 (1996)
49. J. W. Stelmokas, A. G. Zink, J. L. Loferski: Image Correlation Analysis of Multiple Bolted Wood Connections, Wood Fiber Sci. **29**, 210 (1997)
50. M. A. Sutton, W. Zhao, S. R. McNeill, J. D. Helm, W. Riddell, R. S. Piascik: Local Crack Closure Measurements; Development of a Measurement System Using Computer Vision and a Far-Field Microscope, in: 2nd International Symposium on Advances in Fatigue Crack Closure, San Diego, CA (1997)
51. W. Riddell, R. S. Piascik, M. A. Sutton, W. Zhao, S. R. McNeill, J. D. Helm: Local Crack Closure Measurements: Determining Fatigue Crack Opening Loads from Near-crack-tip Displacement Measurements, in: 2nd International Symposium on Advances in Fatigue Crack Closure, San Diego, CA (1997)
52. H. Lu, G. Vendroux, W. G. Knauss: Surface Deformation Measurements of a Cylindrical Specimen by Digital Image Correlation, Exp. Mech. **37**, 433 (1997)
53. J. L. Thorpe, D. Choi, R. B. Hanna: Role Cockle, Following its Evolution, TAPPI J. **74**, 204 (1991)
54. S. R. McNeill, M. Paquette: Initial Studies of Stereo Vision for Use in 3-D Deformation Measurements, unpublished internal report, Univ. South Carolina, SC (1988)
55. P. F. Luo, Y. J. Chao, M. A. Sutton: Application of Stereo Vision to 3-D Deformation Analysis in Fracture Mechanics, Opt. Eng. **33**, 981 (1994)
56. P. F. Luo, Y. J. Chao, M. A. Sutton: Accurate Measurement of Three-Dimensional Deformations in Deformable and Rigid Bodies Using Computer Vision, Exp. Mech. **33**, 123 (1993)
57. P. F. Luo, Y. J. Chao, M. A. Sutton: Computer Vision Methods for Surface Deformation Measurements in Fracture Mechanics, ASME-AMD Novel Experimental Methods in Fracture **176**, 123 (1993)
58. J. D. Helm, S. R. McNeill, M. A. Sutton: Improved 3-D Image Correlation for Surface Displacement Measurement, Opt. Eng. **35**, 1911 (1996)
59. J. D. Helm, M. A. Sutton, D. S. Dawicke, G. Hanna: Three-Dimensional Computer Vision Applications for Aircraft Fuselage Materials and Structures, in: 1st Joint DoD/FAA/NASA Conference on Aging Aircraft, Ogden, Utah (1997)
60. C. R. Dohrmann, H. R. Busby: Spline Function Smoothing and Differentiation of Noisy Data on a Rectangular Grid, in: Proc. SEM Spring Conference Albuquerque, NM (1990) p. 76
61. Z. Feng, R. E. Rowlands: Continuous Full-Field Representation and Differentiation of Three-Dimensional Experimental Vector Data, Comput. Struct. **26**, 979 (1987)
62. J. D. Foley, A. van Dam, S. K. Feiner, J. F. Hughes: *Computer Graphics: Principles and Practice*, 2nd edn. Addison-Wesley, (1996) p. 1174
63. M. R. James, W. L. Morris, B. N. Cox, M. S. Dadkhah: Description and Application of Displacement Measurements Based on Digital Image Processing, ASME-AMD, Micromechanics: Experimental Techniques **102**, 89 (1989)
64. D. Post: Moire' Interferometry: Its Capabilities for Micromechanics, ASME-AMD, Micromechanics: Experimental Techniques **102**, 15 (1989)
65. J. McKelvie: Experimental Photomicromechanics: Fundamental Physics and Mathematical Contrivances, ASME-AMD, Micromechanics: Experimental Techniques **102**, 1 (1989)

66. J. R. Maclin: Commercial Airplane Perspective on Multi Site Damage, in: International Conference on Aging Aircraft and Structural Airworthiness (1991)
67. C. E. Harris: NASA Aircraft Structural Integrity Program, NASA Report-No. TM-102637 (1990)
68. B. E. Amstutz, M. A. Sutton, D. S. Dawicke, M. L. Boone: Effects of Mixed ModeI/II Loading and Grain Orientation on Crack Initiation and Stable Tearing in 2024-T3 Aluminum, ASTM STP 1296 on Fatigue and Fracture Mechanics **27**, 105 (1997)
69. M. A. Sutton, W. Zhao, M. L. Boone, A. P. Reynolds, D. S. Dawicke: Prediction of crack growth direction for mode I/II loading unsing small-scale yielding and void initiation/growth concepts, Int. J. Fract. **83**, 275 (1997)
70. W. N. Sharpe Jr.: Crack-Tip Opening Displacement Measurement Techniques, in J. S. Epstein (Ed.): *Experimental Techniques in Fracture*, VCH Weinheim, (1993) p. 219
71. D. Post, B. Han, P. Ifju: *High Sensitivity Moire*, Springer, Berlin, Heidelberg (1994)
72. M. A. Sutton, W. Zhao, S. R. McNeill, J. D. Helm, R. S. Piascik, W. Riddell: Local Crack Closure Measurements: Development of a Measurement System Using Computer Vision and a Far-Field Microscope, ASTM STP on Advances in Fatigue Crack Closure Measurements and Analysis, in press
73. W. Riddell, R. S. Piascik, M. A. Sutton, W. Zhao, S. R. McNeill, J. D. Helm: Local Crack Closure Measurements: Interpretation of Fatigue Crack Closure Measurement Techniques, ASTM STP on Advances in Fatigue Crack Closure Measurements and Analysis, in press

Laser Doppler and Pulsed Laser Velocimetry in Fluid Mechanics

Jeremy M. Coupland

Department of Mechanical Engineering, Loughborough University,
Ashby Road, Loughborough, Leicestershire, LE11 3TU, UK
j.m.coupland@lboro.ac.uk

Abstract. Since the introduction of the laser in the late 1960s, optical metrology has made a major impact in many branches of engineering. This is nowhere more apparent than in the field of fluid mechanics where laser technology has revolutionised the way in which fluid flows are studied. The light scattered from small seeding particles following the flow contains information relating to the particle position and velocity. The coherence characteristics and high power densities achievable with a laser source allow well-defined regions of flow to be investigated in a largely non-intrusive manner and on a spatial and temporal scale commensurate with he flow field of interest. This review outlines the laser-based methods of velocimetry that are now available to the fluid dynamicist and discusses their practical application. Laser Doppler velocimetry provides a means to produce time-resolved measurements of fluid velocity at a single point in the flow. The optical design of instruments of this type is addressed with reference to spatial resolution and light gathering performance. Typical Doppler signals produced at both high and low particle concentrations are analysed and signal processing techniques are briefly discussed. Pulsed laser velocimeters use imaging optics to record the position of seeding particles at two or more instants and provide information concerning the instantaneous structure of the flow field. The optical configurations and analysis procedures used for planar velocity measurements are described and whole-field three-dimensional velocity measurements using holographic techniques are introduced.

1 Introduction

The prediction of fluid-flow characteristics is one of the most profound and interesting problems in engineering mechanics. Although straightforward in concept, the equations of fluid motion cannot be solved analytically except in the simplest of cases. Until recently the design and control of fluid processes was largely an art based on intuition and experience. As in other areas of engineering the use of computers has radically transformed the field and realistic flow phenomena can now be modelled numerically to a greater accuracy and detail than was previously thought possible. Perhaps paradoxically, the emergence of computational fluid dynamics (CFD) has promoted the role of the experimental fluid dynamicist somewhat, to provide validation of flow models.

It has been recognised for many years that optical methods provide an attractive means to studing fluid flow. For example, with suitable lighting, the introduction of dye, smoke or small particles into a flow can be used to investigate streamlines around a projectile. These and other similar methods are usually referred to as *flow visualisation methods* [1] since in general they give *qualitative* insight into flow phenomena. The development of *quantitative* optical methods in experimental fluid dynamics largely followed the introduction of the laser providing a source of high intensity with well-defined spatial and temporal characteristics.

The basis of all optical velocimeters is to derive velocity information from light scattered by the fluid flow. Although some light is scattered by the fluid itself, it is generally necessary to introduce particulate matter or seeding to the flow of interest to increase the scattering efficiency. The scattering process is illustrated schematically in Fig. 1 which shows consecutive wavefronts scattered from a single particle as it moves along the path AB. It can be seen that the wave fronts are irregularly spaced and the wavelength and frequency of the scattered radiation is therefore a function of angle. In addition, it should be noted that the scattered wavefronts are essentially spherical. The first of these observations is the well-known Doppler shift, and its measurement forms the basis of the techniques which we will refer to as laser Doppler velocimeters. The second property effectively locates the position of the particle. Measurement of the particle location at two or more times forms the basis of the techniques which are generally referred to as pulsed laser velocimeters.

Laser Doppler and pulsed laser velocimetry are now well-established methods which are routinely used to provide flow data. Extensive texts and review articles concerning each of these techniques have been written by many authors, although, they are rarely discussed together. In particular, the texts by *Durst* et al. [2], *Durrani* and *Greated* [3] and *Adrian* [4] provide an introduction to the development of laser Doppler instrumentation, and the articles and texts by *Adrian* [5], *Hinsch* [6], and *Grant* [7], review pulsed laser velocimetry. It is not the intention of this chapter to provide a detailed discussion of all the work covered in this vast subject area, but rather to introduce the fundamental optical principles, and demonstrate the capability of each

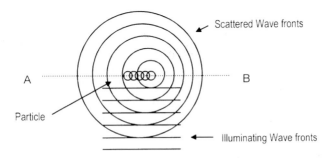

Fig. 1. Scattered wavefronts from a moving particle

of the measurement techniques. By discussing laser Doppler and pulsed laser velocimetry in a single chapter, it is hoped that similarities and differences between the techniques will be more apparent and as such the chapter will be of interest to both new and established engineers working in the field of experimental fluid mechanics. The chapter begins with a brief discussion of the scattering and dynamic properties of seeding particles.

1.1 The Scattering and Dynamic Properties of Seeding Particles

In laser Doppler and pulsed laser velocimeters the seeding particles themselves can be considered to be individual velocity probes. It is clear that the particles must be sufficiently small to track the flow accurately, yet be sufficiently large for the scattered light to be detectable, and a compromise must be achieved in practice. In most cases *linear* (elastic) scattering is dominant, and for the case of near-spherical particles can be estimated using Mie theory [8]. It is found that the scattering efficiency reduces dramatically if the particles are significantly smaller than the wavelength of illumination and the majority of the light is generally scattered in the forward direction. For these reasons, particles of comparable size to the wavelength of illumination are used and forward-scatter optical geometries are preferred provided that they are practically feasible.

Fluorescent seeding particles have also been used in optical velocimetry [9,10]. In this case light is scattered at a wavelength which differs from the incident radiation due to a *non-linear* scattering mechanism. The chief advantage of fluorescent seeding is that light scattered from the seeding material can be filtered from extraneous light scattered by the flow apparatus, thereby increasing the signal-to-noise ratio. It is worth noting that many materials can be made to fluoresce, and this is the basis of spectroscopic techniques known as laser-induced fluorescence (LIF) [11]. Although LIF is usually used to map the species of combustion products, fluid velocity measurements are also possible using a similar method to the planar Doppler velocimeters described in this chapter and a great deal of research is currently underway in this area. In general, however, linear scattering mechanisms are generally significantly more efficient than their non-linear counterpart and the majority of practical measurements are made in this way.

The dynamic properties of spherical seeding particles in homogeneous, sub-sonic fluid flows have been examined by a number of researchers [2]. These treatments relate the particle to the fluid motion in terms of their response to an oscillating infinite fluid. In general the highest frequency at which particles of a given size will faithfully follow the fluid flow is a complex function of the fluid viscosity and the relative density ratio of the particles to the fluid. However, the behaviour of the seeding particles is greatly simplified in two special cases, which, rather conveniently, most frequently occur in practice.

If the seeding particles are neutrally buoyant, that is to say, they are of a material which has density equal to that of the fluid, then the fluid motion is followed precisely by small particles. With care this condition is relatively easy to satisfy in liquid flows, but cannot be achieved in gaseous flows, where either solid or liquid seeding is usually employed. In this case the bulk density of the seeding material is considerably greater than that of the fluid and the dominant force on the particles is due to Stokes drag, F_D, given by

$$F_D = 3\pi\mu d_p (V_p - V_f) \,, \tag{1}$$

where μ is the fluid viscosity, d_p is the particle diameter and V_f and V_p are the fluid and particle velocities, respectively. In this case the equation of particle motion can be written in the standard differential form

$$\frac{\rho_p d_p^2}{18\mu} \frac{dV_p}{dt} + V_p = V_f \,, \tag{2}$$

where ρ_p is the density of the particle material. By analogy with communications theory, the particle inertia can be considered as a linear filter with a time constant, τ, given by

$$\tau = \frac{\rho_p d_p^2}{18\mu} \,. \tag{3}$$

For example water droplets, of 1 μm diameter suspended in air, are found to have a time constant of approximately 3×10^{-6} s, (taking $\rho_p = 1000\,\mathrm{kg\,m^{-3}}$, $d_p = 10^{-6}$ m and $\mu = 1.8 \times 10^{-5}$ Pa s) and are therefore capable of tracking most flows of interest. It is worth noting that by identifying and measuring the velocity of particles differing size it is possible to measure the fluid acceleration [12], and the practical generation of particles of well defined size has been recently reviewed by *Melling* [13].

2 Laser Doppler Velocimetry (LDV)

Laser Doppler velocimeters measure a change in frequency, the Doppler shift, which is observed when radiation is scattered from a moving particle. In all cases of practical interest the Doppler shift is relatively small compared with the frequency of the unscattered light and thus no apparent change in colour is observed by the eye. The Doppler shift is most easily measured using interferometeric methods, and this is the basis of the majority of laser Doppler velocimeters. The following section outlines the fundamental principles of this technique.

2.1 Fundamentals of LDV

This basic principle is illustrated in Fig. 2 which shows a classical Mach-Zehnder interferometer. Let us consider the light scattered by a single particle

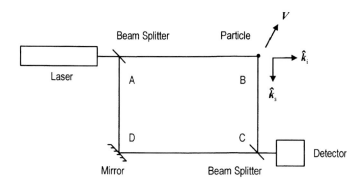

Fig. 2. Mach-Zehnder LDV configuration

denoted by the subscript m. A proportion of this light propagates around the path ABC to form the object arm of the interferometer and is received at the detector. If the wavelength and frequency of the illuminating laser are λ and ν, respectively, the complex amplitude, U_m, incident on the detector due to this component can be written in the form

$$U_m = A_m \exp\left[\mathrm{j}\,2\pi(\nu t - \frac{x}{\lambda})\right], \qquad (4)$$

where x is the path length and A_m is the amplitude of the scattered radiation at the detector. If the particle is moving with constant velocity, V, for small excursions from this basic geometry the path length can be written in the form,

$$x = x_0 - n(\hat{k}_s - \hat{k}_i).\mathbf{V}t, \qquad (5)$$

where $\hat{k}_s$ and $\hat{k}_i$ are unit vectors in the directions of the illuminating and scattering directions respectively, n is the refractive index of the flow medium, and x_0 is a constant. The complex amplitude is then,

$$U_m = A_m \exp\left[\mathrm{j}\,2\pi\left(\nu t + \frac{n(\hat{k}_s - \hat{k}_i).\mathbf{V}t}{\lambda} - \frac{x_0}{\lambda}\right)\right]. \qquad (6)$$

It is clear from this equation that an additional term is added to the frequency of the scattered light. This is the Doppler shift, $\Delta \nu_m$, given by

$$\Delta \nu_m = \frac{n(\hat{k}_s - \hat{k}_i).\mathbf{V}}{\lambda}. \qquad (7)$$

The reference beam propagates around the path ADC and the complex amplitude, U_R, of this beam at the detector can be written

$$U_R = A_R \exp\left[\mathrm{j}\,2\pi\left(\nu t - \frac{x_R}{\lambda}\right)\right], \qquad (8)$$

where A_R is the reference beam amplitude and x_R is the path length. The signal output by the detector is proportional to the incident intensity, I, which can be written

$$I = |U_m + U_R|^2 = A_m^2 + A_R^2 + 2A_m A_R \cos(2\pi \Delta \nu_m t + \phi_m), \qquad (9)$$

where the phase constant, ϕ_m, is given by $\phi_m = 2\pi(x_R - x_0)/\lambda$. If the frequency of this signal is measured directly, a sign ambiguity in the Doppler frequency is observed. The ambiguity is most easily resolved by introducing a frequency shift, $\Delta \nu_R$, to the reference beam such that the detector output can be written

$$I = |U_m + U_R|^2 = A_m^2 + A_R^2 + 2A_m A_R \cos[2\pi(\Delta \nu_m - \Delta \nu_R)t + \phi_m]. \qquad (10)$$

In this way, a carrier frequency is introduced and, provided that its magnitude is greater than the Doppler shift, it allows the particle velocity to be determined unambiguously. It is worth noting that the method also has the advantage of separating the Doppler signal from low frequency noise generated by both optical and electronic sources.

In practice, the complex amplitude at the detector is often the sum of individual particle contributions such that the intensity is given by

$$I = \left| \sum_m U_m + U_R \right|^2,$$

$$= \left| \sum_m U_m \right|^2 + A_R^2 + 2A_R \sum_m A_m \cos[2\pi(\Delta \nu_m - \Delta \nu_R) + \phi_m]\}. \qquad (11)$$

It is important to note that the time-varying part of this equation is simply the sum of the Doppler components due to each particle.

A practical measurement of fluid velocity by the technique we now refer to as laser Doppler velocimetry was first reported by *Yeh* and *Cummins* in 1964 [14] using the configuration shown in Fig. 3. Light scattered from the region of interest was imaged onto the entrance aperture of a photomultiplier tube and the output was fed to a radio-frequency spectrum analyser. This pioneering experiment showed that a Doppler shift was present in the scattered radiation and is linearly related to the component of the flow velocity in the direction $(\hat{k}_s - \hat{k}_i)$.

The optical configuration proposed by Yeh and Cummins is generally referred to as a *reference beam velocimeter* and is still of general use in the study of atmospheric turbulence where large distances are involved such that small-aperture back-scattered geometries are necessary. Most single point LDV measurements are now performed, however, using the so-called *cross beam, fringe, dual scatter* or *differential velocimeter* configuration. The cross-beam velocimeter was proposed in the late 1960s in almost equivalent forms by *Bond* [15], *Rudd* [16] and *Mayo* [17]. The basic concept is to illuminate

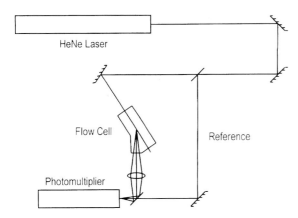

Fig. 3. The reference beam LDV of *Yeh* and *Cummins* [14]

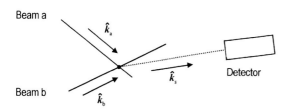

Fig. 4. Cross-beam configuration

the particle with two probe beams as shown in Fig. 4. With reference to the previous discussion, if a differential frequency shift $\Delta\nu_R$ is applied between the beams the complex amplitude at the detector due to a single particle, U_m, can be written in the form

$$U_m = A_m \Big(\exp\{j\,[2\pi\,(\nu + \Delta\nu_a - \Delta\nu_R/2)\,t + \phi_a)]\} \\ + \exp\{j\,[2\pi\,(\nu + \Delta\nu_b + \Delta\nu_R/2)\,t + \phi_b]\} \Big) . \qquad (12)$$

where ϕ_a and ϕ_b are the initial phases of beams a and b respectively and $\Delta\nu_a$ and $\Delta\nu_b$ are the Doppler shifts in each component given by

$$\Delta\nu_a = \frac{n(\hat{k}_s - \hat{k}_b).V}{\lambda}, \qquad (13)$$

$$\Delta\nu_b = \frac{n(\hat{k}_s - \hat{k}_a).V}{\lambda}. \qquad (14)$$

The output of the detector is proportional to the intensity, I, given by

$$I = |U_m|^2 = 2A_m^2\{1 + \cos[2\pi(\Delta\nu_m - \Delta\nu_R)t + \phi_{ab}]\}, \qquad (15)$$

where $\phi_{ab} = \phi_a - \phi_b$, and $\Delta\nu_m$ is the differential Doppler shift given by

$$\Delta\nu_m = \Delta\nu_a - \Delta\nu_b = \frac{n(\hat{k}_b - \hat{k}_a).V}{\lambda}. \qquad (16)$$

Clearly, the differential Doppler shift is independent of observation angle, and thus imaging optics of arbitrarily large aperture can be used to collect scattered light. In addition, since the cross-beam geometry performs a differential measurement, the optical path length changes due to air currents and vibrations of the optical components are largely common to both beams and do not adversely affect the result. A practical cross-beam LDV configuration is shown in Fig. 5. This compact design uses largely the same optics to focus the probe beams into the measurement volume and subsequently collect the backscattered light. An acousto-optic modulator (Bragg cell) is used to give a reference frequency shift.

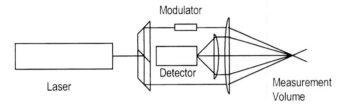

Fig. 5. Backscatter cross-beam LDV

The performance of the cross-beam geometry is significantly different from the reference beam velocimeter, however, when light is received simultaneously from more than one particle. In this case the intensity at the detector can be written

$$I = \left|\sum_m U_m\right|^2 = \sum_m |U_m|^2 + \sum_{m,n}\sum_{m\neq n} U_m U_n^*$$
$$= 2\sum_m A_m^2 \{1 + \cos[2\pi(\Delta\nu_m - \Delta\nu_R)t + \phi_{ab}]\} + \sum_{m,n:m\neq n}\sum U_m U_n^*. \qquad (17)$$

Thus the output of a cross-beam velocimeter is the sum of the Doppler signals due to individual particles plus a cross-particle term. The effect of this term depends on the *relative* velocity of the particles. If the particles move at the same velocity, the relative phases of the particle contributions will not change and the term can be considered constant. If there is a spatial velocity gradient, however, the relative phases of the particle contributions will change with time and the term can be considered as (highly correlated) noise. In practice, the significance of the cross-particle terms is often negligible since the seeding concentration is such that multiple particles are unlikely to be present in the measurement volume. In addition, by using an imaging lens of large numerical aperture to collect the scattered light the significance of the cross-particle terms is reduced. These details are discussed with reference to the optical geometry in the following section.

2.2 Fourier Optics Model of LDV

In the previous section the basic principles of reference beam and cross beam velocimeters were outlined. It is evident, however, that in general a seeded fluid will contain particles throughout the flow and for this reason care must be taken to ensure that the Doppler signal can be attributed to a well-defined region of the flow. Essentially the optical design of an LDV system consists of a transmitter which illuminates the flow and a receiver which collects the scattered light and outputs the Doppler signal to the signal processor. The designs of both the transmitter and receiver are very important in defining the measurement volume and can be investigated using the analysis of Fourier optics [18]. This approach was first applied to the analysis of LDV systems by *Rudd* [19], although much of the treatment presented here follows the later treatment by Lading [20].

A simplified model of an LDV system, which through minor changes can be configured to either a reference or a cross-beam geometry, is shown in Fig. 6. The transmitter consists of a pair of laser beams which simultaneously cross and are brought to focus in the fluid flow. The receiver is a system of lenses that spatially filters and collects the light scattered from this region at the photodetector. In order to simplify the analysis, the lens elements used in both the transmitter and receiver are of equal focal length, f, and the optical axes are coincident. In this way the instrument is configured to collect light in forward scatter, although it is relatively straightforward to mathematically reconfigure the system for the other scattering modes [20].

The complex amplitude in the front focal plane of the transmitter lens is due to two parallel laser beams. If it is assumed that these beams are of

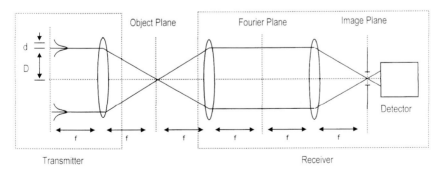

Fig. 6. Fourier model of LDV

Gaussian profile with a peak amplitude, $U_o/2$, and separated by a distance $2D$, the complex amplitude in the front focal plane, $U_a(x_t, y_t)$, can be written,

$$U_a(x_t, y_t) = \frac{U_o}{2}\left[\exp\left(\frac{-[(x_t - D)^2 - y_t^2]}{d^2}\right) + \exp\left(\frac{-[(x_t + D)^2 - y_t^2]}{d^2}\right)\right], \quad (18)$$

where the beam radius, d, is measured at the point where the amplitude is diminished by a factor of $1/e$ (or equivalently the intensity is diminished by a factor of $1/e^2$). For simplicity, no frequency shift has been included and in accordance with the usual notation of Fourier optics the common time factor $e^{j\,2\pi\nu t}$ has been dropped. The complex amplitude in the back focal plane of the transmitter, $U_T(x_o, y_o)$, is then given by

$$U_T(x_o, y_o) = \frac{1}{j\lambda f} \tilde{U}_a\left(\frac{x_o}{\lambda f}, \frac{y_o}{\lambda f}\right), \quad (19)$$

where $\tilde{U}_a(f_x, f_y)$ is the Fourier transform of $U_a(x, y)$ such that,

$$U_T(x_o, y_o) = U_0 \frac{\pi d^2}{j\lambda f}\exp\left(\frac{-\pi^2 d^2(x_o^2 + y_o^2)}{\lambda^2 f^2}\right)\cos\left(\frac{2\pi D x_o}{\lambda f}\right). \quad (20)$$

The intensity in the focal plane of the transmitter, $I_T(x_o, Y_o)$, is often referred to as the *code* of the velocimeter and for the cross-beam geometry is given by

$$I_T(x_o, y_o) = |U_T(x_o, y_o)|^2,$$
$$= I_o \frac{\pi^2 d^4}{2\lambda^2 f^2}\exp\left(\frac{-2\pi^2 d^2(x_o^2 + y_o^2)}{\lambda^2 f^2}\right)\left[1 + \cos\left(\frac{4\pi D x_o}{\lambda f}\right)\right]. \quad (21)$$

Hence the code of a cross-beam velocimeter consists of a set of parallel fringes inside a Gaussian envelope. It is straightforward to show that the number of fringes, N, observed within the diameter of the envelope (measured between the $1/e^2$ points) is given by $N = 4D/\pi d \cong D/d$.

In the idealised LDV system of Fig. 6, the receiver is a unit-magnification imaging system, in the form of a classical $4f$ optical processor (since the object and image planes are separated by a distance equal to $4f$) [18]. As such, the complex amplitude in the object plane $U_o(x_o, y_o)$ is related to that in the image plane $U_i(x_i, y_i)$ by the convolution integral

$$U_i(x_i, y_i) = \int_{-\infty}^{\infty}\int_{-\infty}^{\infty} h(x_o, y_o) U_0(x_i - x_o, y_i - y_o) dx_o dy_o,$$
$$= h(x_i, y_i) \otimes U_o(x_i, y_i), \quad (22)$$

where $h(x_o, y_o)$ is the impulse response which is defined by the pupil function, $P(x_f, y_f)$, describing the amplitude transmittance in the Fourier plane such

that
$$h(x_o, y_o) = \frac{1}{\lambda^2 f^2} \tilde{P}\left(\frac{x_o}{\lambda f}, \frac{y_o}{\lambda f}\right), \tag{23}$$

where $\tilde{P}(f_x, f_y)$ is the Fourier transform of $P(x, y)$. In general, the light scattered by the seeding particles is small compared to the illuminating field, the multiple scattering can be neglected, and to a good approximation we can write

$$U_o(x_o, y_o) = U_T(x_o, y_o) + U_S(x_o, y_o), \tag{24}$$

where $U_S(x_o, y_o)$ represents the field scattered from the seeding particles. If we assume that the particles scatter isotropically and for the moment consider only particles which are within the depth of field of the imaging system we can model the scattered field as

$$U_S(x_o, y_o) = \sum_m a_m \exp\left(\frac{\mathrm{j} 2\pi z_m}{\lambda}\right) U_T(x_o, y_o) \delta(x_o - x_m, y_o - y_m), \tag{25}$$

where a_m is a constant which defines the scattering efficiency and x_m, y_m, z_m are the co-ordinates of the m-th particle. The output of the detector, i, is proportional to the incident intensity, $I_i(x_i, y_i)$, integrated over the aperture, A, and is given by,

$$i \propto \iint_A I_i(x_i, y_i) \mathrm{d}x_i \mathrm{d}y_i = \iint_A |U_i(x_i, y_i)|^2 \mathrm{d}x_i \mathrm{d}y_i. \tag{26}$$

Making appropriate substitutions and expanding, we find that the detector output consists of three terms

$$i \propto \iint_A \{|h \otimes U_T|^2 + |h \otimes U_S|^2$$
$$+ [(h \otimes U_T)(h \otimes U_S)^* + (h \otimes U_T)^*(h \otimes U_S)]\} \mathrm{d}x_i \mathrm{d}y_i, \tag{27}$$

where the constant of proportionality depends on the detector sensitivity. The relative importance of these terms depends on the position and size of the aperture in the Fourier plane of the receiver and determines whether the configuration will behave as a cross-beam or reference beam velocimeter. A typical cross-beam arrangement is shown in Fig. 7. The configuration is characterised by the fact that both of the directly transmitted beams are blocked by the receiver aperture, and this is referred to as the *homodyne mode*. In effect these components are spatially filtered from the complex amplitude distribution in the object plane and terms 1 and 3 in (27) vanish such that the photodetector output can be written

$$i \propto \iint_A |h \otimes U_S|^2 \mathrm{d}x_i \mathrm{d}y_i$$
$$\propto \iint_A |h(x_i, y_i) \otimes \sum_m a_m \exp\left(\frac{\mathrm{j} 2\pi z_m}{\lambda}\right)$$
$$\times U_T(x_i, y_i) \delta(x_i - x_m, y_i - y_m)|^2 \mathrm{d}x_i \mathrm{d}y_i. \tag{28}$$

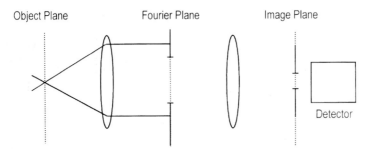

Fig. 7. Cross-beam configuration

If the spatial extent of the impulse response is less than the particle separation then the particle images will not overlap and we can write,

$$i \propto \iint_A \sum_m |a_m U_T(x_i, y_i) h(x_i - x_m, y_i - y_m)|^2 \mathrm{d}x_i \mathrm{d}y_i$$
$$\propto \sum_m |a_m|^2 I_T(x_i, y_i) , \qquad (29)$$

where the last expression follows since $\iint h(x_i, y_i)\mathrm{d}x_i\mathrm{d}y_i = \text{const}$. Thus the photodetector output is proportional to the sum of the intensity of the illuminating field at the positions defined by that of each particle. In the context of LDV, this is called *incoherent* detection. If the spatial extent of the impulse response is larger than the detector aperture then the detected light is the *coherent* sum of contributions from each particle, and the photodetector output can be written

$$i \propto \left| \sum_m a_m U_T(x_m, y_m) \right|^2 . \qquad (30)$$

Hence *coherent detection* implies that cross particle terms appear in the Doppler signal, as discussed in the previous section. In a reference beam velocimeter the Doppler signal is produced by mixing the scattered light with a coherent reference wave (normally frequency shifted) and this is generally referred to as the *heterodyne mode* of operation. The generalised LDV configuration of Fig. 6 can be made to operate in this mode with the Fourier plane aperture shown in Fig. 8. The dominant signal is due to light scattered by beam a interfering with the directly transmitted beam b, and the second term of (27) vanishes. The dynamic part of the signal is given by the last term, and it is found that

$$i \propto \sum_m a_m I_T(x_m, y_m) . \qquad (31)$$

Hence, as in the case of incoherent detection in the homodyne mode, no cross-particle terms appear in the output of the reference beam LDV. It is

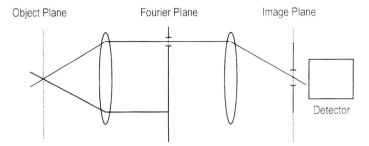

Fig. 8. Reference beam configuration

worth noting that both the reference beam and cross-beam configurations discussed in this section are sensitive to particle movement perpendicular to the optical axis. Provided that no cross-particle terms appear, we have incoherent detection and the photodetector output is due to the weighted sum of the intensity scattered by each of the particles. If a frequency shift is present, the dynamic part of the output is due to a change in particle position relative to the frame of reference defined by a translating fringe pattern illuminating the region.

The discussion above assumes that the particles are in focus. It is clear that this not always true and some consideration of the three-dimensional nature of the problem is therefore necessary. In general, three-dimensional wave propagation and spatial-filtering operations are rather awkward to calculate and for this reason the procedure is only outlined briefly below. The starting point for this calculation is a better approximation to the scattered field in the object plane of the receiver. Using the Fresnel approximation we can write the scattered field as the sum of quadratic wave fronts, such that

$$U_S(x_o, y_o) = \sum_m \frac{A_m}{j \lambda z_m} \exp\left(\frac{j 2\pi z_m}{\lambda}\right)$$
$$\times \exp\left\{\frac{j 2\pi}{\lambda z_m}[(x_o - x_m)^2 + (x_o - x_m)^2]\right\}, \quad (32)$$

where A_m is given by the Fresnel transformation

$$A_m = a_m \int_{-\infty}^{\infty} \int_{-\infty}^{\infty} \frac{U_T(x_o, y_o)}{j \lambda z_m} \exp\left(\frac{-j 2\pi z_m}{\lambda}\right)$$
$$\times \exp\left\{\frac{j 2\pi}{\lambda z_m}[(x_o - x_m)^2 + (x_o - x_m)^2]\right\} dx_o dy_o. \quad (33)$$

The photodetector output can then be calculated in a similar manner to the two-dimensional case presented above. The results of this analysis can be summarized as follows. Provided that the aperture of the receiver is sufficient to resolve separated particle images, incoherent detection is observed regardless of defocus. Interestingly this is also true if other aberrations are present in the receiver optics. Secondly, the analysis shows that the illuminating field can be considered as a fringe pattern consisting of very nearly equally spaced planes of constructive and destructive interference in the region of overlap. The error describing the departure from this case is inversely proportional to the number of fringes, N, and can reasonably be neglected in most cases of interest. Finally, the measurement volume is found to be defined by a function which is approximately an elongated Gaussian function, as shown in Fig. 9. From (21), the width, W, of the Gaussian function (measured at the $1/e^2$ point) in the object plane is $W = 2\lambda f/\pi d$, while the full three-dimensional analysis shows that the depth of the measurement volume, L, is effectively increased by a factor f/D such that $L = 2\lambda f^2/\pi Dd$. It is worth noting, however, that an approximately spherically symmetric measurement volume can be configured by arranging the optical axes of the transmitter and receiver to be orthogonal such that the light is collected in side scatter.

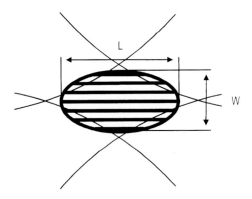

Fig. 9. LDV measurement volume

2.3 The Doppler Signal and Signal Processing

Much of the work in the area of LDV in the past 30 years has concerned the development of dedicated electronic processors for the extraction of velocity data from Doppler signals. Although a detailed discussion of this work is beyond the scope of this review, it is worth considering some of the general features of the Doppler signal and the data therein. With reference to (21), a single particle moving through the centre of the measurement volume at a time, t_m, with constant velocity, V_x, will result in a Doppler signal, $i_m(t)$, which has the form

$$i_m(t) = i_{0m} \exp\left(\frac{-V_x^2(t-t_m)^2}{a^2}\right)$$

$$+ \exp\left(\frac{-V_x^2(t-t_m)^2}{a^2}\right)\cos\{2\pi[bV_x(t-t_m)+\Delta V_R t]\}\right), \quad (34)$$

where $i_m(t)$ is a constant and a and b are given by $a = \lambda f/\sqrt{2}\pi d$ and $b = 2D/\lambda f$, respectively. This signal, which includes a frequency shift, $\Delta \nu_R$, is generally referred to as a *Doppler burst* and contains two terms, each with a Gaussian envelope. The first term is known as the *pedestal function* and is superposed on a signal with frequency defined by the sum of the carrier frequency and Doppler shift and phase defined by the random variable, t_m. Using values of $D = 10$ mm, $f = 100$ mm and $\lambda = 500$ nm, the mean Doppler frequency corresponding to a particle traversing the measurement volume at $1\,\mathrm{ms}^{-1}$ is found to be 400 kHz, and it is not uncommon for Doppler signals to have a bandwidth of several MHz. In the early days of LDV, instrumentation capable of processing signals of this frequency was expensive and limited in capability. For this reason a great deal of effort was directed towards the development of methods to reduce the processing burden. For example, several methods to remove the pedestal from the Doppler signal using additional detectors [2] were proposed. With modern electronics this task is relatively straightforward using a large frequency shift (around 40 MHz) and standard filtering techniques.

In the temporal-frequency domain the power spectrum corresponding to a single Doppler burst (with the pedestal removed) referred to as the *burst spectrum* can be written

$$S_m(\nu) = \left|\int_{-\infty}^{\infty} i_m(t)\exp(-2\pi\mathrm{j}\nu t)\mathrm{d}t\right|^2,$$

$$= \frac{i_{0m}\pi a^2}{\nu_x^2}\exp\left[-\frac{2\pi^2 a^2}{\nu_x^2}(\nu - bV_x - \Delta\nu_R)^2\right]. \quad (35)$$

It can be seen that the burst spectrum corresponding to a single particle traversing the measurement volume at constant velocity can be modeled as a single Gaussian peak with position linearly related to the particle velocity. Hence, in the absence of noise, the burst spectrum is a deterministic function of the fluid velocity, and the latter can be retrieved exactly, for example, using a curve-fitting routine to locate the peak position.

In practice the Doppler signal is generally due to the sum of individual bursts, and typical signals are shown in Fig. 10 for (a) low seeding concen-

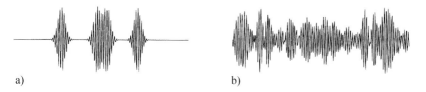

Fig. 10. Doppler signal at (**a**) low and (**b**) high seeding concentration

trations and (b) high concentrations. In the former case separate bursts are clearly distinguishable and can be analyzed in the manner described above. When the expected number of particles within the volume exceeds unity, a nearly continuous signal is observed. It is clear that a continuous signal is advantageous in that it provides a complete time history of the fluid velocity; however, there is an inherent uncertainty in the data obtained from this signal. This arises because the Doppler signal is a function of the random variables, t_m, which define the relative phases of the individual Doppler bursts. The spectrum obtained from the analysis of a finite portion of the Doppler signal will average to the form of the burst spectrum of (35) as the data length is increased. For a finite data length, an uncertainty in the measured velocity therefore results which decreases as the measurement bandwidth is reduced.

Several methods have been applied to the recovery of fluid velocity data from the Doppler signal. Until recently signals obtained at high seeding concentration were analyzed using analogue demodulation methods including zero-crossing ratemeters, phase-locked loops and frequency-tracking filters [2]. The statistical performance and robustness of these methods depends critically on the signal-to-noise ratio, the frequency-tracking filter generally being superior to the others in this respect. The *burst spectrum analyser* (BSA) is a modern instrument capable of extracting data from signals obtained at low or high seeding concentrations. In essence this instrument consists of a burst detector which provides a validation signal to gate the output of a spectrum analyser. At low seeding concentrations a BSA provides estimates of the individual particle velocities. At high seeding concentrations the burst detector can be adjusted to provide a compromise between measurement bandwidth and uncertainty.

It is worth noting that some care should be exercised in the interpretation of the data obtained using LDV. For example, a BSA can be operated in a mode that collects the statistics of the *particle* velocity derived from the analysis of individual burst spectra. It should be noted that these statistics differ slightly from those of the *fluid* velocity, since for a uniformly seeded fluid the number of particles traversing the measurement volume per unit time is directly proportional to the fluid velocity and hence must be weighted accordingly. In addition to this, velocity gradients within the measurement volume and other mechanisms can lead to bias in the estimates of fluid velocity statistics. In most cases the bias is negligible, but these mechanisms can be important in the characterisation of low-intensity turbulence, and further discussion of this topic can be found in the work by *Adrian* [4].

Finally, it is also possible to measure the Doppler shift *directly* from the light scattered from a moving particle. In essence this method utilises an optical filter with a sharp roll-off across the bandwidth of the Doppler signal such that a change in frequency is accompanied by a corresponding change in amplitude. Although direct demodulation was first proposed in the context of LDV by *James* et al. [21], it is difficult to achieve the frequency resolution required to study low-speed flows and the methods discussed above are generally preferred. A significant advantage of this method, however, lies in its ability to demodulate many Doppler signals in parallel and is the principle underlying the planar techniques discussed in Sect. 3.

2.4 LDV Measurements in Practice

LDV is now a mature measurement technique and several companies specialise in equipment for this purpose. It is not the intention of this section to describe the detail of these systems, since this is best done by sales representatives. However, it is worth explaining some of the general features of modern equipment including instrumentation capable of multiple component measurements and scanning LDV systems.

In order to increase the data rates of velocity measurements it is generally useful to have instrumentation capable of producing valid data from small seeding particles. In order to maximise this capability it is clearly necessary to use a combination of powerful illumination and sensitive detection equipment. An argon-ion laser is usually used to illuminate the flow giving several watts of multiple line output and is often used in conjunction with fibre optic delivery to facilitate accurate positioning of the LDV optics relative to the flow. Most commercial LDV systems employ a frequency shift derived from a Bragg cell (acousto-optic modulator). These cells induce a travelling wave which modulates the refractive index of a transparent solid such as quartz. Light diffracted from these devices is Doppler-shifted at a frequency of approximately 40 MHz. Early LDV systems often used two Bragg cells operating at slightly different frequencies in each of the beams, such that the difference frequency was of the order of a few MHz and was suitable for further processing. Modern systems generally use one cell and, if necessary, the carrier frequency can be moved to an intermediate frequency electronically. The light collected by the LDV system is usually detected by a photomultiplier tube. Although these devices require high-voltage drive-circuitry, they are still used due to their high-bandwidth, and low-noise characteristics. It is likely, however, that in the future solid-state detectors such as avalanche photodiodes will play an increasing role as LDV systems are miniaturised by using laser diodes and fibre optics together with using modern photonic engineering methods [22].

LDV systems capable of simultaneously measuring more than one component of the fluid velocity are of major interest in the study of complex three-dimensional flow fields. The most straightforward extension of the single component LDV geometry of Fig. 5 is to replicate the basic geometry in an orthogonal plane such that the transmitters and receivers of both channels have common optical axes and the measurement volumes are coincident. In this way, the imaging optics are common to both channels and a very compact two-component system can be realised. It is apparent, however, that a means is required to separate the light propagating in each channel. The multiple-line argon-ion laser is useful in this respect, since the output can be split into the dominant lines at 488 nm and 512 nm wavelengths. The channels can be separated using a dichroic beamsplitter to divide the laser output, and a similar splitter to direct the collected light to separate photomultipliers.

In general, three-component LDV systems are much more complex than one- or two-dimensional systems, since more than one observation or viewing angle is necessary and this often precludes their use in practice. The most straightforward method to implement three-component LDV is to augment a two-component system with a separate single component unit. In this way the channels can be separated by wavelength, for example by using a HeNe laser or by ensuring that orthogonal polarisations are maintained between channels of common wavelength. Clearly care must be taken to ensure that the measurement volumes are coincident to avoid bias in the collection of the flow statistics.

LDV systems can be used to examine the spatial variation of the flow field by scanning the measurement volume through the fluid. For one and two-component systems this can be achieved by placing the LDV on a traverse and the collecting data point by point. Care must be taken, particularly if windows are present, to ensure that the probe beams cross precisely at each measurement point. In practice, the added complexity of three-component systems often means that for small flows (for example the fuel sprays discussed below) it is experimentally more efficient to scan the flow past the LDV system.

As an example of a scanning LDV measurement, Fig. 11 shows the flow field in a high-pressure, Gasoline Direct-Injection (GDI) fuel spray from a study by *Hargraves* et al. [23]. The study was performed using a two-component LDV system which sampled the flow at 1 mm intervals in the radial direction and at 5 mm intervals in the axial direction by incrementally repositioning the injector. The top half of the figure shows an ensemble average of the data obtained approximately 5 ms after the initiation, and the bottom half is the corresponding frame from a high-speed video sequence. To the right of the figure, a toroidal vortex can be identified which, from a time sequence of similar data, is found to form at the jet boundary and propagates downstream during the injection event. This result clearly shows the ability of LDV to measure small-scale flow phenomena and in this case provides valuable data to further the understanding of fuel mixing.

Scanning LDV systems collect time-resolved data describing the fluid velocity from sequentially sampled measurement volumes within the flow field. This data can be processed to give the time-averaged flow structure provided that the flow statistics do not change on a timescale longer than that used to collect data from a single measurement volume, i.e. the flow statistics must be stationary. Many parameters of interest to the fluid dynamicist, such as vorticity and strain rates, require knowledge of the flow structure, which can

Fig. 11. LDV study of a GDI fuel spray (after *Hargraves* et al. [23])

only be gained from equipment capable of simultaneous multiple-point velocity measurement. Clearly, multiple, probe LDV systems are prohibitively expensive in all but the simplest cases; however, simultaneous measurements are possible in a plane using optical methods to demodulate the Doppler signal. This approach is called planar Doppler velocimetry and is outlined in the following section.

3 Planar Doppler Velocimetry (PDV)

Planar Doppler velocimetry was first reported by in 1991 *Komine* and *Brosnan* [24], who utilised direct optical demodulation to process the light scattered from a laser light sheet in a simultaneous, *parallel* operation. The configuration of Komine and Brosnan is shown in Fig. 12 and consists of a system of relay lenses to image the light scattered from the light sheet onto two CCD cameras. A frequency discriminating filter in the form of an optical cell filled with iodine vapour and a neutral density attenuator is placed before the first and second cameras, respectively.

The spectral characteristics of iodine vapour include a system of lines which have peak absorption approximately 2 GHz above the centre frequency of the 514 nm argon-ion line. Figure 13 shows the intensity transmittance of an iodine cell determined by *Chan* et al. [25], as a function of frequency above the 514 nm line at several temperatures. It can be seen that the position of the absorption peak remains approximately constant but the feature deepens and widens with temperature, which is attributed to collision broadening

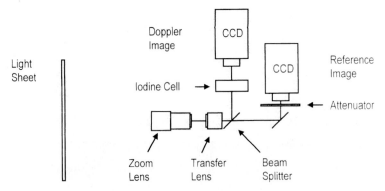

Fig. 12. PDV configuration of *Komine* and *Brosnan* [24]

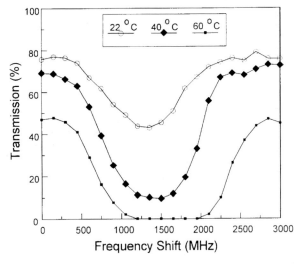

Fig. 13. Iodine cell transmission (after *Chan* et al. [25])

due to an increase in vapour pressure and Doppler broadening, respectively. At 40 °C the linear region of the absorption edge covers a bandwidth of approximately 400 MHz and has a gradient such that a 10 MHz frequency change corresponds to approximately 1% change in intensity transmittance.

The 514 nm line of argon-ion laser can be tuned to coincide with a point midway along the absorption edge of the iodine cell. The Doppler shift associated with the particle velocity will cause a proportional change in the attenuation of the cell and the intensity of the image will be modulated accordingly. However, this intensity distribution is essentially a coherent sum of the light which has been scattered from the random distribution of particles within the light sheet and collected by the imaging system. Hence the

Doppler image is modulated by laser speckle and must be normalised with respect to an unfiltered reference image.

The resolution of PDV is strongly dependent on the quality of the Doppler and reference images. From the transmission characteristics of the iodine cell, outlined above, if the intensity transmittance can be estimated to an accuracy of $\pm 1\,\%$, then, in principle, the Doppler shift can be estimated to an accuracy of $\pm 10\,\mathrm{MHz}$. For a side-scatter geometry, it is straightforward to show from (7) that this corresponds to a random error in the velocity of approximately $3.6\,\mathrm{ms}^{-1}$. The upper limit to the velocity is determined by the $400\,\mathrm{MHz}$ bandwidth of the iodine cell, which for this geometry corresponds to a velocity of $144\,\mathrm{ms}^{-1}$. Hence PDV is particularly suited to high-speed flow measurement and has been successfully demonstrated in transonic flows.

In principle, it is possible to measure velocity to a higher resolution than this; however, there are many experimental factors which confound the measurement. Since, in the absence of particle motion, the intensity of the normalised image depends on the frequency of the laser source relative to the absorption edge of the iodine cell, these components must be stabilised. By temperature-controlling the cell to a precision of $\pm 0.1\,°\mathrm{C}$, *Ainswoth* and *Thorpe* [26] report that the position of the absorption edge can be maintained to an accuracy of approximately $\pm 2\,\mathrm{MHz}$, while *Roehle* and *Schodl* [27] describe a stabilised argon-ion laser which is actively locked to a reference iodine cell. The most significant limit to the resolution of PDV are errors introduced by misalignment of the Doppler and reference images and the noise generated by the CCD detectors. *Ainsworth* et al. [28] describe an alignment method using a warping procedure to compensate for differential distortion of the two images. The effect of CCD noise depends on the bandwidth and consequently restricts the dynamic range. *McKenzie* [29] has shown that the output of typical CCD cameras with 8-bit resolution leads to an uncertainty of about 1% in the intensity transmittance over a bandwidth of around $200\,\mathrm{MHz}$ and the use of cooled CCD sensors with 12-bit resolution increases the velocity resolution to around $1\,\mathrm{ms}^{-1}$.

The extension of PDV to measure the three-components of the fluid velocity within a plane is straightforward using additional imaging systems. Figure 14 shows one possible configuration to record the Doppler shift from three perspectives. At the centre of this configuration (the point where the optical axes of the imaging systems cross), using (7) it can be shown that the sensitivity vectors of the three imaging systems are orthogonal. Consequently, the three-components of the fluid velocity can be determined with equal accuracy. Instrumentation of this kind has been implemented in the NASA Ames wind tunnel facility by *Meyers* [30]. Figure 15 is a result taken from a simultaneous, three-component study of a transonic jet flow operating at temperatures of around $700\,°\mathrm{C}$. In this case, the light sheet was positioned at a distance of $0.25\,\mathrm{m}$ from the jet exit, and the out-of-plane (axial) component was derived from the three-component data. Despite the necessarily

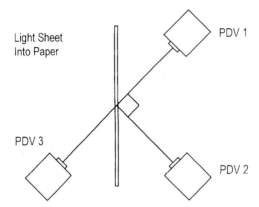

Fig. 14. Three-component PDV

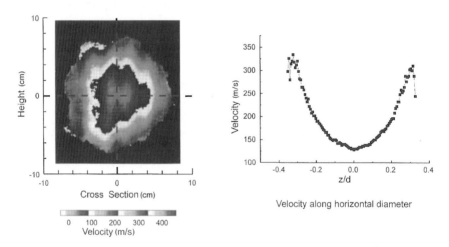

Fig. 15. PDV study of a transonic jet (after Meyers [30])

large stand-off distance (15.5 m), a spatial resolution of 1.25 mm was obtained and the velocity deficit caused by a conical core at the centre of the jet can be clearly seen.

The size of the measurement volume(s) in PDV is defined by the thickness of the light sheet and the resolution of the camera. Phase curvature within the light sheet is not critical since the broadening of the Doppler shift by this mechanism is much less than the bandwidth of the iodine cell; however, it is clear that the sheet should be of uniform thickness. Typically a light sheet thickness of around 250 μm is achievable over an area of 100 mm × 100 mm and high resolution video cameras are able to resolve regions measuring 100 μm × 100 μm within this area. It is worth noting that, provided that sufficient light is available, PDV can be used to investigate large-scale flow structure and this is a major advantage of the technique. Although much research effort is being directed towards increasing the resolution of PDV, multiple-point measurements of comparatively low speed-flows are undertaken using imaging techniques which are insensitive to the Doppler frequency shift. Since these techniques typically require a pulsed laser light source, they are often referred to collectively as pulsed laser velocimetry and are discussed in the following section.

4 Pulsed Laser Velocimetry

Pulsed laser sources used in conjunction with photographic, video or holographic recording techniques can be used to freeze the motion and record the position of seeding particles suspended in a flow field. By using a double-or multiple-pulsed laser, the particle locations at two or more instants in time can be recorded and this approach forms the basis of the techniques which we will refer to as pulsed laser velocimeters. Although the ability to record images which are clearly identified as individual particles is not a strict requirement of these techniques, it is generally beneficial to subsequent analysis, and for this reason the techniques described in this chapter are most often referred to as particle image velocimetry.

4.1 Particle Image Velocimetry (PIV)

In its simplest form PIV uses double-exposure photography to record the positions of particles illuminated by a light sheet at two instants in time. Although a coherent source is not strictly necessary, in practice a pulsed laser is most often used as an intense light source which is easily fashioned into a thin light sheet. The basic configuration is shown in Fig. 16. The light sheet is usually generated using cylindrical optics to illuminate a plane of interest within the flow field. The camera is positioned such that the axis is perpendicular to the light sheet and the shutter is opened while the laser is firing. The resulting photographic transparency is then developed and analysed.

Historically, PIV was developed from the work of *Dudderar* and *Simpkins* [31], who in 1978 applied laser speckle photography (LSP) to the measurement of fluid flow. In contrast to the study of solid-surface deformation, in a fluid-measurement situation the control of the concentration of the scattering particles is possible and better results are obtained if individual particle images can be identified [32]. With reference to the LDV analysis presented in Sect. 2, this is analogous to the case of *incoherent* detection, where the images do not overlap on the detector. If the camera system is treated in a similar manner to the LDV receiver, the intensity in the image plane, $I_i(x_i, y_i)$, for the case of incoherent detection, can be written as

$$I_i(x_i, y_i) = h_I(x_i, y_i) \otimes \sum_m \delta(x_i - x_m, y_i - y_m) , \qquad (36)$$

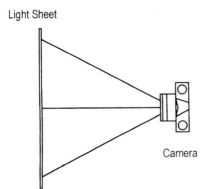

Fig. 16. PIV configuration

where $h_I(x_i, y_i)$ is the incoherent impulse response of the system given by $h_I(x_i, y_i) = |h(x_i, y_i)|^2$. Piecewise analysis of PIV transparencies involves the interrogation of small regions to identify and measure the displacement of a repeated pattern of particle images. Clearly, each interrogation region will contain the paired images of particles located within a well-defined measurement volume. The dimensions of each measurement volume are defined by the size of the interrogation region and the thickness of the light sheet. As in the case of PDV, the thickness of the light sheet depends strongly on the required propagation length, and hence the depth of each measurement volume is around 250 µm for a light sheet measuring 100 mm × 100 mm.

In early studies the analysis of PIV transparencies was attempted using the method of *Young's fringe analysis* which was introduced in the context of LSP by *Burch* and *Tokaski* [33]. In essence the transparency is probed using an unexpanded laser beam and a fringe pattern is generated in the back focal plane of a convex lens, as shown in Fig. 17. A simplified yet useful model of this method of analysis can be formulated by considering only paired particle images. In this way the transmission function of the (positive) transparency is proportional to the sum of the intensity distributions at the time of the first and second exposures, and the complex amplitude transmitted by the

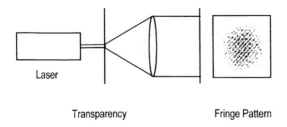

Fig. 17. Young's fringe PIV analysis

region can be written in the form

$$U(x_i, y_i) = U_0 h_I(x_i, y_i) \otimes \sum_m \delta(x_i - x_m, y_i - y_m)$$
$$\otimes \left[\delta(x_i, y_i) + (x_i - \Delta x, y_i - \Delta y)\right], \quad (37)$$

where $\Delta x, \Delta y$ are the components of the particle image displacement and U_0 is a constant. The intensity distribution in the back focal plane of the convex lens is proportional to the spatial power spectrum, $W(f_x, f_y)$, of this field, which is given by

$$W(f_x, f_y) = \left| \int_{-\infty}^{\infty} \int_{-\infty}^{\infty} U(x_i, y_i) \exp(j\, 2\pi[f_x x_i + f_y y_i]) \mathrm{d}x_i \mathrm{d}y_i \right|^2,$$
$$= H_I(f_x, f_y) \times \{1 + \cos[2\pi(f_x \Delta x + f_y \Delta y)]\}$$
$$\times \sum_m \sum_n \left(1 + \cos\{2\pi[f_x(x_m - x_n) + f_y(y_m - y_n)]\}\right). \quad (38)$$

In this expression, the function $H_\mathrm{I}(f_x, f_y)$ is the spatial power spectrum of the incoherent impulse response function and represents a broad envelope. The second factor in this expression represents a cosinusoidal modulation and consists of parallel fringes analogous to the Young's fringes formed by two slits. The spatial frequency of these fringes is proportional to the particle velocity. The final factor represents a further modulation in the form of a set of cosinusoiudal fringes, each generated by the cross particle pairing of the images within the interrogation region, and can be considered as multiplicative noise.

If one image pair is present within the interrogation region, there are no cross-particle terms and the fringes have a form which is directly analogous to a two-dimensional version of a single-particle burst spectrum in LDV. For this reason much of the early work in PIV concerned the development of techniques to measure the spatial frequency of the fringes using methods which can be considered analogous to the ratemeters in LDV [34]. However, the highly correlated multiplicative noise, which is largely absent in an LDV signal, makes this type of analysis highly problematic in PIV. In order to separate the noise, further spectral analysis is necessary. The spatial Fourier transform of the spatial power spectrum is the spatial autocorrelation function [18], and is given by

$$R_{UU}(u, \nu) = \int_{-\infty}^{\infty} \int_{-\infty}^{\infty} U(x, y) U^*(x - u, y - \nu) \mathrm{d}x \mathrm{d}y,$$
$$= R_{h_I h_I}(u, \nu)$$
$$\otimes [\delta(u - \Delta x, \nu - \Delta y) + 2\delta(u, \nu) + \delta(u + \Delta x, \nu + \Delta y)]$$
$$\otimes \sum_m \sum_n \delta[u - (x_m - x_n), \nu - (x_m - x_n)], \quad (39)$$

where $R_{h_I h_I}(u, \nu)$ is the autocorrelation function of the incoherent impulse response function. Expanding and rearranging this expression gives

$$R_{UU}(u,\nu) = 2NR_{h_Ih_I}(u,\nu)$$
$$+NR_{h_Ih_I}(u,\nu) \otimes [\delta(u-\Delta x, \nu-\Delta y) + \delta(u+\Delta x, \nu+\Delta y)]$$
$$+R_{h_Ih_I}(u,\nu) \otimes \begin{bmatrix} [\delta(u-\Delta x, \nu-\Delta y) + 2\delta(u,\nu) + \delta(u+\Delta x, \nu+\Delta y)] \\ \otimes \sum_{m,n}\sum_{m\neq n} \delta[u-(x_m-x_n), \nu-(x_m-x_n)] \end{bmatrix} \quad (40)$$

The first two terms are the *central* and *signal* peaks, representing the zero frequency and fringe frequency in the Young's fringe pattern, respectively. The final term is the noise which can be considered as the relative uncertainty in the particle pairing. Both the spatial power spectrum and the autocorrelation function are second-order statistics of the distribution of light transmitted by a PIV transparency and, due to their Fourier transform relationship, contain identical information. It is clear, however, that the particle displacement can be retrieved more easily from the autocorrelation function using relatively straightforward peak-location algorithms. The methods of analysis are illustrated in Fig. 18, which shows typical realisations of (a) the modeled interrogation region, (b) the spatial power spectrum, and (c) the spatial autocorrelation. It can be seen that, although the spatial power spectrum contains several distinct spatial-frequency components, the dominant frequency can be identified in the autocorrelation function and the particle displacement can be found using a peak-finding algorithm based on the centre of intensity [5].

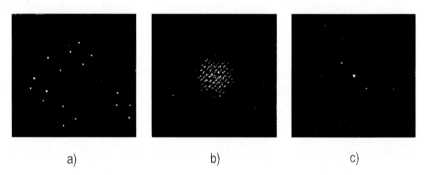

Fig. 18. PIV analysis: (**a**) signal, (**b**) Youngs fringe, and (**c**) autocorrelation

Correlation analysis of PIV transparencies can be accomplished using digital image processing to perform spectral processing of the Young's fringe distribution or by direct calculation of the autocorrelation function of the transmitted intensity distribution [5]. Typically, digital processing systems are capable of analysing transparencies at rates of a few vectors per second. However, it is also possible to analyse PIV transparencies at considerably faster rates using parallel optical processing techniques [35,36].

Clearly, the expected number of particle image pairs within each interrogation region has a strong influence on the ability of the analysis system to accurately identify and measure the position of the signal peaks. Using models similar to that described above, *Keane* and *Adrian* [37] have investigated the random and systematic bias errors introduced by the method of analysis. In essence their work showed that ideally each interrogation region should contain around 10 particle image pairs and should be digitised to a resolution of at least 128×128 pixels, although more particle image pairs are necessary if strong velocity gradients are present within the region. If this is the case, the measured velocity is a good estimate of the two-dimensional fluid at the *centre of the measurement volume*, at the time *midway between the exposures*.

Although the autocorrelation analysis described above provides a good estimate of the fluid velocity it should be noted that an ambiguity exists in the sign of the velocity since the order of the exposures is not recorded. In addition, it is evident that the minimum measurable particle image displacement must be greater than the image diameter in order for the signal peaks to be separated from the zero-order peak. These deficiencies are overcome by the methods discussed in the following section.

4.2 Removal of Directional Ambiguity

The directional ambiguity problem in PIV is analogous to that observed when no frequency shift is present in an LDV system and can be solved in a similar manner. In PIV an *image shift*, rather than frequency shift, can be introduced at the recording stage using either a rotating mirror [38] or a birefringent plate [39]. In addition to these techniques it is also possible to label the images corresponding to each exposure. If relatively low speed flows are of interest, it is straightforward to record the images on separate frames, although care must be taken to ensure registration. *Image labelling* can also be accomplished using a colour-sensitive emulsion and two spectrally separated laser sources to record each exposure [40]. Dual reference beam, image plane holography and encoding methods using adaptive and polarisation-sensitive pupil functions have also been investigated for this purpose [41,42].

In addition to the removal of directional ambiguity, image shifting and labelling techniques also offer greater dynamic range. With reference to the analysis presented in the previous section, the first and second exposure images can be modelled as

$$A(x,y) = h_\mathrm{I}(x,y) \otimes \sum_m \delta(x - x_m, y - y_m) , \tag{41}$$

and

$$B(x,y) = h_\mathrm{I}(x - \Delta x, y - \Delta y) \otimes \sum_m \delta(x - x_m, y - y_m) . \tag{42}$$

The corresponding cross-correlation function, $R_{AB}(u,\nu)$, is given by,

$$R_{AB}(u,\nu) = R_{h_\mathrm{I} h_\mathrm{I}}(u,\nu) \otimes \delta(u - \Delta x, \nu - \Delta y)$$

$$\otimes \sum_m \sum_n \delta[u - (x_m - x_n), v - (x_m - x_n)] ,$$
$$= N R_{h_I h_I}(u - \Delta x, \nu - \Delta y) + R_{h_I h_I}(u, \nu)$$
$$\otimes \sum_{m,n} \sum_{m \neq n} \delta[u - (x_m - x_n), v - (x_m - x_n)] . \tag{43}$$

Comparison with (40) shows that the cross-correlation of image-labelled data does not contain a central peak and thus displacement measurements of less than one particle diameter are possible. In addition it should be noted that the noise contribution is considerably reduced.

4.3 PIV Measurements in Practice

It is clear that for the methods of correlation analysis outlined above the PIV recordings should ideally be of high spatial resolution. For this reason the majority of studies have used wet-film processing methods, usually using a large-or medium-format camera. In contrast with the idealised imaging geometry assumed in the previous section, it is worth pointing out that most cameras produce a slightly distorted image of the light sheet, particularly if a wide-angle objective lens is used. In effect these lenses record the *angular* position of the particles and the measured velocity components are therefore those perpendicular to the viewing direction. This effect can be used to advantage in stereoscopic systems to estimate the out-of-plane component of velocity but is usually ignored for typical recording geometries using relatively long-focus objective lenses.

As for LDV, it is a general requirement of a PIV system to provide valid velocity data from the light scattered by small particles. In PIV the problem is exacerbated somewhat by the relative inefficiency of side scatter and the small camera aperture which is often required to ensure that the depth of field is greater than the thickness of the light sheet and that optical aberrations are minimised. A high power laser source is usually used in PIV studies, since the output can be fashioned easily into a light sheet. For low-speed flows the continuous output of an argon-ion laser is often used. An interesting and efficient method for creating a light sheet from a continuous laser is to use a rotating prism or a galvanometer mirror to scan the beam through the flow [43]. For high-speed flows a pulsed laser source is necessary to provide the required exposure. Originally double-pulsed Ruby lasers were used for PIV studies but it is now more common to use a pair of frequency-doubled Nd:YAG lasers, giving increased flexibility.

In early PIV studies, the analysis hardware often consisted of a computer driven translation stage to interrogate the transparency using a low-power laser beam and a CCD array to digitise the Young's fringe pattern or particle image distribution directly. With the advances in computer imaging, however, the process can be accomplished more efficiently using digital scanning technology. Currently, personal-computer-based image-processing hardware is capable of processing a few vectors per second and commercial packages have been written specifically for PIV transparency analysis.

Figure 19 shows a good example of a practical PIV measurement, from an extensive study of the flow within an internal combustion (IC) engine by *Reeves* [44]. Optical access to the combustion chamber was provided through a fused-silica window in the piston crown, and the complete swept-volume could be viewed through the fused-silica cylinder. The result shows the flow ahead of a flame front in a horizontal plane approximately 10 mm below the spark plug. In this study the flame front could be accurately located since the silicone oil used for seeding was consumed by the combustion process. It can be seen that, as expected, the flow is largely radial as the unburned gas is compressed.

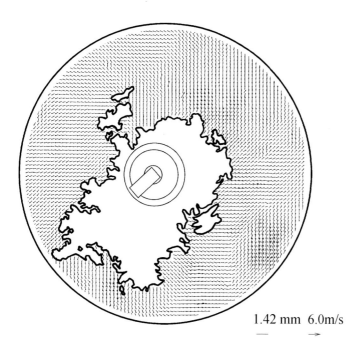

Fig. 19. PIV study of flow in a fired IC engine (after *Reeves* [44])

More recently, complete PIV systems have become commercially available using high-resolution video cameras rather than photographic recordings. Inevitably spatial resolution is sacrificed to photographic recording. However, there are clear benefits in terms of the time required to make a particular flow study and to collect the quantity of flow data necessary for statistical predictions. An interesting feature of video-based systems is the electronic implementation of image shifting using the CCD chip itself to translate the recorded image between exposures. Image labelling is also possible using line-transfer or dual-chip gated CCDs [45]. Video-based PIV analysis is similar to that discussed in the previous section, and commercial systems use optimised processors capable of a few hundred vectors per second for this purpose.

Both the photographic and video-based PIV methods outlined in this section are capable of two-component velocity measurement of fluid flow in a single plane of interest. Recently much attention has concerned the extension of these techniques to provide three-component measurements, and this is discussed in the following section.

4.4 Three-Dimensional PIV

Although the development of the planar two-component PIV techniques described in the previous section represents a significant advance in experimental fluid dynamics, three-component measurements are clearly of benefit in order to provide a more complete description of the flow structure. Several planar and whole-field pulsed laser velocimeters, which measure three-components of fluid velocity within a plane and throughout the flow field respectively, have been proposed in the literature. Three-component planar PIV can be implemented using a colour-sensitive recording medium and overlapping dual-colour light sheets [46]. In this way the depth information is effectively encoded as the colour balance of each of the particle images and can be retrieved by correlation methods. Much research has also concerned the investigation of stereoscopic photogrammetric techniques applied to PIV recordings [47]. In essence, a stereoscopic PIV image of the light sheet is produced from two different perspectives, for example using a pair of cameras fitted with wide-angle objectives. As mentioned in the previous section, these cameras measure the velocity components in a plane which is perpendicular to the viewing direction. If the optical axes are parallel but the cameras are laterally displaced, a different viewing angle is obtained for each recording of the flow region and the out-of-plane component of the velocity can be estimated. Three-component stereoscopic PIV has been studied by *Lawson* and *Wu* [48] who showed that in general the random error in the out-of-plane component is a factor of 2-4 times worse than that of the in-plane components, depending on the configuration used. Simultaneous measurement of the three components of fluid velocity throughout the flow region has also been investigated holographically, and this method is discussed in the following section.

4.5 Holographic Particle Image Velocimetry (HPIV)

Holographic images contain information concerning the three-dimensional distribution of recorded objects. Pioneering work by *Trolinger* [49] and *Thompson* [50] within the field that is now referred to as particulate holography demonstrated that, amongst other parameters, the velocity could be measured by focusing into a double-or multiple-pulsed holographic image of a seeded flow using a travelling microscope. This type of analysis is usually referred to as *particle tracking* since individual particles are identified and their position is effectively tracked through the flow volume. As in the analysis of conventional photographic PIV, identification of individual particles is computationally intensive and detailed analysis of holographic records in this way is laborious. Despite this a number of impressive studies have been performed in this way including an interesting study of flow fields in the microgravity environment aboard a space-borne laboratory [12].

Methods of extracting the velocity using the correlation-based techniques of PIV have also been investigated as means of efficiently analyzing holographic records. This form of velocimetry, usually referred to as holographic particle image velocimetry, does not rely on identification of individual particles and in essence allows a group of particles to be identified and their displacement measured. To date the majority of work has considered the correlation of the *intensity distribution* captured on a CCD array image when a travelling microscope is focused into the reconstructed image. In particular *Barnhart* et al. [51] have demonstrated measurements of this type using the stereoscopic holographic recording and reconstruction geometry shown in Fig. 20. In essence, using a dual-reference-beam geometry two labelled images from different perspectives are recorded. The three components of the flow velocity are then deduced from cross-correlation of the labelled images obtained from each perspective in a similar manner to the planar stereoscopic methods outlined above. It is worth noting that, if an extended flow volume is recorded in this way, out-of-focus images are present in each reconstruction. However, the phase-conjugate reconstruction process allows a relatively large aperture recording to be made without severe aberrations such that particles rapidly defocus with depth. Using state-of-the-art digital processing hardware full-field, three-component data sets containing approximately 0.5 million velocity vectors have been extracted from a fluid volume in this way. A second approach to the analysis of holographic records is to

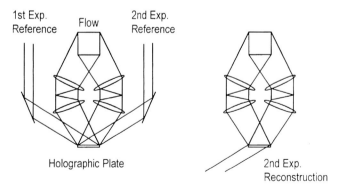

Fig. 20. The stereoscopic HPIV system of *Barnhart* et al. [51]

correlate the *complex amplitude* describing the particle image distribution, which can be accomplished conveniently and at high speed using basically the same optical-correlation techniques used for the analysis of photographic PIV transparencies [36]. This approach differs fundamentally from the intensity based correlation analysis and for this reason it will now explained in some detail.

A configuration capable of processing the complex amplitude distribution recorded by a particulate hologram is shown in Fig. 21. In essence, a phase conjugate reference wave is used to reconstruct the double-pulsed hologram and an aperture is placed in the real image. The complex amplitude transmitted by the aperture, $U(x,y)$, can be considered as the sum of the contributions, $A(x,y)$ and $B(x,y)$, which result from the scattered light recorded by each of the two exposures such that

$$U(x,y) = A(x,y) + B(x,y) \ . \tag{44}$$

With reference to (32), the complex amplitude due to a plane wave scattered by small particles can be written as the sum of quadratic wavefronts such that

$$A(x,y) = \sum_m \frac{A_m}{j\lambda z_m} \exp\left(\frac{j 2\pi z_m}{\lambda}\right) \exp\left\{\frac{j 2\pi}{\lambda z_m}\left[(x-x_m)^2 + (y-y_m)^2\right]\right\}. \tag{45}$$

It is assumed that these particles move in unison a distance $\Delta s = (\Delta x^2 + \Delta y^2 + \Delta z^2)^{\frac{1}{2}}$, then using the Fresnel approximation once again the optical field as a result of the second exposure recording can be written

$$B(x,y) = A(x,y) \otimes \frac{1}{j\lambda \Delta z} \exp\left\{\frac{-j\pi}{\lambda \Delta z}\left[(x-\Delta x)^2 + (y-\Delta y)^2\right]\right\} \ , \tag{46}$$

where constant phase factors have been omitted for simplicity. Hence the

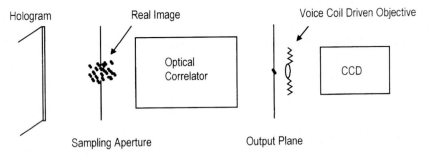

Fig. 21. Optical correlation of holographic recordings

complex amplitude describing the double-pulsed reconstruction can be written

$$U(x,y) = A(x,y) \otimes \left(\delta(x,y) + \frac{1}{j\lambda\Delta z} \right.$$
$$\left. \exp\left\{ \frac{-j\pi}{\lambda\Delta z} [(x-\Delta x)^2 + (y-\Delta y)^2] \right\} \right). \quad (47)$$

It is interesting to consider the diffraction pattern observed in the far focal plane of a convex lens, which is proportional to the spatial power spectrum and is given by

$$W(f_x, f_y) = |\tilde{U}(f_x, f_y)|^2 ,$$
$$= |\tilde{A}(f_x, f_y)|^2 \times \{2 + 2\cos\left[\pi\lambda\Delta z(f_x^2 + f_y^2)\right.$$
$$\left. - 2\pi(f_x\Delta x + f_y\Delta y) - \pi/2 \right]\}. \quad (48)$$

It can be seen that the far-field pattern corresponds to the spatial power spectrum of the first exposure images modulated by a fringe pattern which is a function of the displacement. The loci of the minima in this fringe pattern are defined by the set of equation

$$\lambda\Delta z(f_x^2 + f_y^2) - 2(\Delta x f_x + \Delta y f_y) - 1/2 = 2n + 1 , \quad (49)$$

where n is an integer. In contrast with the Young's type fringes of (37), the fringes are a family of concentric circles, as shown in Fig. 22, centred at $(\Delta x/\lambda\Delta z, \Delta y/\lambda\Delta z)$ with radii, r_n, given by,

$$r_n = \left[(\Delta x^2 + \Delta y^2)/\lambda\Delta z^2 + (2n + 3/2)\lambda\Delta z\right]^{\frac{1}{2}} . \quad (50)$$

Although in principle the three-dimensional particle displacement could be extracted from the fringe pattern as in the two-dimensional case, the

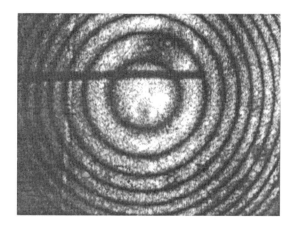

Fig. 22. Circular fringes formed in the back focal plane of a convex lens

finite scattering population results in noise which is manifest in the $|\tilde{A}(x,y)|^2$ term in (48). In a similar manner to the two-dimensional case, the dominant fringe pattern can be identified from the complex autocorrelation function, $R_{UU}(u,\nu)$, given by

$$R_{UU}(u,\nu) = R_{AA}(u,\nu)$$
$$\otimes \left(\frac{-1}{j\lambda \Delta z} \exp\left\{ \frac{j\pi}{\lambda \Delta z} \left[(u - \Delta x)^2 + (\nu - \Delta y)^2 \right] \right\} + 2\delta(x,y) \right.$$
$$\left. + \frac{1}{j\lambda \Delta z} \exp\left\{ \frac{-j\pi}{\lambda \Delta z} \left[(u + \Delta x)^2 + (\nu + \Delta y)^2 \right] \right\} \right), \quad (51)$$

where $R_{AA}(u,y)$ is the complex autocorrelation function of $A(x,y)$. Expanding this function gives

$$R_{AA}(u,\nu) = \sum_m |A_i|^2 \delta(u,\nu)$$
$$+ \sum_{m,n} \sum_{m \neq n} A_m A_n^* \frac{-1}{j\lambda(z_m - z_n)}$$
$$\exp\left\{ \frac{-\pi j}{\lambda(z_m - z_n)} \left[(x_m - x_n - u)^2 + (y_m - y_n - \nu)^2 \right] \right\}. \quad (52)$$

The first term in this equation describes a peak at the origin with magnitude proportional to the sum of the scattered intensity from each particle. The second term consists of randomly phased quadratic wavefronts resulting from uncorrelated particle images and can be considered as noise. Thus the complex autocorrelation can be written,

$$R_{UU}(u,\nu) = \frac{-\sum_m |A_m|^2}{j\lambda \Delta z} \exp\left\{ \frac{j\pi}{\lambda \Delta z} \left[(u - dx)^2 + (\nu - \Delta y)^2 \right] \right\}$$
$$+ 2 \sum_m |A_m|^2 \delta(u,\nu)$$
$$+ \frac{\sum_m |A_m|^2}{j\lambda \Delta z} \exp\left\{ \frac{-j\pi}{\lambda \Delta z} \left[(u + \Delta x)^2 + (\nu + \Delta y)^2 \right] \right\}$$
$$+ \text{noise terms} \quad (53)$$

It can be seen from comparison with (40) that the complex autocorrelation function consists essentially of a central peak and two signal peaks; however, in this case the signal peaks *represent* quadratic wavefronts. If optical-correlation techniques are used to compute the complex correlation the signal peaks are *physically* quadratic wavefronts. In this way the wavefronts can be brought to focus at a distance proportional to Δz in front of and behind the correlation plane, as shown in Fig. 23. The quadratic wavefronts corresponding to the noise terms in (52) are generally out of focus and can be removed from an image formed on a CCD camera using a suitable threshold.

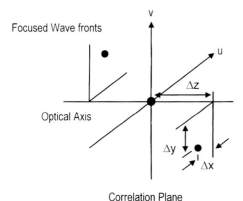

Fig. 23. Output of an optical correlator

In practice, the depth information can be retrieved from the output of an optical correlator by examining the image captured by a high-speed CCD camera equipped with a voice-coil-driven objective lens. The resolution to which the depth component, Δz, can be measured depends on the numerical aperture of the holographic recording. However, it can be shown that the process is tolerant to phase distortion introduced during the holographic recording process and other aberrations which are common to both exposures [52]. For this reason measurements have been made from large-aperture holograms (NA > 0.5) with subwavelength resolution in each of the displacement components [53].

HPIV using optical correlation is a relatively new development and more study is necessary to characterize the method as a practical measurement technique. It is clear that the aperture placed in the real image defines the lateral dimensions of the measurement volume in the plane of the aperture. The light transmitted by the aperture has contributions from particle images recorded throughout the depth of the illuminated region of the flow. In general, these images will have uncorrelated displacements and will largely be removed by the correlation process. However, it should be noted that this is not always the case and, although the relative intensity of these additional contributions will diminish as the inverse square of the distance to the aperture, this is a relatively slowly varying function compared to the Gaussian forms which define the measurement volumes in LDV and conventional PIV.

Finally, it is worth pointing out that the basic HPIV method outlined above is subject to directional ambiguity in the same way as conventional PIV. The problem can be solved in a similar manner, however, using variations of image shifting [53], or multiplexed holographic image-labelling techniques [41]. In addition, *Hinsch* et al. [54] proposed a time-of-flight multiplexed holographic configuration which utilizes a partially coherent laser to record multiple light sheets in the fluid. The practical implementation of these techniques in a challenging measurement situation has yet to be achieved.

5 Conclusions, Discussion and Future Development

In this chapter the fundamentals of laser Doppler and pulsed laser velocimetry have been introduced with particular reference to optical configurations and signal analysis methods. There are clearly many similarities between the techniques discussed and some clear distinctions to be made.

The Doppler signal output from an LDV system is essentially a sum of individual Doppler bursts resulting from the transitions of particles through the measurement volume. This volume is well defined by the configuration of the transmitter optics and in theory can be a region as small as a few wavelengths in diameter. Since the measurement volume is finite, both the mean frequency and bandwidth of the Doppler burst are proportional to the particle velocity and a *proportional error* is therefore expected in a given measurement. For a cross-beam velocimeter this error is inversely proportional to the number of fringes within the measurement volume. Clearly there is a compromise between velocity and spatial resolution and the majority of current research within the field of LDV is concerned with improving the estimation of burst frequency, thereby increasing spatial resolution [55]. In practical LDV systems, the upper limit to the measurement range is ultimately limited by the applied frequency shift that is necessary to resolve directional ambiguity. Using modern burst spectrum analysis methods and digital electronics there is effectively no lower limit to the velocity which can be measured using LDV, which gives the technique unsurpassed dynamic range.

The PDV technique utilizes an iodine cell to demodulate the Doppler signal directly. The advantage of this technique is that the light scattered from many regions in the flow can be analysed in parallel. In order to define separate measurement volumes it is necessary to illuminate the flow with a laser light sheet. The ability to produce a light sheet of near constant thickness across the flow volume provides a lower limit to the achievable depth of the measurement volume, and this is typically 1–2 orders of magnitude greater than that achievable using LDV. The precision to which the Doppler frequency can be measured in a PDV set up is determined by the experimental parameters. These include the relative stability of the iodine cell and laser source together with limits imposed by the requirement of simultaneously recorded, and aligned, Doppler and reference images. These factors result in an *absolute error* in velocity measurement and provide a lower limit to the measurement range. The upper limit to the measurement range of PDV is set by the useful bandwidth of the iodine cell. These two considerations mean that at present PDV is ideally suited to the measurement of large-scale transonic flows, although much research is directed toward reducing the lower velocity limit [28].

Particle-image velocimetry differs fundamentally from laser Doppler velocimetry in that particle displacement rather than velocity is measured. The signals produced in PIV are in the form of photographic, video or holographic records of particle images at two or more times and the velocity measurement range is readily adjusted by changing the pulse separation. Spatial correlation analysis of small interrogation regions taken from these records produces signals which are similar to the spectrum produced from a Doppler burst in LDV; however, additional peaks are present which reflect the uncertainty in particle pairing. An *absolute error* in the measurement in the particle location, which is proportional to the size of the particle images, is present, and this provides a lower limit to the measurement range. An upper limit to the measurement range is set by the size of the interrogation region, which also dictates the spatial resolution, and a compromise between these requirements must be achieved in practice. It is worth noting that HPIV is able to measure displacements which are smaller than the wavelength of light and has the potential to provide data of superior spatial resolution [56].

Recent developments in high-resolution video and digital processing technology has recently led to the commercialization of specialist PIV instrumentation. Although the video format provides less resolution than photographic (or holographic) recording [57], no wet processing is necessary and the time taken to collect data is therefore reduced dramatically. At present the analysis of video data is usually performed using correlation-based techniques, which are fundamentally the same as those applied to photographic records. It is worth noting that these methods effectively fit a linear particle-displacement model to the set of particle images recorded within each interrogation region. In effect, rather than producing a velocity vector from each individual particle-image pair, a single vector is produced which represents the most likely particle-image displacement in that particular region. This apparent loss of information can be attributed to the fact that the analysis technique is required to identify the pairing at each location, taking no account of the results obtained from adjacent regions. If this requirement is relaxed, then it is possible to utilize the information present in a given record more efficiently by fitting a non-linear displacement model to the data. More efficient data analysis is chiefly of interest in the analysis of lower resolution video records but is also applicable to photographic transparencies as a way of increasing the tolerance to spatial velocity gradients. The search for optimal analysis procedures has been a part of PIV development from its very conception and remains a major topic of current research [58,59,60].

References

1. W. Merzkirch: *Flow Visualisation*, Academic, New York (1987)
2. F. Durst, A. Melling, J. H. Whitelaw: *Principles and Practice of Laser Doppler Anemometry*, Academic, New York (1976)
3. T. S. Durrani, C. A. Greated: *Laser Systems for Flow Measurement*, Plenum, New York (1977)
4. R. J. Adrian: Laser Velocimetry, in R. J. Goldstein (Ed.): *Fluid Mechanics Measurements*, Springer, Berlin, Heidelberg (1983)
5. R. J. Adrian: Particle-Imaging Techniques for Experimental Fluid Dynamics, Annu. Rev. Fluid Mech. **23**, 261 (1991)
6. K. D. Hinsch: Particle Image Velocimetry, in R. S. Sirohi (Ed.): *Speckle Metrology*, Dekker, New York (1993)
7. I. Grant: Particle Image Velocimetry, in P. K. Rastogi (Ed.): *Optical Measurement Techniques and Applications*, Artech House (1997)
8. H. C. Van der Hulst: *Light Scattering by Small Particles*, Wiley, New York (1957)
9. M. A. Northrup, T. J. Kulp, S. M. Angel: Fluorescent Particle Image Velocimetry, Appl. Opt. **30**, 3034 (1991)
10. D. P. Towers, C. E. Towers, C. H. Buckberry, M. Reeves: Directionally Resolved Two-Phase Flow Measurements Using PIV with Fluorescent Particles and Colour Recording, 9th Int. Symp. Laser Applications in Fluid Mechanics, Lisbon (1998)
11. B. Hiller, R. K. Hanson: Simultaneous Planar Measurements of Velocity and Pressure Fields in Gas Flows using Laser Induced Fluorescence, Appl. Opt. **27**, 33 (1988)
12. J. D. Trollinger, M. Rottenkolber, F. Elandalouissi: Development and Application of Holographic Particle Image Velocimetry Techniques for Microgravity Applications, Meas. Sci. Technol. **8**, 1573 (1997)
13. A. Melling: Tracer Particles and Seeding for Particle Image Velocimetry, Meas. Sci. Technol. **8**, 1406 (1997)
14. Y. Yeh, H. Cummins: Localised Fluid Measurements with a HeNe Laser Spectrometer, Appl. Phys. Lett. **4**, 176 (1964)
15. R. L. Bond: Contributions of System Parameters in the Doppler Method of Fluid Velocity Measurement, PhD thesis, Univ. of Arkansas (1968)
16. M. J. Rudd: A Laser Doppler Velocimeter Employing the Laser as a Mixing Oscillator, J. Phys. E **1**, 723 (1968)
17. W. T. Mayo: Laser Doppler Flow Meters - A Spectral Analysis, PhD thesis, Georgia Inst. of Technology (1969)
18. J. W. Goodman: *Introduction to Fourier Optics*, McGraw-Hill, New York (1968)
19. M. J. Rudd: A New Theoretical Model for the Laser Doppler Velocimeter J. Phys. E **2**, 55 (1969)
20. L. Lading: A Fourier Optical Model for the Laser Doppler Velocimeter, Opto-Electron. **4**, 385 (1972)
21. R. N. James, W. S. Babcock, H. S. Seifert: A Laser Doppler Technique for the Measurement of Particle Velocity, AIAA J. **6**, 160 (1968)
22. J. D. C. Jones: New Opto-Electronic Technologies for Laser Anemometers, in L. Lading, G. Wigley, P. B. Buchave (Ed.): *Optical Diagnostics for Flow Processes*, Plenum, New York (1994)

23. G. K. Hargraves, G. Wigley, J. Allen, A. Bacon: Optical Diagnostics and Direct Injection of Liquid Fuel Sprays, in: *Optical Methods and Data Processing in Heat and Fluid Flow*, City University, London (1998) pp. 121
24. H. Komine, S. J. Brosnan: Real Time Doppler Global Velocimetry, in: AIAA 29th Aerospace Sciences Meeting, Reno, N. V. (1991), Tech. Dig. paper 91-0337
25. V. S. S. Chan, A. L. Heyes, D. I. Robinson, J. T. Turner: Iodine Absorption Filters for Doppler Global Velocimetry, Meas. Sci. Technol. **6**, 784 (1995)
26. R. W. Ainsworth, S. J. Thorpe: The Development of a Doppler Global Velocimeter for Transonic Turbine Applications, in: ASME Gas Turbine and Aero Engine Congress and Exposition, The Hague, (1994), Tech. Dig. paper 94-GT-146
27. I. Roehle, R. Schodl: Evaluation of the Accuracy of the Doppler Global Technique,in: Optical Methods and Data Processing in Heat and Fluid Flow, City University, London (1994) pp. 121
28. R. W. Ainsworth, S. J. Thorpe, R. J. Manners: A New Approach to Flow-Field Measurement-A View of Doppler Global Velocimetry Techniques, Int. J. Heat Fluid Flow **18**, 116 (1997)
29. R. L. McKenzie: Measurement Capabilities of Planar Doppler Velocimetry Using Pulsed Lasers, Appl. Opt. **35**, 948 (1996)
30. J. F. Meyers: Development of Doppler Global Velocimetry as a Flow Diagnostics Tool, Meas. Sci. Technol. **6**, 769 (1995)
31. T. D. Dudderar, P. G. Simpkins: Laser Speckle Photography in a Fluid Medium, Nature **270**, 45 (1977)
32. C. J. D. Pickering, N. A. Halliwell: Speckle Photography in Fluid Flows: Signal Recovery with Two-Step Processing, Appl. Opt. **23**, 1129 (1984)
33. J. M. Burch, J. M. J. Tokarski: Production of Multiple Beam Fringes from Photographic Scatterers, Opt. Acta **15**, 101 (1968)
34. R. Meynart: Digital Signal Processing for Speckle Flow Velocimetry, Rev. Sci. Instrum. **53**, 110 (1982)
35. J. M. Coupland, N. A. Halliwell: Particle Image Velocimetry: Rapid Transparency Analysis Using Optical Correlation, Appl. Opt. **27**, 1919 (1988)
36. Z. Q. Mao, N. A. Halliwell, J. M. Coupland: High Speed Analogue Correlation for PIV Transparency Analysis Using a Ferroelectric Liquid Crystal Spatial Light Modulator, Opt. Lasers Eng. **24**, 301 (1996)
37. R. D. Keane, R. J. Adrian: Optimisation of Particle Image Velocimeters. Part I: Double Pulsed Systems, Meas. Sci. Technol. **1**, 1202 (1990)
38. R. J. Adrian: Image Shifting Technique to Resolve Directional Ambiguity in Double Pulsed Velocimetry, Appl. Opt. **23**, 3855 (1986)
39. C. C. Landreth, R. J. Adrian: Electro-Optic Image Shifting for Particle Image Velocimetry, Appl. Opt. **29**, 4216 (1988)
40. A. Cenedese, G. P. Romano: Colours in PIV, Atlas Vis. **3**, 83 (1995)
41. J. M. Coupland, C. J. D. Pickering, N. A. Halliwell: Particle Image Velocimetry: Theory of Directional Ambiguity Removal Using Holographic Image Separation, Appl. Opt. **26**, 1576 (1987)
42. M. Reeves, N. J. Lawson, N. A. Halliwell, J. M. Coupland: Particle Image Velocimetry: Image Labelling by Encoding of the Point Spread Function by Application of a Polarisation-Sensitive Pupil Mask, Appl. Opt. **34**, 194 (1995)
43. C. Gray, C. A. Greated, D. R. McCluskey, W. J. Eason: An Analysis of the Scanning Beam PIV Illumination System, Meas. Sci. Technol. **2**, 461 (1991)

44. M. Reeves: Particle Image Velocimetry applied to Internal Combustion Engine In-Cylinder Flows, PhD thesis, University Loughborough (1995)
45. D. R. McClusky, T. Jacobsen: Instrumentation for Real Time PIV Measurements, in: Joint ASME, JSME, and EALA Conference, Hilton Head Island, SC (1995)
46. S. Arndt, C. Heinen, M. Hubel, K. Reymann: Multi-Colour Laser Light Sheet Tomography (MLT) for Recording and Evaluation of Three-Dimensional Turbulent Flow Structures, in: *Optical Methods and Data Processing in Heat and Fluid Flow*, City University, London (1998) pp. 481
47. M. P. Arroyo, C. A. Greated: Stereoscopic Particle Image Velocimetry, Meas. Sci. Technol. **2**, 1181 (1991)
48. N. J. Lawson, J. Wu: Three-Dimensional Particle Image Velocimetry: Experimental Error Analysis of a Digital Angular Stereoscopic System, Meas. Sci. Technol **8**, 1455 (1997)
49. J. D. Trolinger: Particle Field Holography, Opt. Eng. **14**, 383 (1975)
50. B. J. Thompson: Holographic Methods for Particle Size and Velocity Measurement - Recent Advances, in: Holographic Optics II: Principles and Practice, Proc. SPIE **1136**, 308 (1989)
51. D. H. Barnhart, R. J. Adrian, G. C. Papen: Phase Conjugate Holographic System for High Resolution Particle Image Velocimetry, Appl. Opt. **33**, 7159 (1994)
52. J. M. Coupland, N. A. Halliwell: Holographic Displacement Measurements in Fluid and Solid Mechanics: Imunity to Aberrations by Optical Correlation Processing, Proc. R. Soc. **453**, 1053 (1997)
53. D. H. Barnhart, V. S. S. Chan, C. P. Garner, N. A. Halliwell, J. M. Coupland: Volumetric Three-Dimensional Flow Measurement in IC Engines Using Holographic Recording and Optical Correlation Analysis, in: Optical Methods and Data Processing in Heat and Fluid Flow, City University, London (1998) pp. 51
54. K. Hinsch, H. Hinrichs, G. Kuhfahl, P. Meinlschmidt: Holographic Recording of 3-D Flow Configurations for Particle Image Velocimetry (PIV), in: ICALEO'90, Laser Inst. America, Orlando, FL (1991) pp. 121–130
55. C. Tropea: Laser Doppler Anemometry: Recent Developments and Future Challenges, Meas. Sci. Technol. **6**, 605 (1995)
56. H. Royer: Holography and Particle Image Velocimetry, Meas. Sci. Technol. **8**, 1562 (1997)
57. R. J. Adrian: Dynamic Ranges of Velocity and Spatial Resolution of Particle Image Velocimetry, Meas. Sci. Technol. **8**, 1393 (1997)
58. I. Grant, X. Pan: An Investigation of the Performance of Multi-Layer Neural Networks Aplied to the Analysis of PIV Images, Exp. Fluids **19**, 159 (1995)
59. F. Carasone, A. Cenedese, G. Querzoli: Recognition of Partially Overlapped Particle Images Using Kohonen Neural Network, Exp. Fluids **19**, 225 (1995)
60. J. Ko, A. J. Kurdila, J. L. Gilaranz, O. K. Rediniotis: Particle Image Velocimetry via Wavelet Analysis, AIAA J. **36**, 1451 (1998)

Surface Characterization and Roughness Measurement in Engineering

David J. Whitehouse

University of Warwick, UK and Metrology Consultant, Taylor Hobson Ltd., UK
esdjw@eng.warwick.ac.uk

Abstract. Surfaces are becoming more important especially as miniaturization progresses and with the advent of micro mechanics. The role of the surface is explored with respect to control of manufacture and prediction of performances at interfaces and rubbing bodies. This is developed to include some aspects of designer surfaces. Instruments for measuring surfaces are critically examined. These include the well-known stylus method, optical methods and the new generation of scanning probe methods: Atomic Force Microscopy (AFM) and Scanning Tunneling Microscopy (STM). Finally, the ways in which surfaces are characterized from profile traces and over an area are described.

1 General

1.1 Historical

The importance of surfaces has been known for many years. Some of the great names in science suspected the surface to be important in everyday phenomena. Friction is a good example. Leonardo, Amonton and Coulomb all explored friction with regard to loads and surface areas. It took many years to formulate the relationship between friction and the surface [1]. The same is true for lubrication [2] and wear [3].

However, the driving force for measuring surfaces was industrial and not research. In particular mass production, especially in the automotive industry in the 1920s and 1930s, provided the economic need. In the USA car giants such as General Motors played a big part in surface measurement. Around 1930 there was an awareness, perhaps for the first time, that a very smooth surface did not necessarily mean that it was a good surface. The Bentley racing-car company in England around 1926 found that their engines produced in-house to exacting tolerances and roughnesses performed worse than bought-in engines.

It is now known that this shortfall in expected performance was due to the fact that the cylinders were too smooth to retain oil and as a result they were prone to seize during high speed running. At the time it was not so obvious.

In the late 1920s and early 1930s, three important engineering areas were evolving. These were the use of instrumentation for surface measurement, the establishment of national standards for surface roughness, and the use of quality control. As usual these themes did not develop along logical lines but haphazardly with many false starts. Also development was not the same either in rate or discipline for different countries. The countries involved split fairly evenly into two groups: the USA and UK on the one hand and Germany and the Soviet Union on the other.

The Anglo-Saxon approach tended to be more pragmatic than that of the German and Soviet industrialists. In the UK and in the USA immediate problems were attacked with a view to getting solutions quickly. In Germany a longer-term view was held, especially in instrumentation, and the Soviet Union became very involved in tribological aspects of surface behaviour, albeit on a mainly empirical basis.

The two different approaches still permeate the National and International Standards although more in a complementary role than adversarial, as was once the case. An appreciation of this dichotomy helps in understanding the content of the ISO standards in use today.

One simple example given below illustrates this point. It is now fairly well accepted that the surface has a dual role. One aspect is concerned with the influence of the surface functionally, e.g. in terms of friction and wear. The other aspect is in terms of use of the surface as a control and performance monitor of the manufacturing process. In the 1930s the German and Soviet approach was mainly concerned with the surface function; the UK and USA approach tended more towards manufacturing control. The former produced an emphasis on peak parameters and the latter on average parameters. Unfortunately the standards have to cater for both points of view simultaneously, and with the advent of much international subcontracting and multinational organizations the mixture of parameters has proved to be frustrating in some cases and costly, especially to instrument makers.

In instrument terms there has been a similar international split. Generally, but not exclusively, the German instrument makers [4] at least at first, concentrated their efforts on non-contacting methods such as optical methods and capacitance methods, whereas the UK and the US developed stylus methods.

1.2 Nature and Importance of Surfaces

This is shown schematically in Fig. 1 and 2. These figures bring together what has been said above. It is important to realize that these figures are quite general and relate to any method of measurement whether optical or otherwise.

Figure 1 shows the central role of surface measurement from the manufacturing point of view. It has been discovered recently that the manufacture can not only be controlled quite effectively by looking at the surface which has been generated on the workpiece, but with suitable instrumentation and data processing the machine tool can be monitored for vibration, slideway

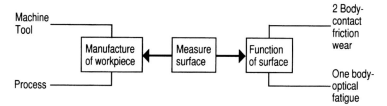

Fig. 1. Usefulness of surfaces

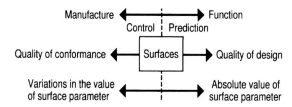

Fig. 2. General scheme for measurement

error, chatter and so on. The philosophy is simple. It is based on the fact that the surface is extremely sensitive to the method of production. If anything changes either in the process or within the machine tool it will reveal itself on the surface: it may not be a large change but it will be present and significant [5].

The emphasis from right to left in Fig. 2 [6] therefore is one of process control and machine tool diagnostics. This is quite a different emphasis to that regarding the surface function. Here the quality of the surface can be absolutely fundamental when performance is concerned. Hence the general scheme for measurement is shown in Fig. 2. In this figure it should be made clear that it is not necessarily the same surface parameter which is referred to regarding control and prediction.

It can be seen that the surface and hence its measurement are important across the spectrum of quality control. Although the two roles of surface finish have now been clearly identified (Fig. 1 and 2), the measurement of surface texture has not become easier. In fact, as more manufacturing processes are used and more applications for surface texture have been found, the measurement has become more difficult. This is because the texture is extremely complex. It has more variation in size and shape than practically any other type of signal. In height, for example, surfaces can have an amplitude of millimetres in, say, shaping down to nanometres or less in polishing: a range of a million to one or more. In length the range of values is not so large but is still in the tens of thousands. Surfaces tend to have spacing parameters which are larger than the height parameters by an order of about one hundred as a result of the tool shape and its cut in the surface. In addition to the scale of size problem, surface profiles can have a wide range in shape, as illustrated

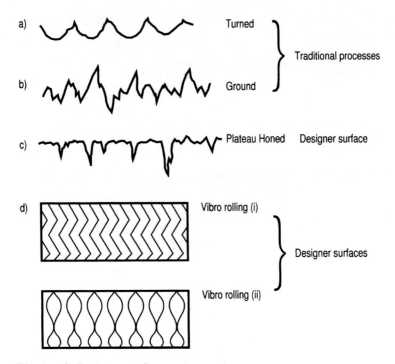

Fig. 3a–d. Designer surfaces

in Fig. 3. Also, there is a question of how to deal with the areal patterns, the lay, generated by tool movement or other means (see Fig. 3d). The patterns developed in face milling, for example, are very difficult to assess because the general pattern changes from point to point. Every method of measurement, whether optical or not, runs into this problem.

It has to be stated that as a general rule profiles of the surface and their associated parameters ranging from average values to random processes are sufficient to control the manufacturing processes, as will be seen in Sect. 5. However, for machine tool problems, and to an even greater degree for performance, it is the areal information which is the most useful.

1.3 Trends

1.3.1 Scale of Size Arguments

Traditionally the surface texture has been within the micron scale of size superimposed on macro-scale workpieces. As an example, for a journal bearing shaft of 25 mm diameter a surface texture of 0.2 µm roughness is typical. These dimensions have been chosen to satisfy a given dynamic need. It is well known that there are three elements in the dynamic force and energy equations: inertial effects, damping effects, and stiffness effects, respectively.

Their importance is determined to a large extent by their physical dimensions: volume, area, and length. This is illustrated in Table 1. It can be seen from the table that at micrometre and nanometre scales, volume effects become small - at a rate of $1/l^3$ or $1/l^5$ depending on the nature of the function, i.e. translation or rotation.

Table 1. Trend in surfaces with scale of size [4,6]

(A) Energy and Force	Components of Force and Dimension		
Scale	Inertia (Volume)	Damping (Area)	Elastic (Length)
(a) Macro scale - Effectiveness	Large	Medium	Small
(b) Micro scale - Effectiveness	Small	Medium	Large
(c) Nano scale - Effectiveness	Negligible	Large	Very Large
(B) Information	Components of Dimension		
	Volume	Area	Length
Effectiveness	Large but difficult	Large	Small

Obviously, in nanotechnology, stiffness or elastic effects should be dominant. They are in principle, but they depend mainly on material properties, and at this scale high stiffness is difficult to achieve. The greatest potential is in area effects, which include damping, surface tension, etc. It is these effects which depend so much on the roughness of the surface. Unfortunately, although the reduction in the general size of the object is straightforward, maintaining the same scale reduction of the roughness is not generally possible. This has had the result of increasing the ratio of roughness to linear dimension to the detriment of performance control. This is one reason why the measurement and control of roughness is so important. It is well known that frictional effects due to roughness are one of the limits to the performance of micro-dynamic systems and MEMS (micro-electro-mechanical systems). Also in the same table, the other driving force behind today's technological advances, namely information transfer and storage, is becoming increasingly dependent on areal effects. In this case it is the lateral areal characteristics such as defects and grain size which are important, rather than the amplitude characteristics of the roughness.

1.3.2 Designer Surfaces

In the earlier days of surface measurement the workpiece was usually turned or milled down to size at a relatively crude level. These cutting tools produced surface texture values of microns in height. For example, planing produced $2-3\,\mu$m, milling $1-2\,\mu$m, and turning $0.5-1.0\,\mu$m, where the number stood for the arithmetic average (AA, CLA, or R_a, which are all the same). If a finer finish were needed, for example in optical applications, then abrasive methods like grinding, polishing and lapping were used, resulting in roughness values of $0.1\,\mu$mand less.

The surface produced traditionally had the geometric form determined by the last process. This meant that the waveform of the surface profile actually could be related to the shape of the cutting tool or the grain producing it. In Fig. 3, profiles are shown of a turned part and a ground part, representing a periodic and random waveform, respectively. However, the fact that these somewhat arbitrary surfaces have been used in various functional applications is more a result of the availability of suitable machine tools rather than the process's relevance to performance.

The turning point in the specification of the process came about in the automotive industry. The traditional way to optimize the performance of a car engine was to run it in for a given distance, usually in the thousands of miles or kilometres. During running-in the engine revolutions had to be carefully controlled. After this period of running-in, it was noticed that the surface finish looked as if the asperities of the roughness had been rubbed off leaving in effect two components: (1) the original texture containing the middle and lower parts of the original process marks, and (2) a much smoother part for the upper part of the profile. This upper part of the surface had the characteristics of abrasive wear and no longer resembled the original process marks. Once having identified this trend it became clear that the whole running-in process could be bypassed by using two different processes, one producing the top part of the profile, and the other the lower part. Such a profile is shown in Fig. 3c under the title of Plateau Honing. This represents a designer surface - the upper part of the surface profile supporting the load and the lower part retaining the oil and trapping any wear debris. The former needs low surface curvatures and long plateaux, whereas the latter needs deep wide valleys.

A word of caution is needed here because it is only the surface geometry which has been matched to the requirement for high performance. Sub-surface properties have not necessarily been considered.

In the second example (Fig. 3d(i)) [7,8] the lay of the surface rather than the roughness amplitude can be structured to optimize the performance. For example, the surface marks can be isolated from each other (Fig. 3d(i)), which could be needed in the case of mechanical seals, or the marks can interact, as in Fig. 3d(ii), which could be important in lubrication. The examples shown in Fig. 3d are made from Vibro rolling, which allows a very good regular pattern to be produced. It can, however, be more abusive metallurgically than more conventional processes [8].

Perhaps the most straightforward of designer surfaces is the aspheric form used to correct spherical abrasion in optical instruments. In the past, spherical shapes have been generally produced, but since the advent of accurate diamond turning and grinding machines aspheric shapes have been growing in importance.

2 Instrumentation

2.1 General Points

The instrument used to measure surface texture depends on the application. It could be a very simple instrument used to control the manufacture using a limited set of parameters near to or on the machine or it could be a comprehensive instrument for use in a laboratory. In both cases the instrument could be based on an optical technique or a tactile method using a stylus. In what follows the comprehensive instrument will be described rather than the simple instrument. The former type of instrument has high fidelity but is usually slow. The latter is quick in operation but restricted in application. All the important features of optical and stylus methods are discussed in [1,2].

There are three aspects to instrumentation, whether by optical or stylus methods:

i) the pick-up and the reference;
ii) filtering and data pre-processing; and
iii) parameters.

These points will be addressed in turn.

2.1.1 Reference

All types of surface measuring are a calliper in one form or another. The principle is to compare the test surface with a reference [9]. There are a number of ways of producing the reference from which to measure. Figure 4a shows the usual form in which the test surface is compared mechanically or optically with a true linear reference held somewhere within the surface instrument. Usually the reference has to be levelled relative to the test surface in order that they are both in the same general direction. This method is basically that of a calliper in which one arm is guided by the reference and one by the test piece. The calliper is moved through the gap between them and the variations in the angle or the linear distance between the two styli are monitored. The actual value of the distance between the two arms of the calliper is not important in surface metrology because it is only variations which are important. The way in which the calliper is shown below is the usual configuration because it allows the calliper to be inserted into holes. This side-acting configuration is easy to achieve mechanically but not easy optically.

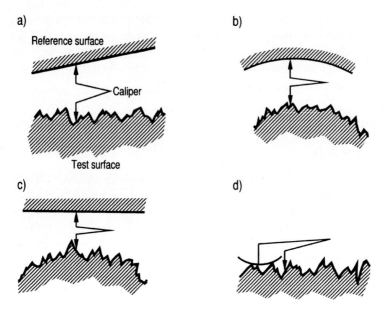

Fig. 4a–d. Mechanical references

If the test surface is curved or shaped in any way, it is usual to try to match the shape of the reference, as shown in Fig. 4b. If this cannot be done, for example in the case of epitrochoids for Wankel engines or aspheric surfaces, the signal has to be obtained conventionally with a straight datum (Fig. 4c), and the actual reference shape computed from the data by using a computer. This reference shape is then positioned relative to the raw data. This is an example of a two-step shape reference. There is another type of reference called a skid datum in which the movement of both arms of the calliper is determined by the test surface (Fig. 4d). However, in the tactile case one of the arms ends in a blunt stylus rather than a sharp one. In the optical case one path is focussed, corresponding to the sharp stylus, and a non-focussed optical path is used for the skid. The philosophy is that the blunt stylus integrates the surface by spanning the gaps between peaks [10]. The resultant signal therefore is the difference between the sharp stylus vertical movement and that of the skid or shoe, which has a much more limited spatial bandwidth. This skid method has the advantage that the signal from the skid is automatically in the same direction as that of the test surface so that no levelling is needed. Neither is there a need for a curved datum.

The skid method is convenient in terms of set-up and so is faster than the conventional method, but care has to be taken on two counts. The first is that the skid has to be blunt enough to adequately span the peaks. It should not drop significantly into the valleys, although it will always drop to a small extent (Fig. 5a). It should not however, be as bad as is shown in

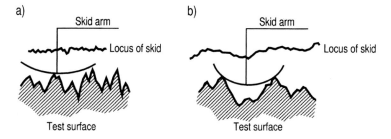

Fig. 5a,b. Surface integration using skid as reference

Fig. 5b. The other possible source of error can be seen in Fig. 6, which shows problems associated with the lateral spacing between the skid and the sharp stylus. In the case of Fig. 6a the spacing is such that the skid and stylus move in phase, with the result that hardly any roughness is recognized. In Fig. 6b the opposite occurs: the skid and stylus move in antiphase because the skid stylus lateral separation is half the wavelength of the surface. This gives an output of twice the true value. It can be seen that when the situation shown in Fig. 6c holds it is not acceptable to use a skid. So before a skid type instrument is used the peak spacings should be checked for both problems

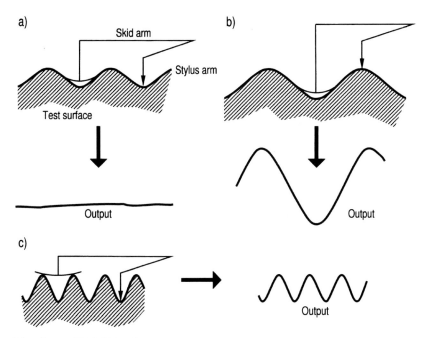

Fig. 6a–c. Skid distortion

outlined above [11]. In optical instruments the reference focussed arm and the skid arm usually more concentrically.

It should be noted that the configurations between the test surface and reference and the problems of the skid instrument exist for optical, capacitative, and any other method, although the errors induced may be more difficult to quantify.

It should should be noted that the skid-stylus configuration is not the same as differential optical methods which use two focussed beams.

2.2 The Stylus

2.2.1 Mechanical Stylus

The mechanical stylus contacts the surface. It is usually made of diamond because of the very hard surfaces, such as the ceramics, which have to be measured. There are a variety of shapes and tip dimensions allowed by international standards. The most conventional one is conical in shape with a 2 μm radius and a cone angle of 60° (Fig. 7a). An alternative shape is the pyramid shown in Fig. 7b.

This is usually asymmetrical, as shown in Fig. 7c, primarily to give the stylus added strength against shock but also to lower the pressure. However, it should be remembered that the direction of traverse should be across the 'lay' of the roughness marks as the figure shows, with the smaller dimension 'a' at right angles to the movement. If the surface is isotropic then the effective dimension of the tip is 'b' not 'a' and is more likely to be 7 or 8 μm which integrates out a lot of detail on fine surfaces. The minimum dimension of the stylus is usually not specified.

It has been argued that the precise tip shape is not important because it always ends up slightly curved at the edges and flat in the middle. If the original shape is that of Fig. 7a the bottom will get a flat due to wear, and it if is that of Fig. 7b the edges on the corners will be smoothed over with usage. For very fine surfaces, even if the stylus is infinitely sharp, there will be some integration due to the slope of the flank. Figure 8 shows the effect for a groove.

In principle this effect can be minimized by using a stylus with a more acute angle slope. This can be produced by polishing. In tactile measurements

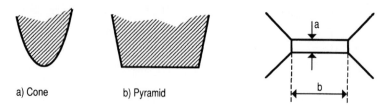

a) Cone   b) Pyramid

Fig. 7a,b. Stylus shape

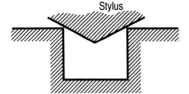

Fig. 8. Stylus slope effect

the slope can be produced completely independent of the tip dimension but the stylus has to be used with care if the angle becomes less than 45°. It is worth reiterating that the tip dimension and angle are independent for the mechanical stylus.

2.2.2 Optical Stylus

The lateral resolution of the optical stylus, which is used in optical follower techniques, is dependent on the numerical aperture of the objective lens viewing it (Fig. 9a,b). The spacing $d \propto \lambda/\mathrm{NA}$, where λ is the wavelength of the light being used, NA is the numerical aperture and is approximately given by S/f where S is the size of the objective lens and f is its focal length. The effective angle is $\tan^{-1}(S/2f)$. But, as can be seen, it is not independent of the resolution: both depend on the numerical aperture of the objective. The stylus method, therefore, is more flexible than the optical method, e.g. it is possible to get a tactile stylus with very small lateral resolution at the same time as leaving a very acute angle. The lateral resolution of the stylus method can be as small as required simply by polishing mechanically. Tips of 0.1 μm have been made [B1]. The optical lateral resolution is limited by the wavelength of light.

2.2.3 Comparison Between the Mechanical and Optical Stylus

The obvious point is that the optical stylus does not contact the workpiece whereas the mechanical stylus does. For soft surfaces such as copper this constitutes an advantage for the optical method because the mechanical stylus has to have a finite pressure in order to follow the surface. This pressure usually indents the surface elastically according to Hertz's equations [11], but if the tip is broken or badly worn it can sometimes cause plastic deformation and hence damage the surface, which obviously impairs fidelity. Usually, however, the damage occurs rather less often than expected, because the outer 'skin' of the test piece, of whatever material, is usually harder than the bulk (Fig. 10), the dislocations which allow plastic deformation having a spacing larger than the area of contact of the stylus. This effect increases the effective hardness of the surface by factors of as much as five [12]. Mechanical

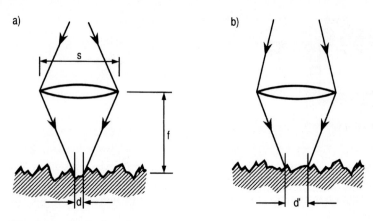

Fig. 9a,b. Optical probe

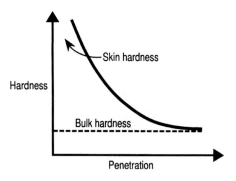

Fig. 10. Skin hardness

stylus methods can be made to measure other features of the surface such as hardness and friction, neither of which is possible with optics.

Optical methods also have their problems because it is not the true geometry which is measured but the optical path length or difference. Figure 11 shows a perfectly smooth surface with equal thickness films of refractive indices n_1 and n_2 on the top. The interface between the two films will show an edge with optical methods which does not resemble the actual geometry of the surface.

Diffraction effects can also result if the local curvature on the surface is high ($\approx 10\,\mu m$) (Fig. 12) [B6]. The corner can act as a secondary source of light that produces an extraneous peak, which in the case above highlights the edge. This may be good visually but it will give a distorted value for the step height. For rough surfaces, e.g. those with a surface roughness of the order of $1\,\mu m$ down to $0.25\,\mu m$, the above differences can usually be ignored, but for fine surfaces they cannot. Unfortunately, fine surfaces are now the norm,

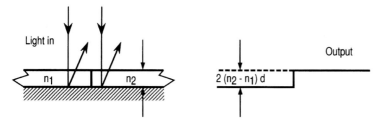

Fig. 11. Optical path length

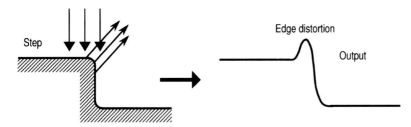

Fig. 12. Optical edge enhancement

so care has to be taken in interpreting optical results. Similar problems arise with SEM.

This discrepancy between the two techniques is based on physical laws and does not mean that either the mechanical or optical stylus produces the 'true' result. Both are valid for what they do. For fine surfaces the two methods should not agree. Optical methods tend to differentiate the surface, e.g. produce spikes, whereas mechanical stylus methods tend to integrate the signal. For molecularly smooth surfaces the optical result is invariably bigger - sometimes by a factor of two or more than the mechanical stylus result. This effect means that it is practically impossible to get fine surface standards to read the same for both methods. In fact each method should have its own standard!

One final point is that the optical method does not suffer from damage to the stylus, as can happen to the mechanical stylus if used improperly. The real danger is not being aware of a chipped or otherwise damaged stylus. For this reason special calibration standards are used to detect for breakage. Figure 13 shows such a standard. It consists of a set of grooves of different widths. The size and sometimes shape of the stylus can be inferred from a graph of penetration versus groove width.

2.3 Basic Instrument

Figure 14 shows a method of comparing the relative merits of various instruments. This particular graph uses the dynamic range of the instrument

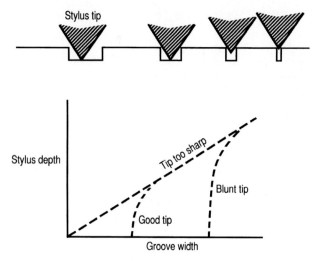

Fig. 13. Calibration of stylus tip

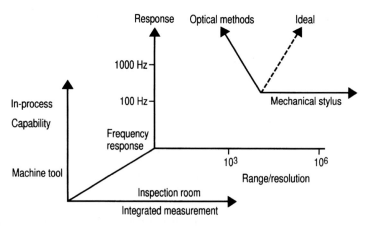

Fig. 14. Measurement trends

(range/resolution) as one axis and the response time as the other (or maximum frequency response). This merit chart is basically for industrial use because it highlights the two factors which are most important in practice. One is the frequency response at which a reading can be taken off the surface and the other is the number of actual factors which can be measured at any one time. The former decides how close to the manufacture the measurement can be made whereas the latter determines the number of calibrations and set-up operations that are needed. Moving positively along both axes determines the speed and fidelity of the measurement.

There are other merit graphs in use such as that of Stedman, which plots log amplitude versus wavelength (see, for example, page 570 of [B1]), but these are mainly used to assess the range of surface over which measurements can be made and so are very important in research. However, they do not lend themselves to the reality of industrial usage because of the lack of information about response time.

In what follows in Sect. 4 and 5, the most common instruments are examined, namely the mechanical stylus and optical methods. Both of these have a good pedigree, having been devised at the same time in recognizable form by *Schmalz* in 1929 [13]. The mechanical stylus technique was intended to replace the human nail for assessing roughness and the optical method was intended to replace the eye; both were intended in effect to take the element of subjectivity out of surface assessment.

2.4 The Stylus Method (Mechanical)

As previously mentioned, this configuration allows the probe to be put into holes to measure roughness: this is necessary because well over half of components made have an axis of revolution. The problem with this is that the pick-up assembly shown in Fig. 15 necessarily has a high moment of inertia and hence a relatively low frequency response. At one time it was thought that the system could only be used effectively with surface speeds in which the dominant surface periodicities occurred well below the resonant frequency of the pick-up. However, it is possible to work close to the resonant frequency if a damping factor of 0.59 is used [14]. This optimizes the measurement of surfaces having a wide spectral content such as abrasive surfaces and also, incidentally, sharp edges such as are now found in VLSI and lithographic processes.

An alternative scheme is shown in Fig. 16. This is in effect a cantilever mode of operation in which the transducer is directly above the probe. This basic system is used not only in traditional fast-scanning methods but also

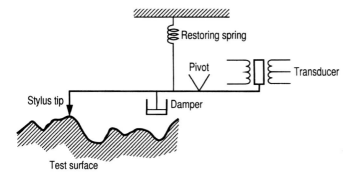

Fig. 15. Side acting gauge

in the new generation of scanning microscopes, including STM and AFM. In the cantilever mode the frequency response can be many times higher than in the conventional mode, but it is restricted to nominally flat surfaces.

The transducer system is now usually optical for the scanning methods. In the engineering applications the optical system is a flag system in which the end of the stylus shank blocks off a portion of the light. For the much smaller movements of the nanometer scanning instruments it is usual to use an interferometer. In both of these the idea is not to impose any significant force on the stylus shank. Stylus methods for the serious measurement of surfaces started with *Schmalz*'s work [13]. It is not well known that he was also the pioneer in the other principal method of measuring surfaces, namely optical techniques. These will be outlined below.

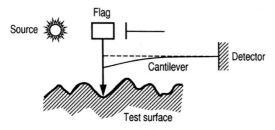

Fig. 16. Cantilever mode

2.5 The Optical Methods

Such is the versatility of the optical method that many possibilities exist (15), as shown in Fig. 17. On the extreme-left hand side are those optical instruments mimicking the mechanical stylus. Under this heading is the optical follower. Further to the right brings less focussed light until on the extreme right the surface is floodlit. As a rule of thumb, methods on the right are fast but not very accurate. They are also cheap. Those on the left are slower but more accurate. Also, methods to the right tend to give indirect measurement of the surface whereas those on the left are more absolute and usually more costly.

Basically, all optical methods project light onto the surface, either focussed or not. Also, the light can pass by an obstruction on the way to the surface or after scattering. One simple method referred to on the extreme-right hand side of Fig. 17 is the gloss meter (Fig. 18) [16]. In this method two detectors pick up the scattered light. Detector A is positioned at the specular angle and detector B at some other angle. The surface roughness can be inferred from the ratio of the light received at these detectors. For a fine surface most of the light is detected at A, whereas for a rough surface it is evenly

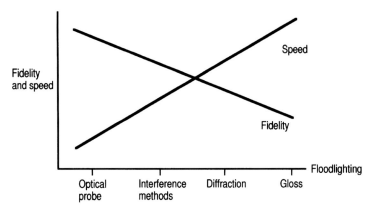

Fig. 17. Comparison optical methods

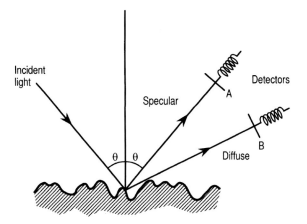

Fig. 18. Glossmeter

spread between A and B. The degree of scatter is determined by the ratio of the wavelength of light to the roughness value (rms R_q). This technique can only be used as a comparator because the law which determines the scatter depends on the manufacturing process.

In a more refined version of the scatter method coherent light is used (Fig. 19) [17]. In this form the light source is imaged in the Fourier transform plane. The image in the transform plane is modulated by the surface roughness, in particular the surface slopes. If the surface roughness is fine - usually less than $\lambda/4$ - then the intensity pattern received at the back focal plane of the lens can be interpreted as the power spectrum of the surface and as such can be used very effectively in diagnostic monitoring of the machine tool and the process. Also, under certain circumstances the rms roughness, the curvature, and the slope information can be derived from the moments

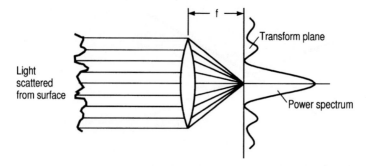

Fig. 19. Diffractometer

of the spectrum [B1]. This method of surface assessment is unique because the smaller the lateral detail on the surface, the bigger the scattered angle, which is very convenient for fine surfaces.

Figure 20 shows a typical modern interferometric method [18], which replaces the conventional methods due to *Tolansky* [19]. In this method a Mireau interferometer is used [20]. The reference mirror and the test surface are positioned relative to the objective. The fringe pattern produced is viewed from above by a CCD camera. This is then scanned and stored in a computer. Then some subsequent movements are made between the objective lens and the test surface. This movement is repeated a number of times up to one wavelength shift, the fringe pattern being stored each time. The subsequent area picture of the surface, either in contour form or straight scans, is unravelled by the computer and displayed. This method gives an overall picture of the surface, from which it is possible to get numerical information. However, it works best with flat surfaces such as those which occur in the semiconductor industry, e.g. wafer technology. Great care has to be taken when the objective lens is moved, otherwise the computer can get confused! This refers particularly to a pitch and yaw movement of the lens introduced when the lens is moved axially. Such movements can impose angular shifts into the image which cause overlapping problems.

One large group of optical methods uses scanning as part of the technique. The obvious starting point is the flying-spot microscope [21]. In this instruments very high spatial information is obtained by restricting the illumination on the surface to a very small spot. Light collected at the same time can obviously only emanate from the illuminated spot by scattering from the surface. The extraneous noise is therefore kept to a minimum.

A variation on this is the confocal microscope shown in Fig. 21 [22]. In this the imaging and receiving optics are exactly the same. As, above the optics produces a spot on the surface. The reflected light is picked up via a pinhole by a detector. All light from an out-of-focus plane is blocked and cannot reach the detector. The result is a dramatically sharp image. Resolution of 1 µm

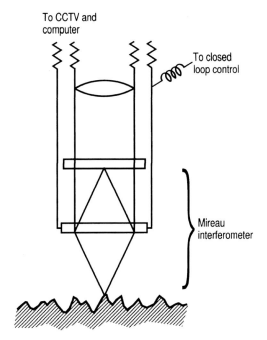

Fig. 20. Mireau interferometer

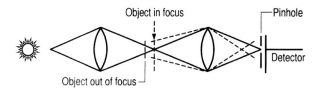

Fig. 21. A confocal microscope

in the x-direction and 0.1 μm in height are readily obtained. The absolute profile is obtained by measuring the axial movement needed to keep the surface in focus during the scan. This is in effect an optical follower method in which the objective lens is moved to establish a null on the detectors. As the scan proceeds across the surface the movement of the lens constitutes the topography of the surface. Other earlier methods [23] utilize the Foucault method of optical testing in which a knife-edge is situated at the back focal plane of the system

Amongst the most successful optical methods is the so-called heterodyne technique, in which either two wavelengths or two polarizations are used as in the method shown in Fig. 22 [24]. This clever system actually uses one polarized path as a skid and the other path as a sharp stylus. These follower

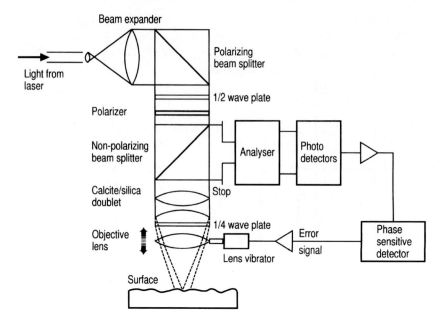

Fig. 22. A heterodyne follower

methods are not as fast as the floodlit methods because the heavy lens system has to be moved in order to maintain the null signal.

There are many other optical techniques: one using differential interference microscopy [25]; Fringes of Equal Chromatic Order (FECO) methods [26]; and speckle and holography [27]. However, they are very restricted in application. In no way can speckle or holography be considered to be general purpose methods.

The fact that there are so many optical methods is a sign of the potential of the method. Unfortunately the problems with spikes and the difficulty of calibration have meant that they have not enjoyed as much popularity as the mechanical stylus method.

Historically, the mechanical stylus gave more attention to the vertical detail on the surface. This was an engineering approach in which the tribological function was considered the most important; contact and similar phenomena are dominated by the heights of asperities. Optical methods, on the other hand, paid more attention to lateral detail because the principal use was to look for areal structure. This sort of information appeared at the time to be more relevant in metallurgy and chemistry. In truth neither method, mechanical stylus or optical, give the full story. Optical methods such as in the follower method are switching to height information and relying on the scan to get structural detail. Similarly, mechanical stylus methods are now incorporating scanning mechanisms to complement the height information. Perhaps the best way to decide on the most suitable method is to find which one is closer to the functional requirement of the test piece. For example, it is better to use an optical method if the test piece is a mirror.

2.6 Other Conventional Methods

Pneumatic methods [28] and capacitative methods [29] have been used mainly in Germany by Perthen and others, but they have not enjoyed the popularity of the mechanical stylus or the optical methods described above and so will not be addressed here. They all have in common the fact that they measure directly over an area and give an integrated measure of the surface. Both methods require either the electrode or the skirt to have the same nominal shape as the surface under examination. Failure to do so gives incorrect result.

2.7 Non-conventional Methods

Under this heading are the new scanning methods, which include Scanning Electron Microscopy (SEM) [30], Transmission Electron Microscopy (TEM,) Scanning Tunneling Microscopy (STM) [31], Atomic Force Microscopy (AFM) etc. These methods are in the general realm of nanotechnology rather than conventional engineering. It should suffice here to say that they (AFM, STM) represent stylus methods in which the point touches or nearly touches the surface. Many new details of the surfaces are being explored by these methods other than topography. Indeed they can explore charge density, force, chemical constitution, and many other physical contact phenomena such as friction and hardness. These scanning methods are excellent for picking up surface structure and for pin-pointing defects but still have problems in the calibration of height. For a detailed review of these nanometrology instruments see [B1].

3 Pre-Processing and Filtering

3.1 Levelling

This can mean simply levelling the data by removing a best fit or similar line from the collected data. What this exercise does in effect is to position the reference surface actually within the profile so as to enable the roughness to be measured from it (Fig. 23).

3.2 Curve Fitting for Form

Any curve can be fitted to the raw data so as to enable the roughness to be measured. If the instrument has a sufficiently long word length and if the calculations do not introduce bias then the roughness and the radius of curvature, which is a form of dimensional metrology, can be obtained simultaneously. There are many types of curve which can be fitted, it just depends

Fig. 23. Establishing a reference

on the subsequent application. For example, in Fig. 24 if the workpiece has to be assembled into a system then tolerances are important, and a Chebychev zone could be fitted to the data which would be more meaningful than fitting a best-fit least-squares curve. This is because the Chebychev functions can take up all of the tolerance zone by oscillating between the two limits of the zone. This in effect gives equal weight to all of the zone instead of the limits just being touched twice in best-fit methods.

The advantage of fitting functions or just polynomials to the data is that nothing is lost. The disadvantage is that for best results some idea of the shape of the workpiece is necessary otherwise odd effects can occur.

The alternative approach is to use filtering to isolate the roughness from the general curve. This has the advantage that the filter needs no prior knowledge, but it does need data at the beginning or end or both to enable the filter to get rid of transients [32].

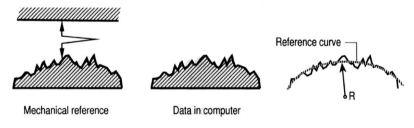

Fig. 24. Curve fitting

3.3 Filtering for Waviness

3.3.1 Reason for Filtering

The removal of waviness is a contentious subject which attracts great attention. First, waviness is that signal left on the surface by faults in the machine tool. If the machine is perfect then no waviness will occur. On the other hand, roughness will always be present because it is the mark of the process. Waviness is usually of longer wavelength than roughness but it is not necessarily periodic as the name implies; it could just be a narrow-band random waveform (Fig. 25). If the ultimate use of the workpiece is single bodied, i.e. it is being painted or used as a mirror then the total geometry is all that is needed - waviness should not be precluded. However, if the ultimate use requires contact, such as in a bearing, then the roughness and waviness should be isolated. The reason is given as follows.

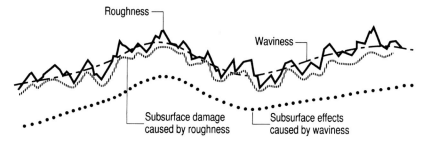

Fig. 25. Need to separated waviness

It is often not just the geometry of the waviness and roughness which can cause trouble; when the workpiece is functioning, they could also have effects under the surface. The subsurface effects of the roughness are quite severe because they are produced by the processing, in which a lot of energy is expended (Fig. 26). The waviness, on the other hand, is produced relatively slowly - the subsurface effects are lower and less abusive than for the roughness. So the waviness is likely to produce different functional effects than the roughness, apart from the effects due to geometry alone.

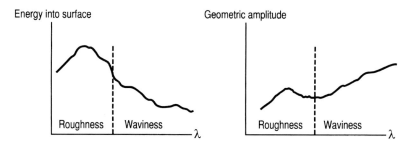

Fig. 26. Energy versus geometry (function of wavelength)

3.3.2 Type and Value of Filter

The traditional filter for extracting the roughness from the profile signal is shown in Fig. 27 [32]. It consists essentially of two cascaded CR filters arranged so that their cut-off frequency (or wavelength) acts at 75% transmission. This value is historical and is an amalgam of the British 80% and the US 70.7%. The parameter λ_c is called the cut-off, the filter cut-off, or sometimes the meter cut-off. The current name is the sampling length. They are all the same. There are two more lengths which have general usage in surface metrology. One is called the assessment length and is the length of surface over which the measurement of the roughness takes place. It is usually long enough to encompass five sampling lengths end to end. This number, five, is arbitrary but it is the length generally accepted by instrument makers. The second length the traverse length, is the length of surface actually traced. It is obviously greater than the cut-off length and the assessment length.

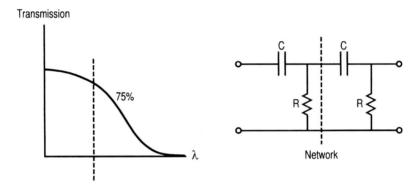

Fig. 27. Characteristic of standard filter (2CR)

Apart from the arbitrary 75% for the cut-off, the standard 2CR roughness filter distorts the signal as shown in Fig. 28. It can be seen that there is a delay of the mean line representing the waviness and form relative to the profile which also contains the roughness. As the roughness is the difference between the profile and the mean line representing the waviness and form it is obvious from the figure that some peak parameters could get distorted, and in particular be higher than their true value. This is a consequence of the fact that a CR filter is a form of differentiator. Average parameters like rms R_q are not affected. True averages like R_a are not affected significantly, e.g. below 3%.

To correct for the distortion, alternative filters have been tried [33]. The most suitable filters belong to the class called phase-corrected filters (strictly, linear phase filters with a shift introduced). These do not distort the roughness (Fig. 29 and 30) [27]. The transmission curve now used is derived from

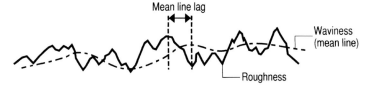

Fig. 28. 2CR distortion

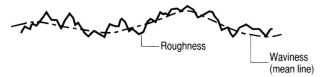

Fig. 29. Phase corrected filter

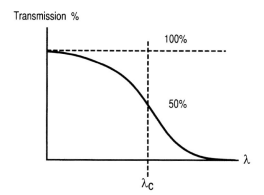

Fig. 30. Phase corrected filter transmission

the Gaussian distribution and is now included in the National and International Standards. This shape is not perfect but it is acceptable [34]. Note that there is no fundamental functional reason why the Gaussian filter should be used [B1]. The big advantage of phase-corrected filters is that at every x point on the profile the mean-line value and the roughness add up to unity. The value of 50% for the transmission characteristic of the cut-off evenly separates the roughness and the waviness. No distortion occurs with this type of filter and is a consequence of the filter having an impulse response (weighting function) with an axis of symmetry, as shown in Fig. 31.

At one time it would have been very difficult, if not impossible, to use a phase-corrected filter in an instrument, hence the use of electronic elements such as capacitors and resistors in the earliest filters. However, now that computers are in general use there is no longer a problem.

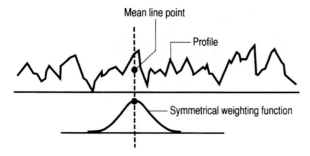

Fig. 31. Weighting function

Filtering, as discussed above, is a very straightforward procedure with mechanical stylus instruments because the signal from the transducer is electrical and usually analogue, which can easily be converted into a digital equivalent. It is not so easy for optical systems which are not of the "follower" type. The "follower" has a suitable output but other optical methods do not. For example it is difficult to provide a waviness filter in a glossmeter or with conventional fringe methods unless some considerable work is carried out on the image. The exception to this is in diffraction, where quite effective filtering can take place in the transform plane (Fig. 32) by means of an opaque stop with graded transmission, which can easily be made to be Gaussian in shape, which is the same as the function in Fig. 31.

Although this optical filtering takes place at the speed of light, parameters have to be obtained via moments of the power spectrum available in the transform plane. About the only convenient parameter available is the root-

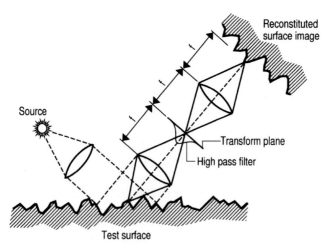

Fig. 32. Optical filter

mean-square of the surface, an estimate of which can be obtained from the light transmitted. It is to be expected that such filtering systems will only work for fine surfaces, where the true power spectral density is available in the transform plane. Such a technique has been used successfully in biology applications [35].

There has been an attempt to provide enough information to allow the bandwidth to be decided. This is shown in Table 2.

Table 2. Metrological characterization of phase correct filters and transmission bands for use in contact (Stylus) instruments (ISO 3274)

Gaussian filter. Short wavelength cutoff λ_s;
Long wavelength cutoff λ_c

λ_c (mm)	λ_s (μm)	$\lambda_c : \lambda_s$	Max stylus tip (μm)	Max sampling spacing (μm)	Max number of points per cutoff
0.08	2.5	30	2	0.5	160
0.25	2.5	100	2	0.5	500
0.80	2.5	300	2	0.5	1600
2.50	8.0	300	5	1.5	1666
8.00	25.0	300	10	5	1600

4 Parameters

4.1 General

It is well known that the surface texture has two basic roles in engineering today. One is as a means of controlling the manufacturing process and the other is a recognition of the fact that the texture can affect the performance of the workpiece. Often, just looking at the surface or feeling it is sufficient to tell a skilled operator whether the texture is typical of a process or if it is suitable for an application. Unfortunately it is difficult to communicate these judgments. Using a profile trace of the surface is also not very helpful because it still represents a tremendous amount of information. A primary task of the surface metrologist is therefore one of condensing the data on the profile or the surface itself to a minimum set of parameters. For process control even two parameters would be considered to be one too many. More would never be used in practice. There is a little more scope when trying to predict the suitability of a surface for functional use, but not much. It is clear, therefore, that the task of condensing the surface data into one or at most a few parameters is very difficult and that any effort expended in trying to get the best parameters is justifiable. In what follows, some examples will be given. These fall into a number of categories:

i) height parameters
ii) length parameters
iii) hybrid parameters
iv) areal parameters - the lay.

4.2 Height Parameters (see [36] for example)

The simple height parameters are either averages or extremes.

4.2.1 Average Parameters

The most used is the R_a value - the roughness average, originally called the CLA in the UK and AA in the USA. If the height distribution of the profile is $p(y-\bar{y})$ where y is the height from an arbitrary datum and $\bar{y}$ is the mean-line height both at x,

$$R_a = \int_0^\infty |(y-\bar{y})|p(y-\bar{y})dy, \quad \text{or}$$
$$= \frac{1}{L}\int_0^L |(y-\bar{y})|dx,$$

where $\bar{y}$ is the mean-line height. Although this is relatively easy to measure from a graph [10], it is not easy to obtain using optics.

The next obvious parameter is the R_q the root-mean-square value given by

$$R_q = \int_{-\infty}^\infty (y-\bar{y})^2 p(y-\bar{y})dy, \quad \text{or}$$
$$= \sqrt{\frac{1}{L}\int_0^L (y-\bar{y})dx}.$$

The R_q value is the parameter which can be measured optically via the Fourier transform plane. It is, however, difficult to measure from a graph but not from computed data.

There are two other parameters which can be obtained as averages from the height distribution. These are the skew and the kurtosis. They are not strictly height parameters, but they give information about the shape of the process marks on the surfaces. They also follow on the series starting with the R_a. Thus R_{sk} and R_{ku} given by

$$R_{sk} = \frac{1}{R_q^3}\int_{-\infty}^\infty y^3 p(y-\bar{y})dy, \quad \text{or}$$
$$= \frac{1}{R_q^3}\frac{1}{L}\int_0^L (y-\bar{y})^3 dx,$$

$$R_{\mathrm{ku}} = \frac{1}{R_q^4} \int_{-\infty}^{\infty} y^4 p(y - \bar{y}) \mathrm{d}y, \quad \text{or}$$

$$= \frac{1}{R_q^4} \frac{1}{L} \int_0^L (y - \bar{y})^4 \mathrm{d}x,$$

R_{sk} indicates the extent to which the profile is symmetrical about the mean line. The kurtosis or excess is a measure of the spikiness or bumpiness of the surface. Some examples are shown in Fig. 33. It can be seen at a glance that R_a, R_q, R_{sk}, and R_{ku} are all moments of $p(y)$. The value of 3 for kurtosis represents the value which is obtained for a surface with a Gaussian height distribution and is sometimes subtracted from R_{ku}.

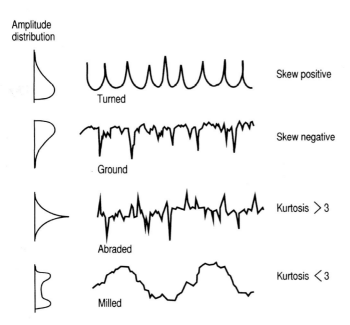

Fig. 33. Examples of amplitude distribution moments

4.3 Peak Parameters

4.3.1 Peak Parameters (General)

These parameters follow the German tradition and were meant to be the parameters of most use in tribology. There are many in use today: R_y, R_z, R_{3z}, R_{tm} e.g. Fig. 34. R_y is the maximum peak-to-valley over the whole assessment. R_z is the average difference between the five highest peaks and five deepest valleys, within one sampling length. R_{tm} is the average height of the biggest peak-to-valley reading in each of the five sampling lengths. R_{pm} is

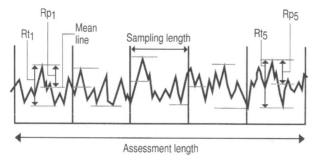

Fig. 34. Some peak parameters

similar but only refers to the highest peak R_p measured from the mean line. Note that peak-to-valley measures do not as a rule depend on the generation of a mean line. This can be useful on small parts, where R_z is usually used.

The problem with peak, and peak-to-valley parameters is that they are not convergent parameters; they are divergent. This means that the more readings that are taken on a surface then the bigger they will get! For average readings the more readings taken the better the stability. In the early days R_y or, as it was then, R_t was advocated, but it was so variable that an average of 5 was made to stabilize the reading to get R_{tm}. In Japan R_{3z} was used for a peak-to-valley measure. It is the height difference between the third highest peak and the third lowest valley. It represents a clever way of getting rid of rogue peaks or valleys

4.4 Spacing Parameters

The spacing parameters have not until recently had much importance attached to them by engineers although spacings have always been important to microscopists. The most used parameter is S_m, which is the average distance between crossings of the mean line, usually over one cut-off length. S_m identifies major peaks and S another spacing parameter is the average distance between local (small peaks) on the profile (Fig. 35).

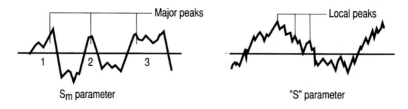

Fig. 35. Peak spacing parameters S and S_m

4.5 Peak Parameters (Statistical)

There is one parameter or family of parameters which has grown up with the subject. This was originally called the Abbott Firestone curve. It has had many names since its inception in 1936 in General Motors by Dr. Abbott. One popular name was the bearing area curve, which became the bearing ratio curve, and now is called the material ratio (MR) curve.

The MR curve is basically a measure of the percentage of material to air at any given level in the surface (Fig. 36):

$$\mathrm{MR}(y) = \int_y^\infty p(y)\mathrm{d}y .$$

For a Gaussian distribution this becomes $\mathrm{MR}(y) = [1 - \mathrm{erf}(y)]$.

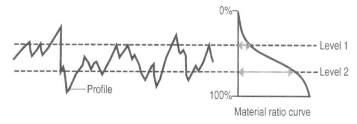

Fig. 36. Material ratio curve

In Fig. 36 two lines have been drawn through the profile, one at level 1 and one at level 2. It can be seen that the amount of material encountered as it cuts through the profile is much higher for level 2 than it is for level 1. The actual percentage of the profile which is material is plotted horizontally on the MR curve. Obviously, at the highest peak MR(%) is zero and at the deepest valley it is 100%. How it changes in going from 0 to 100% is the MR curve, and this shape contains the information which hopefully describes how much material a lapping plate would contact as a function of height.

Because of its "S" shape, investigators in Germany split the curve up into three parts. Each part of the curve according to the designers described a different aspect of the function of the surface when used as part of, say, an engine. The breakdown is shown simply in Fig. 37. It was decided that the three components of the curve could be designated R_{pk}, R_{k}, and R_{vk}, and that each had a different behavioral property. R_{k}, the kernel (or core) part of the profile, was ascribed to the load-carrying part. R_{pk} described the function of the peaks, and R_{vk} described the function [37] of the valleys (Fig. 37a). This is a sensible breakdown but it is slightly ill-conceived because the "S" shape only appears if the abscissa for a random surface is plotted linearly. A random curve would normally be plotted on probability paper, in which

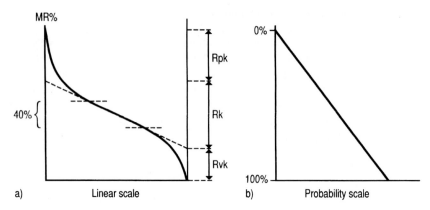

Fig. 37a,b. Parameters derived from the material ratio curve

case it would be a single line (Fig. 37b) showing no simple regions like the topmost curve in Fig. 33 even though it is the same surface.

The mechanisms for establishing where the $R_{\rm pk}$, $R_{\rm k}$, and $R_{\rm vk}$ regions lie on the MR cure have been complicated, but one accepted method is to start with the centre of the MR curve and then fit a window of width 40% such that the curve enclosed is the straightest. This straight line, once found, is then extrapolated to the coordinates at both ends of the curve. Where the intersections occur determins where $R_{\rm pk}$ and $R_{\rm vk}$ start.

Despite this odd justification, this breakdown of the curve can be meaningful for multi-processes like plateau honing described in this chapter. $R_{\rm k}$ is the load-carrying part of the profile and $R_{\rm vk}$ represents the oil-retention part.

Mention of the material ratio curve will also be made in the section on three-dimensional (areal) parameters. There are various other spacing parameters, such as the average wavelength $\lambda_{\rm a}$, which is closely related to $S_{\rm m}$. Also the crossings can be taken at any height of the profile (Fig. 38) and are not restricted to the mean line. This is called the high spot count. Figure 38a shows how the high spot count level can be used to isolate (a) high peaks and (b) deep valleys. These parameters are obviously important in lubrication and corrosion or fatigue [38].

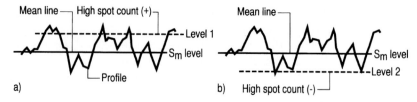

Fig. 38a,b. High spot count

4.6 Hybrid Parameters

4.6.1 Slope

These parameters use both the height and also the spacing information. Probably one of the most important is Δ_a, the average slope of the surface or Δ_q, the rms value:

$$\Delta_a = \frac{1}{L}\int_0^L \left|\frac{dy}{dx}\right| dx ,$$

$$\Delta_q = \sqrt{\frac{1}{L}\int_0^L \left(\frac{dy}{dx}\right)^2 dx} .$$

Δ_q is very important in optics. If $R_q \ll \lambda$ (the wavelength of light) the surface scatters light by diffraction rather than geometrically; in the latter case the value of Δ_q can be found directly from the distribution of light in the Fourier plane.

4.6.2 Curvature

This is usually taken to mean the average curvature at the peaks (or valleys) rather than the modulus of the curvature of the whole profile. For the peak curvature the actual curvature value is calculated from d^2y/dx^2, i.e. y'':

$$P_{curv} = \frac{1}{L}\int_0^L \left(\frac{d^2y}{dx}\right)_{y'=0} dx .$$

This parameter is very important in all aspects of contact, friction and wear because to a large extent it determines whether or not a peak when compressed in contact will deform elastically or plastically. If only elastic deformation occurs, the wear will be negligible. If the deformation is plastic, then wear will occur.

4.7 Effects of Filtering on Parameter Values

These two parameters, slope and curvature are undeniably functionally important. However, they are very difficult to work with in practice because of noise. Both parameters are differentiators and as a result can corrupt the already high values of slope and particularly curvature obtained from the high frequencies. The corruption is brought about usually by numerical noise in algorithms or word length problems in the digital form of the signal. Any other extraneous noise, e.g. electrical noise, can also interfere with the real signal values. This makes the effect of filtering very profound; a small shift of the cut-off can cause a dramatic effect in the value of the parameter [B1]. So sensitive are these and some of the peak parameters that it is meaningless to use them unless the filters are also specified. There is no such thing as a unique value of a surface parameter for a given surface: it all depends on the bandwidth, i.e. the high and low cuts of filters used in the instrumentation. Also it makes a lot of sense to use an upper and lower filter cut-off to match parameters in the manufacture or the application. For instance a ball-race might have for its short wavelength limit the grain size of the grinder which produced it. Also, the long wavelength limit could be the width of the elastic contact of the balls in the bearing!

Unless some experience is brought to bear on the choice of cut-off, any answer is possible. This was one of the reasons why there was a switch in the 1960s to more fundamental parameters (of which incidentally the material ratio is one). This new approach was and is called random process analysis. A simple examination of this type of analysis will be given next.

5 Random Process Analysis in Surface Metrology

5.1 General

Random process analysis [39] can be used to investigate three aspects of surfaces: the height information, the spacings and the areal information such as the lay. At present it is the first two which will be considered [40].

5.2 Height Information

The main tool for investigating heights is the height probability density function already met with early on in Sect. 4.2. It is in fact the function designated $p(y-\bar{y})$. This was the essential part of the formulae for R_a, R_q, R_{sk}, and R_{ku}. It is not coincidental that these different parameters could all be derived from the same base. In fact one factor is missing and this is $\bar{y}$, the position of the mean of the function, i.e. the mean line of the profile.

What is reflected by all these parameters is three factors of $p(y-\bar{y})$: (a) the position, (b) the size, and (c) the shape. The problem is that although $p(y-\bar{y})$ contains all the height information in the profile, the information is not very accessible. In fact it is hardly better than having the profile itself. It is replacing one curve by another, albeit shorter and simpler in shape. The engineer wants to characterize the surface with as few numbers as possible. Luckily, of the three factors above, the actual position is not required because it is concerned with position not deviations of shape. This leaves size and shape to be specified.

5.2.1 Size from the Probability Density Function

In principle, any of the height parameters will suffice, whether it is a peak or average value. The main point is that it only requires one of the parameters to give a reasonably stable estimate of surface height. The choice of parameter is arbitrary, although in practice R_a is most often used because it is easy to measure and is quite reliable.

5.2.2 The Shape of the Probability Density

This is difficult because the functional use of the workpiece could require many shapes. In fact sometimes the Fourier transform of the curve (called the characteristic function) is taken to simplify the characterization and manipulation of the function. Because of the need to keep the number of parameters to a minimum it is usual to have one or two: the skew and sometimes the kurtosis which have been described earlier. Parameters of the Beta function have also been used [41].

What is required is a critical measure of how the shape of the profile changes with height. In most applications this has turned out to be the value of the skew. The most dominant factor about the shape of the profile is whether it has very high peaks or very deep valleys. The skew shows this. It can distinguish between good turning and bad turning. It can also be used to tell whether a grinding wheel needs dressing [4]. This aspect will be discussed in the next section.

5.3 Spacing Information

Random process analysis came about because of problems in communication. So the method devised to investigate the signals was extremely powerful in the time domain or in time series. It is unfortunate that much of the information in, say, a contact or wear situation is in height and not lateral position. The $p(y-\bar{y})$ function tends to average out much useful information, for example the peaks are lost, although $p(y-\bar{y})$ is still obviously vital for height parameters. The point here is that contact situations would be better characterized by the height distribution of the peaks rather than the height ordinates

The two functions used to characterize the horizontal spacing information rather than the height information are the autocorrelation function and the power spectral density [40]. Both of these carry the same information, so either can be used in any given situation. These two functions are a Fourier transform pair. Thus if $A(\tau)$ is the autocorrelation function and $P(w)$ the power spectral density

$$A(\tau) = \frac{1}{2\pi} \int_{-\infty}^{\infty} P(w) \exp(+jw\tau) dw ,$$

$$= \frac{1}{\pi} \int_{0}^{\infty} p(w) \cos(w\tau) dw ,$$

because $P(w)$ is an even function about $w = 0$. And

$$P(w) = \int_{-\infty}^{\infty} A(\tau) \exp(-jw\tau) d\tau ,$$

$$= 2 \int_{0}^{\infty} A(\tau) \cos(jw\tau) d\tau .$$

There is a preferred use of $A(\tau)$ and $P(w)$.

5.3.1 The Autocorrelation Function

This is a measure of the similarity of one part of the profile with another a distance from it. It can be expressed simply as $E[y'(x)y'(x+\tau)]$, which in words is the average value of the product of the profile and itself shifted by the distance τ. Here y' is measured from the mean line. This average is over all x, i.e. all the profile. To get the actual function, the value of the shift τ is changed from zero to whatever shift might be considered to be the upper limit. It could be the Hertzian length in bearings for example. This function is widely known so that its evaluation can be achieved using most mathematical software packages. However, a simple example, given later, will show its potential.

These statistical parameters are effective because they provide a large enhancement of the signal over the noise introduced into the system. For example, each point on the autocorrelation function of a profile taken from a surface is a result of a great deal of averaging. Small changes between surfaces became significant. As a general rule the autocorrelation function can best be used to reveal changes in processes such as grinding whereas power spectral analysis can be used to best advantage in processes which are fundamentally periodic or repetitive, such as in turning or milling. Both the autocorrelation function and the power spectrum hunt for the unit machining event. In the case of grinding, the unit event is the impression left on the surface by an average grain on the grinding wheel. In power spectral analysis it is the periodic signal left on the surface by a clean cutting tool on a perfect machine.

Take grinding, for example, as a case in which autocorrelation is useful. Figure 39a shows the impression left on the surface by a sharp grain. Figure 39c is the correlation function. Notice that the correlation length (the distance over which the correlation drops almost to zero) is a direct measure of the effective grain hit width. For a grinding wheel in which the grains are blunt (Fig. 39d) there is a considerable piling up of material as well as the formation of a chip. By examining the correlation function (Fig. 39f) it is apparent that the piling up or ploughing is revealed by lobing in the autocorrelation function. The width of the central lobe is a measure of the amount of material removed. At a glance, therefore, the shape of the autocorrelation function reveals the efficiency of the grinding in all its aspects (Fig. 39g). Notice that this would not be revealed by looking at the profile or by using simple parameters. In Figure 39 any longer waves in the autocorrelation function of the surface show that there are other problems such as the need to dress the wheel.

Another example shows how the power spectrum can be used to identify problems in turning. As the tool wear and the machine tool deteriorates, significant changes occur in the spectrum of the surface, as shown in Fig. 40 and 41.

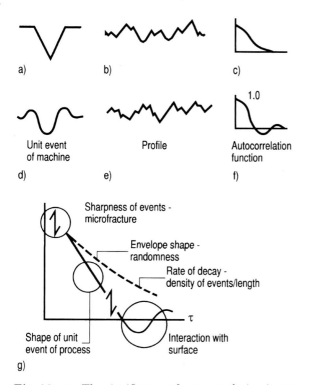

Fig. 39a–g. The significance of autocorrelation in manufacturing

Figure 40 shows a profile of turning produced by a good tool, together with its spectrum. As would be expected for good turning, the spectrum shows some line frequencies, the fundamental corresponding to the feed and a few harmonics due to the shape of the tool.

As the tool wears, the ratio of the harmonic amplitudes to that of the fundamental increases. This is due to random effects of the chip formation and microfracture of the surface. To the left-hand side of the fundamental wavelength there appear periodicities whose wavelengths are much greater than that of the fundamental. These are due to machine-tool problems such as bearing wear, slideway error or even lack of stiffness in the machine tool itself, which may cause chatter. Identifying these effects by using the surface texture is an important first step in remedying the problem.

The spectrum can therefore be split up into two parts, one to the right-hand side of the fundamental frequency (Fig. 40) and one to the left (Fig. 41). On the right-hand side there appear process problems and this has been known for years. The problem is that of knowing how important it is. For many years very simple parameters were used to describe the surface and, it was hoped, its properties. These included the average value R_a, the rms value R_q and various peak height estimates. Investigators in Germany and

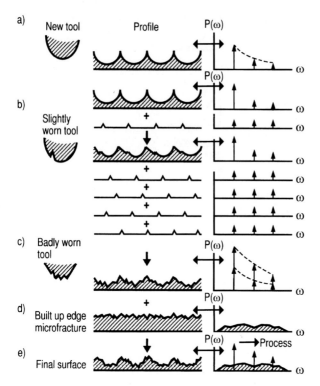

Fig. 40a–e. Power spectral analysis and its use in manufacturing

the USSR used peak measurements rather than average values because they argued that peak measurements correlated more with tribological situations than did average values [5]. Also, the peak measurements of roughness could be measured equally well with optical and stylus methods. This philosophy proved to be practically unsound.

The notion of being able to predict the performance of a workpiece from the geometry of the surface has been attractive for some time. Early investigators used simple models of the surface. These usually involved modelling peaks on the surface as hemispherical spheres scattered on a plane. Then these hemispheres or 'bosses' were assumed to be distributed in a random Gaussian way in height [42,43,44]. This development was closer to real surfaces than previous ones had been but it had the disadvantages that two surface descriptions were needed, one deterministic to describe the shape and size of the hemispherical 'peaks' and one statistical to describe their distribution in space.

This confusing model was eventually replaced by the random-process model mentioned earlier which was based totally on communication theory. This allowed all salient features of the surface to be described with one

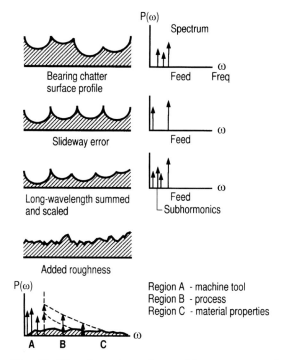

Fig. 41. Use of spectrum for machine, process, and material properties

model. This random-process model was and still is considered to be a big breakthrough in surface characterization.

However, current thinking indicates that even this model needs modifying to reflect better the mechanical situation that occurs, for example in contact where the surfaces contact top down on each other. Another problem which has to be considered is that contact occurs in a parallel mode rather than a serial one. The conclusion has been reached that random-process analysis is adequate for monitoring the manufacturing process, but is rather inadequate for some cases of functional prediction.

Figure 42 shows classification of function and surface features. The classification of function is achieved using the separation of the surfaces and their lateral movement. This classification is an essential element in trying to understand how functional performance is influenced by the surface geometry.

Identifying very specific parameters of the surface geometry with areas in the function map is fraught with problems. In practice only broad categories can be used, as shown in Fig. 42b. Perhaps in the future it will be possible to correlate function and geometry better. However, despite its complexity, Fig. 38 represents a step in the right direction.

Figure 42 shows the type of parameter of the surface that could be useful in various functional situations. The x-axis corresponds to the relative lat-

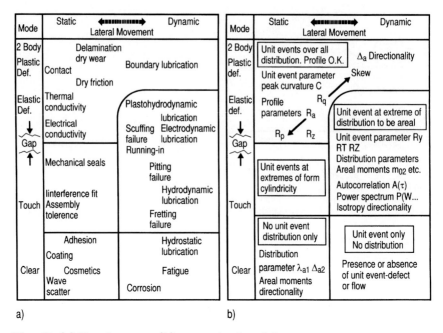

Fig. 42. (a) Function map, (b) parameter template

eral movement between two surfaces and the y-axis to the inverse of their separation. These are (i) unit event characteristics, (ii) profile parameters, (iii) areal parameters, (iv) spatially variable characteristics and extremes of distributions, and (v) defects. It is a fact that often combinations of these parameter types are needed in order to predict the performance of the surface. Figure 42 shows the requirements for surface measurements which in turn decides the specification for the instrumentation in hardware and software. One point to notice is that it is rarely the individual parameter R_a or R_q, for example, which is important but often the <u>type</u> of parameter. Little or no convincing evidence is available to link very specific surface parameters to function.

The important point to notice in Fig. 42 about the surface geometry characteristics important in function is their diversity. They fall into two basic sets. One is the statistical type required, of which there are a number of options. The questions which have to be asked concern whether an average value, an extreme value, the presence or absence of a feature, or perhaps even the spatial variability is required. The other basic issue is that of whether the feature of importance is basically height-orientated, in which case profile information will often suffice, or areal. It should be noticed that areal information reveals structure which is most often important when movement is involved. Under these circumstances it is often extremes of parameters rather than averages which are important. Note that area is something incorrectly called '3D'.

One factor which has been omitted from Fig. 42 is the importance of physical parameters such as nano-hardness and elasticity. Also missing is the presence of thin chemical films on the surface. These factors are known to be important. These non-geometrical properties of the surface should also be measured and included in the surface characterization if a true estimate of the surface performance is to be obtained. There is evidence that multi-disciplinary parameters are now being incorporated into some instruments. However, their use is non-standardized and at present unsatisfactory calibration methods have held back their general implementation.

It should be emphasized that these non-geometrical parameters refer not to bulk properties of the materials but rather to the properties of the outermost skin of the surface, where the skin thickness is taken to be of the same scale as the roughness, if not smaller. The reason for this distinction between bulk and skin properties is primarily that all the action, that is energy transfer, takes place at the surface boundary rather than in the bulk of the material [45].

Whether asperities are crushed or deform elastically is very dependent on the skin property. Apparently soft materials such as copper have a considerably harder surface skin than had previously been thought, which is why stylus measurement is possible despite having stylus pressures greater than the nominal hardness of the material. It is when trying to measure surface parameters which are significant in fatigue or corrosion that the greatest problems arise. This is because in these examples of function it is the isolated deep scratch or defect which often initiates the breakdown. Finding such cracks over a wide area on the surface is a real problem, requiring an instrument to cover a wide area yet with high resolution.

It seems obvious from that which has been reported above that surfaces can be very important. The question is how to measure the surface most effectively.

6 Areal (or Three-Dimensional) Parameters

6.1 General

From what has been said it is clear that the information from a profile or a set of profiles can be very useful in manufacture. It can be used for condition monitoring of the machine tool as well as the process. But if the areal properties are examined rather than the profile, then it becomes more possible to investigate the path of the tool or wheel than it is with only the profile.

Similarly, if there is an aspect of flow of liquid or air the areal properties of the surface will become much more useful than just the profile (Fig. 43). Areal measurement still has no recognized standards. The symbol S is proposed [46]

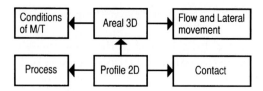

Fig. 43. Importance of areal measurement

rather than R for profiles. Thus the equivalent of R_a is S_a

$$S_a = \frac{1}{L_1 L_2} \int_0^{L_1} \int_6^{L_2} |(y - \bar{y})| \, dx_1 dx_2 , \qquad (1)$$

and so on. For example, the areal skew S_{sk} is given in lateral terms by

$$S_{sk} = \frac{1}{L_1 L_2} S_q^3 \int_0^{L_1} \int_0^{L_2} (y - \bar{y})^3 dx_1 dx_2 . \qquad (2)$$

In areal terms x_1 and x_2 are the two orthogonal spatial axes.

6.2 Comments on Areal Measurement

Obviously, the moment definitions for these areal parameters have to be extended to two dimensions.

6.2.1 Relationship to Profile Data

In general it is not safe to extrapolate areal properties from those of the profile. For example the density of peaks in areal terms is not the square of the profile density, it is slightly smaller, $\approx 20\%$, because of the presence of cols and saddle points which would appear as a maxima in a profile but are not in areally. There are some real discrepancies, for example, the material ratio for a profile is the same as that for an area. The material ratio for area is certainly not the square of the profile value of MR. One major problem is that of peak (or, in areal terms, summit) detection. This is discussed below because it illustrates the difficulty of surface analysis.

6.2.2 Sampling

In addition to having to specify the spacing of digital information for the profile, a comment on the pattern of sampling has to be included (Fig. 44):

(a) trigonal -3 points at $120°$ have to be lower than the central point,
(b) rectangular -4 or 8 surrounding points must be lower than the central point,
(c) Hexagonal, and
(d) as (b).

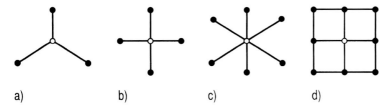

Fig. 44a–d. Sampling patterns

Of these, usually (b) or (d) are chosen. This choice is usually made on practical grounds because it is simple to digitize using conventional tracking carriages. They both have problems because (b) does not sufficiently cover the area; it has relatively few points within the area. The pattern (d) has unequal correlation between the surrounding points and the central one. This is because the surrounding points are not all at the same distance from the central arc. Possibly the best sampling pattern is (c) which does not suffer from the unequal weights and yet does cover the area. It is relatively easy to arrange for a scanning system to enable this pattern to be achieved [47].

There is another point which should be remembered, which is that in principle there should be a very large pattern of digital samples in order to cover the area properly. Because of this lack of complete coverage, digital areal expressions for surface parameters do not necessarily assymptotically converge to the theoretical analogue value as the digital spacing is reduced to zero. For the profile sampling this effect does not happen; results do converge to the analogue. Hence for areal analysis the parameter values do not simply rely on the spacing of ordinates. The values obtained for the parameters also rely on the numerical model used for scanning.

6.2.3 Characterization

There have been many attempts to do this starting from the 1960s [48,49,50]. Some are based on the process of manufacture, the so-called 'hypsometric' method, which although quite revealing was too restricted for general use. Perhaps the methods based on areal correlation functions [51] or power spectral densities are more suited to general use but they have not been adopted. The reason probably is that people versed in manufacture or tribology are not comfortable enough with random process analysis to use them.

One current method of characterization [46] is included here to demonstrate the difficulty of getting even the most obvious and straightforward parameters to work:

1. Density of summits (where a summit is defined as an areal peak)
2. Length of contours as a function of height
3. Areal autocorrelation
4. Areal power spectrum

5. Degree of isotropy
6. Directionality
7. Angular average wavelength
8. Visualization.

We consider these in turn, noting that there is no standard as yet.

6.2.4 Density of Summits

This is designated by S_{ds} where

$$S_{ds} = \frac{\text{Number of summits/unit area}}{(M-1)(N-1)},$$

with M and N being the number of ordinates in the two orthogonal directions.

6.2.5 Areal Autocorrelation and Power Spectra

$A(\tau_x \tau_y)$ and $P(w,v)$ both have formulae which are simple extensions of the profile case, e.g.

$$P(w,v) = \iint f(x,y) \exp[-j(wx + \vartheta y)] \mathrm{d}x \mathrm{d}y .$$

Both of these functions have been used to evaluate such features as summit height, curvature contour lengths etc., first having made some assumptions about the surface model.

6.2.6 Isotropy

This does not exist in the profile. A number of ways of specifying isotropy are possible. One such definition is the isotropy ratio:

$$\text{Isotropy ratio} = \frac{\text{shortest correlation length}}{\text{longest correlation length}},$$

where the correlation length is defined as $\int_0^\infty |A(\tau)| \mathrm{d}\tau$, which for a typical ground surface is $1/e$. Sometimes the correlation length refers to that distance over which the autocorrelation falls to a low level, e.g. 10%.

6.2.7 Directionality

Simply, the directionality is the way in which the isotropy presents itself to the function. See, for example, Fig. 45, which shows that exactly the same profile when reversed in direction can have a completely changed functional significance, for example in a bearing.

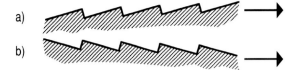

Fig. 45a,b. Directionality

6.2.8 Parameters

One attempt to provide a limited set of areal (three-dimensional) parameters has been set up in Europe (EUR 15178EN) [52].

Amplitude parameters
1. S_q - rms of surface
2. Ten-point height S_z
3. Skew S_{sk}
4. Kurtosis S_{ku}

Spatial parameters
5. Density of summits S_{ds}
6. Texture aspect ratio S_{tr}
7. Texture direction
8. Shortest correlation length S_{al}

Hybrid parameters
9. rms slope S_{Δ_q}
10. Mean summit curvature S_{sc}
11. Material surface area ratio S_{ar}

These above are the purely geometric parameters. In addition to these a number of functional parameters have been suggested. They are quite arbitrary.

Functional Parameters
12. Surface material ratio S_q
13. Void volume ratio S_{vr}
14. Surface bearing index S_{bi} Of these even a basic
15. Core fluid retention S_{ci}
16. Valley fluid retention S_{vi}

set has 14 parameters [52]. The definitions of some of the functional parameters are given below.

Surface material ratio

$$S_{bc} = \frac{S_q}{Z_{0.05}},$$

where $Z_{0.05}$ is the height of the surface at 5% material ratio.

Core fluid retention index

$$S_{ci} = [V_v(0.05) - V_v(0.8)]/S_q \quad \text{(unit area)},$$

where V means valley. If S_{ci} is large then there is good fluid retention. Relationship is $0 < S_{ci} < [0.95 - (h_{0.05} - h_{0.8})]$.

Valley fluid retention index

$$S_{vi}[V_v(h = 0.8)]/S_q \quad \text{(unit area)} \quad 0 < S_{vi} < [0.2 - (h_{0.8} - h_{0.05})].$$

All of these parameters have yet to be standardized and they are quite arbitrary, but they are better than no proposition.

6.2.9 Gap Properties

In many applications, to specify one surface can be relatively meaningless. Such occasions arise in many tribological functions, e.g. contact and friction. Hence there is a severe restraint which should be remembered when dealing with areal parameters, and this is that it should be possible to predict the statistical properties of the functional gap between two surfaces which are to work together. This criteria is usually not possible with peak-type parameters, but fortunately it is for some of the statistical terms such as autocorrelation and power spectrum analysis. These can be added together not forgetting that the parameters only refer to touching surfaces; the physical separation belongs to á Tolerances and Fits' discussion.

The fact that the autocorrelation function is in itself a function of shift τ makes it especially useful in rubbing body investigations. In other words the parameters used should be as close to the functional situation as is possible: the parameter should mimic the function.

7 Conclusions

This review has described in the broadest terms the characterization and measurement of engineering surfaces. The general message is that both the method of measurement and the characterization of the surface should, to get best results, be as close to the application of the surface as possible. In process control, the surface parameters should reveal the tool or grain of machining. In machine tool monitoring, ideally the instrument sensor should follow the path of the cutting tool. In this way only the differences detected are significant. Similarly in functional terms friction, wear and, in particular, contact should be to some extent predictable from the surface parameters, albeit in an indirect way.

It has not been possible to include all the detailed information that is available. For this a bibliography has been provided together with key references. Optical methods have been examined in some detail but not exclusively, because in general terms optical methods have not been as versatile as stylus methods, in engineering at least. Furthermore, optical methods have not been the driving force behind surface processing in engineering. Their total acceptance in industry is still some way off; only recently has there been research into suitable optical standards. Many problems have arisen because tactile standards have been used for optical calibration. It is hoped that this situation is now changing and that optical methods ultimately will stand alongside tactile methods, the former being used for non-contact functions and the latter for contact and general tribology applications.

It is pertinent in summary to make a few comments about the future. Surface texture and surface geometry as a whole are at the crossroads in any discussion between energy transfer or information transfer and storage. Because of the movement towards miniaturization, the characterization and measurement of surfaces will become more important rather than less. Also, any attempt to fully characterize the functional situation using surfaces will have to include other aspects of the surface such as the chemical films on the surface. At the nanoscale more importance will have to be given to the quantum effects, which do not show up when working with the traditional scale of size used by engineers. Already these effects are being seen in instruments. In the future it will be more and more difficult to differentiate between the engineer, the physicist, and the chemist.

General Bibliography

Surface Metrology

B1. D. J. Whitehouse: *Handbook of Surface Metrology*, Inst. of Physics, Bristol (1994)

Optical Surface Metrology

B2. F. G. Bass, I. M.Fuchs: *Wave Scattering of Electromagnetic Waves from Rough Surfaces*, Macmillan, New York (1961)
B3. P. Beckmann, A. Spizzichino: *The Scattering of Electomagnetic Waves from Rough Surfaces*, Pergamon, Oxford (1963)
B4. J. M. Bennet: *Surface Finish and its Measurement, Collected Works in Optics* Vols. 1,2, Opt. Soc. America, Washington, DC (1992)
B5. J. A. Ogilvy: *The Theory of Wave Scattering from Random Surface*, Pergamon, Oxford (1991)
B6. D. J. Whitehouse: *Optical Methods in Surface Metrology*, SPIE Milestone Ser. **129**, (1996)

References

1. J. F. Archard: Proc. R. Soc. Lond. A **243**, 190 (1957)
2. A. E. Norton: *Lubrication*, McGraw Hill, New York (1942)
3. J. T. Burwell: Wear **1**, 119 (1957)

4. J. Perthen: *Prüfen und Messen der Oberflächengestalt*, Hasser, Munich (1949)
5. D. J. Whitehouse: Proc. Inst. Mech. Eng. **192**, 179 (1978)
6. D. J. Whitehouse: J. Inst. Phys. Meas. Sci. Technol. **8**, 955 (1997)
7. D. J. Whitehouse: Proc. Wear Conf. MIT, (1978) p. 17
8. Y. G. Schneider: Prec. Eng. **6**, 219 (1984)
9. R. E. Reason, M. R. Hopkins, R. I. Garrod: *Report on the Measurement of Surface Finish by Stylus Methods*, Rank, Leicester (1944)
10. R. E. Reason: The Measurement of Surface Texture, in Wright-Baker (Ed.): *Modern Workshop Technology*, 2nd edn., Macmillan, London (1970)
11. D. J. Whitehouse: J. Phys. E. **15**, 1337 (1982)
12. J. Prescott: *Applied Elasticity*, Longmans, London (1961)
13. G. Schmalz: Z. VDI, **73**, 144 (1929)
14. D. J. Whitehouse: Proc Inst. Mech. Eng. **202**, 169 (1988)
15. D. J. Whitehouse: *Optical Methods in Surface Metrology*, SPIE Milestone Ser. **129** (1996)
16. A. Guild: J. Sci. Inst. **5**, 209 (1940)
17. J. H. Rakels: J. Phys. E. **19**, 76 (1986)
18. U. Breitmeier: *Optical Follower Optimized*, UBM Tech. Note (1990)
19. S. Tolansky: Verh. Tag. Dtsch. Phys. Ges. **5–7**, Göttingen (1947)
20. J. C. Wyant: Appl. Opt. **11**, 2622 (1975)
21. J. Z. Young, T. Roberts: Nature **167**, 231 (1951)
22. C. J. R. Sheppard, T. Wilson: Opt. Lett. **3**, 115 (1978)
23. M. Dupuy: Proc. Inst. Mech. Eng. **180**, 200 (1968)
24. G. E. Sommarren: Appl. Opt. **20**, 610 (1981)
25. E. L. Church, H. A. Jenkinson, J. M. Zavada: Opt. Eng. **16**, 360 (1977)
26. E. L. Church, T. V. Vorburger, J. M. Zavada: SPIE Proc. **508**, 13 (1985)
27. R. W. Sprague: Pt. Opt. **11**, 2811 (1972)
28. G. Graneck, H. L. Wunsch: Machinery **81**, 707 (1952)
29. K. F. Sherwood, P. M. Cross: Proc. Inst. Mech. Eng. **182** (1968)
30. J. W. S. Hearle, J. T. Sparrow, P. M. Cross: *Scanning Electron Microscope*, Pergamon, Oxford (1972)
31. H. K. Wickramasinghe: AIP Proc. **241** (1991)
32. D. J. Whitehouse, R. E. Reason: *The Equation of the Mean Line of Surface Texture*, Rank, Leicester (1965)
33. D. J. Whitehouse: Proc. Inst. Mech. Eng. **182**, 179 (1968)
34. ISO Standard, No. 3274, Geneva (1996)
35. A. Vander Lugt: Opt. Acta **15**, 1 (1968)
36. ISO Standards Handbook, Applied Metrology, Limits fits & Surface Properties, ISO, Geneva (1988)
37. BS. ISO, No. 13565, parts 1 & 2 (1996)
38. D. J. Whitehouse, M. J. Phillips: Philos. Trans. R. Soc. Lond. A **290**, 267 (1978)
39. M.S Longuet-Higgins: Proc. Roy. Soc. Lond. A **249**, 321 (1957)
40. J. Peklenik: Proc. Inst. Mech. Eng. **182**, 108 (1968)
41. D. J. Whitehouse: Annals CIRP **27**, 491 (1978)
42. J. A. Greenwood: Proc. Roy. Soc. Lond. A **295**, 300 (1966)
43. J. A. Greenwood: Proc. Roy. Soc. Lond. A **393**, 133 (1984)
44. D. J. Whitehouse, J. F. Archard: *ASME Conf. Surface Mechanics*, Los Angeles, CA (1969) p. 36
45. B. F. Gaus: Trans. ASME, J. Tribol. **109**, 427 (1987)

46. K. J. Stout, P. J. Sullivan, W. P. Dong, E. Minsah, K. Subari, and T. Mathia: European Community Document, No. 33, 74/1/0/170/90/2
47. D. J. Whitehouse, M. J. Phillips: Philos. Trans. R. Soc. A **305**, 441 (1982)
48. M. Pesante: Annals CIRP **12**, 61 (1963)
49. M. Ehrenreich: Microtecnik **XII**, 103 (1959)
50. N. Myres: Wear **5**, 182 (1962)
51. J. Peklenik, M. Kubo: Annals CIRP **16**, 235 (1968)
52. European Community Document, No. 15178 EN

Index

4f optical processor, 382

Abbott Firestone curve, 443
Aberrations, 7
Abrasive, 418, 427
Acousto-optic modulator, 380, 389
Acrylate coatings, 255
Adhesive, 157, 158, 250, 257
Adhesives, 158, 257, 258
Aerodynamics, 121
Affine transformation, 332
Airy stress function, 46, 50
Analytic signal, 57
Analyzer, 204, 205, 209, 210, 214, 216, 217, 251, 278, 279, 304, 307
Angular phase shift, 207
Anomalous, 178, 180
Areal
– autocorrelation, 455, 456
– correlation, 455
– parameters, 440, 444, 452–454, 457, 458
– patterns, 416
– power spectrum, 455, 456
Aspect ratio determination, 338
Asperities, 418, 432, 453
Aspheric
– shapes, 419
– surfaces, 4, 420
Astigmatism, 8–11, 22, 25
Autocorrelation, 397–399, 406, 447–449, 456, 458
Automatic digital readout, 131
Average parameters, 414, 436, 440
Axial
– bending, 52
– chromatic aberration, 11
Axis aberration, 8

Background intensity, 110, 184, 332
Bandwidth, 230, 239, 249, 252, 253, 259, 263, 266–268, 289, 387–389, 391, 393, 394, 408, 420, 439, 446
Beam divergence, 241

Beta function, 447
Biaxial test, 351
Bilinear, 328, 331, 332
Binocular vision, 334
Birefringence, 197, 198, 203, 215, 238, 239, 247, 250, 257, 279, 295, 307
Birefringent materials, 200, 203, 304, 307, 319
Bithermal loading, 157, 158
Boundary
– conditions, 33, 41, 42, 46, 49, 51, 52, 88
– constraints, 42
Bragg
– cell, 380, 389
– grating fabrication, 247, 248
– grating reliability, 251
– grating sensor, 233, 246, 247, 250–253, 259, 262, 266–268
– Bragg wavelength, 246, 249–252, 263, 266
Bridge, 239, 273, 275, 279, 283–286, 335
Brillouin, 273, 286–290
Buckling, 151, 185–187, 189, 194, 323, 326, 362
Bulk modulus of elasticity, 41
Burst
– spectrum, 387, 388, 397
– spectrum analyser (BSA), 388

Calibration, 212, 216, 221, 223, 286, 326, 333–335, 338–341, 354, 355, 362, 366, 367, 425, 426, 432, 433, 453, 459
Calliper, 419, 420
Capacitance methods, 414
Carbon coating, 255
Carrier, 27, 57, 58, 77, 78, 81, 156, 160, 191, 277, 378, 387, 389
Carrier fringes, 160–162, 177, 178
Caustics, 295–297, 299, 309–314, 316, 317, 319
Cavity length, 14, 234, 235, 237, 241, 242, 245, 260, 261

CCD camera, 107, 164, 166, 183, 184, 193, 211, 327, 328, 391, 393, 406, 407, 430
Center of curvature, 4, 21, 27
Characteristic diagram, 61–63, 68–71, 95
Chatter, 415, 449
Chebychev zone, 434
Circular polariscope, 200, 207, 209, 210, 213, 216, 225, 307
Circularly polarized light, 198, 202
Coatings, 215, 236, 253–255, 280, 295
Code of the velocimeter, 382
Coherence, 14, 15, 21, 248, 249, 268, 275, 278, 279, 283, 373
– length, 14, 15, 245, 248, 268, 277, 278
Coherent
– detection, 384, 385, 395
– gradient sensor, 299, 300
– shearing interferometer, 295, 296
Color hologram, 30
Coma aberration, 8, 9, 25
Comatic image, 8
Compact shear fixture, 160
Comparative holography interferometry, 117
Comparator, 429
Compatibility, 33, 39, 42–46, 48, 235
Compressive edge-line load, 313
Computational fluid dynamics (CFD), 373
Computer implementation, 68
Confocal microscope, 430, 431
Contour
– interval, 122, 123, 151, 154, 177, 182, 191–193
– planes, 123, 124, 140
– sensitivity, 124
Contouring, 121, 124, 140, 141
Correlation
– fringes, 132–136
– function, 261, 325, 331, 342, 400, 448
– length, 448, 456, 457
Corrosion, 255, 444, 453
Crack closure, 349, 350
Crack-opening displacement, 159, 348, 350, 351

Cross-beam velocimeter, 378, 380, 382, 383, 408
Cross-correlation, 261, 263, 325, 332, 341, 342, 400, 403
– normalized, 330–332
Cross-line grating, 152, 158, 167, 168
Cross-sensitivity, 289
Cryogenic, 158
CTE, 171–173, 191, 237, 238, 242–244
Curvature, 3–5, 7, 10, 19, 21, 27, 49, 50, 117, 167, 172, 173, 190, 213, 284, 394, 418, 424, 429, 433, 445, 456, 457
Curve-fitting, 387
Curved surface, 144, 166, 170, 323, 340

Damping effects, 416
Dark-field
– circular polariscope, 209, 216
– plane polariscope, 205, 218
Defocused speckle photography, 128, 129
Deformation, 23, 27, 28, 33, 37–41, 47, 103, 104, 106, 108–110, 112, 114, 116, 125–129, 132–135, 139, 151, 153, 154, 156–159, 161, 162, 164, 166, 167, 172, 173, 175–178, 184, 187, 189, 191, 273, 275, 279, 280, 285, 286, 295, 297–300, 302, 303, 316, 323–328, 330, 342, 345, 352, 358, 367, 395, 423, 445
Degree of
– isotropy, 456
– scatter, 429
Demodulation, 233, 239, 241, 253, 258, 259, 262, 263, 266, 277, 278, 388, 391
Density
– field, 121
– of summits, 455–457
Designer surfaces, 413, 416, 417, 419
Detector nonlinearity, 59
Deviatoric stress, 41
DFT algorithm, 62, 66, 68, 83, 86, 95
Diamond turning, 419
Differential
– interference microscopy, 432
– velocimeter, 378
Diffraction
– grating, 30, 173, 264, 266
– halo, 126, 128, 130, 133, 139

Digital
- holographic interferometry, 107
- image correlation, 323, 326, 334, 358, 367
- image processing, 134, 197, 211–213, 222, 398

Dipole, see Mnopole, 87
Direction of displacement, 106, 109, 126
Directional ambiguity, 399, 407, 408
Directionality, 456, 457
Dispersion of birefringence, 203
Displacement field, 110, 128, 164, 166, 176, 185, 327, 328, 331, 332, 352, 353
Distortion, 10, 52, 313, 333–335, 337, 339–341, 343, 393, 407, 421, 436, 437

Distributed
- strain, 290
- temperature, 290

Doppler
- burst, 387, 408, 409
- shift, 287, 374, 376–380, 387, 388, 392–394
- signal, 378, 381, 384, 386–388, 391, 408

Double
- aperture, 137
- exposure, 105, 106, 120
- refraction, 197, 198

Dynamic events, 57, 58

Edge detection, 333
Effective stress, 227
Efficiency factor, 76, 86
Elastic effects, 417
Electro-optic holographic interferometry, 110
Electronic packaging, 152
Elliptically polarized light, 198, 202
Engineering strain, 160, 163, 164
Epoxy, 158, 167, 171, 173, 257, 258, 282, 307, 345, 362
Equilibrium, 33, 34, 39, 41–46, 48, 49, 51
ESPI, 133, 134, 136

Extrinsic
- Fabry-Perot (EFP) Sensor, 240, 243
- parameters, 335

Fabry-Perot, 233–236, 238–246, 250, 253, 259, 261–263, 265, 267, 268, 275, 276
Fabry-Perot filter, 262, 263
Far-field microscope, 325, 348
Fatigue, 227, 244, 345, 444, 453
Fermat's principle, 1–3

Fiber-optic
- cable, 255
- coatings, 253
- sensors, 233, 239, 241, 247, 253, 254, 256, 273, 275–277, 279, 281, 283, 285–287, 289, 291, 293

Figure of merit, see Eficiency factor, 76
Filter cut-off, 436, 446
Filters, 183, 219, 262, 381, 388, 436, 437, 439, 445, 446
First-order optics, 3, 5, 7
Fizeau interferometer, 15–22
Flag system, 428
Flaw detection, 118, 135
Floodlit, 428, 432
Flow, 110, 121, 275, 373–375, 377, 378, 381, 389, 390, 393–395, 400–403, 407, 408, 453
Fluorescent seeding, 375
Flying-spot microscope, 430
Focal length, 5, 7, 17, 126, 304, 362, 381, 423
Focused speckle photography, 125, 126, 128
Follower, 423, 428, 431, 432

Foucault
- method, 431
- test, 27

Four-wavefront mixing, 122

Fourier
- plane, 131, 139, 140, 382–384, 445
- transform, 126, 264, 266, 382, 397, 398, 429, 440, 447

Fracture mechanics, 152, 176, 227, 295, 297, 309, 315, 316, 318, 323, 367

Frequency
- response, 426–428
- shift, 287–290, 378, 380, 382, 384, 386, 387, 389, 395, 399, 408

Fringe
- contrast, 12, 14, 15, 110, 120, 151
- control, 106
- counting, 158, 160, 259, 260, 277, 278
- multiplication, 151, 156, 164, 177, 185, 191, 193, 211, 213, 214, 227
- order, 105, 151, 154, 156, 159, 162, 163, 181, 182, 184, 185, 192, 204, 206, 214–222, 226, 227, 229, 236, 303, 314, 317

Functional parameters, 457
Fundamental frequency, 288, 449

Gap properties, 458
Gauge length, 49, 81
Gaussian
- distribution, 437, 443
- equation, 5

Gladstone-Dale constant, 121
Glossmeter, 429, 438
Grain hit width, 448
Gray-level, 212, 214, 221, 222, 228, 229
- transformation, 212

Grid, 58, 59, 78, 137, 138, 144, 171, 173, 191, 333, 335, 338, 339, 348, 353, 354, 362
Grinding, 418, 419, 447, 448
Groove, 422, 425

Half-fringe photoelasticity, 216, 219, 227
Hard surfaces, 422
Harmonics, 449
Heat transfer, 121, 309
Height
- distribution, 440, 441, 447
- parameters, 415, 440, 446, 447

Helium-Neon laser, 14, 15
Hermetic coating, 255
Heterodyne, 128, 384, 431, 432
- mode, 384

Hologram, 29–31, 103, 104, 106, 108, 110, 119, 120, 122–124, 145, 404
Holographic
- interferometry, 31, 103–113, 115, 117–119, 121, 123, 125, 127, 129, 131, 133, 135, 137, 139, 141, 143, 145, 147, 149, 180, 295, 296
- principles, 29

- shearing, 115, 117

Homodyne mode, 383, 384
Hooke's Law, 48, 49
Hybrid parameters, 440, 445, 457
Hydrogen loading, 248

Image
- decorrelation, 120
- height, 8
- labelling, 399, 402
- shift, 399

Immersion interferometer, 164, 165
In-plane
- displacement, 45, 112, 115, 126–128, 136, 151, 152, 325, 343, 352
- maximum shear stress, 226
- principal stress, 226

Incoherent detection, 384, 385, 395
Index of refraction, 200, 203
Inertial effects, 416
Insertion loss, 241
Interference, 1, 12–14, 20, 25, 27, 31, 103–106, 110–112, 115, 121, 122, 125, 126, 132, 134, 135, 153–155, 162, 163, 202, 234, 236, 238, 241, 247, 248, 259, 275, 277, 278, 299, 302, 385, 432
- grating, 103

Interferometric
- demodulation, 266
- sensor, 250

Internal mirror, 234–236
Intrinsic
- Fabry-Perot (IFP) sensor, 238
- parameters, 335

Iodine cell, 391–394, 408
Isoclinic fringe pattern, 206
Isoclinics, 209, 222, 226, 228
Isotropy, 456

Knife edge, 27, 28
Kurtosis, 440, 441, 447, 457

Lapping, 418, 443
Laser, 14, 15, 21, 24, 126, 128, 130, 131, 139, 152, 183, 186, 247–249, 253, 259, 266, 268, 277, 288, 300, 301, 304, 373–379, 381, 383, 385, 387, 389–393, 395–397, 399–403, 405, 407–409, 411
- diode, 15

Index 467

- Doppler velocimetry (LDV), 376
- speckle photography (LSP), 395
- induced fluorescence (LIF), 375

Lateral
- shear interferometer, 23–25
- shearing, 130, 299, 300, 304

Laws of refraction, 2, 3
Length parameters, 440
Levelling, 420, 433
Light
- sources, 14, 187, 198, 362
- field isochromatic fringe pattern, 219

Linear scattering, 375
Live fringes, 120
Lubrication, 413, 418, 444

Mach-Zehnder, 276, 288, 295, 296, 299, 376, 377
Machine tool diagnostics, 415
Macromechanics, 155
Magnification chromatic aberration, 11
Marginal rays, 7
Mask technique, 132, 133, 136
Maximum shear stress, 36, 37, 48, 226
Mechanical Stylus, 422–425, 427, 428, 432, 433, 438
MEMS, 417
Meridional rays, 4, 5
Merit graphs, 427
Michelson, 13, 16, 17, 19, 20, 134, 141, 276, 278, 295, 296, 299
Micro-dynamic systems, 417
Microbending, 273, 280, 281, 284, 286
Microcracks, 115
Microfracture, 449
Micromechanics, 151, 177
Mireau interferometer, 430, 431
Mixed mode I/II, 343
Model verification, 323, 367
Modes of vibration, 119
Mohr's circle, 35, 36
Moiré, 58, 59
Monopole, 87
MR curve, 443, 444
Multiple-angle relation, 214
Multiplexing, 233, 239, 240, 246, 251–253, 262, 265–268
Multiplication factor, 168, 177, 185, 193

Nano-hardness, 453
Nanotechnology, 417, 433
Non-birefringent materials, 295, 297, 309, 311
Non-linear scattering, 375
Nondestructive testing, 125, 128, 135, 145
Nonlinear phase shift, 64, 65
Normal
- strain, 37, 49, 131, 160, 161, 175
- stress, 34, 36, 49–51

Normalized CP, 82
Null signal, 432
Numerical aperture, 380, 407, 423

O/DFM, 166, 167, 177, 185, 186, 189–191, 193, 194
Object wave, 103, 104, 110
Object-plate tandem, 106
Oblique rays, 4
Optical
- axis, 4, 5, 8, 122, 126, 129, 130, 135, 137, 140, 301, 302, 305, 335, 384
- filtering, 126, 438
- metrology, 1, 296, 373
- path, 2, 13–19, 21–23, 105, 109, 157, 198, 200, 234, 261, 280, 297–299, 310, 311, 315, 316, 380, 420, 424, 425
- phase, 107, 109, 115, 121, 122, 144, 236, 237, 241, 242, 246, 259, 266, 297, 299, 302, 303, 310
- stylus, 423, 425
- switch, 239, 240

Optically isotropic materials, 306, 307
Oversampling, 328, 345

Paraxial rays, 5, 7
Particle
- image velocimetry (PIV), 395, 412
- tracking, 403

Peak parameters, 414, 436, 441–443, 445
Pedestal function, 387
Perspective, 267, 326, 343, 368, 403
Petzval surface, 11
Phase distribution, 115, 122, 137, 142
Phase
- corrected filters, 436, 437
- grating, 157

– mask, 249
– origin, 60, 63
– retrieval, 277, 278
– shift, 110, 128, 202, 204, 207, 297, 299, 302, 303
– sign ambiguity, 110
– stepping, 134, 164, 180, 184, 185, 191, 192, 194, 217, 218, 222
– unwrapping, 28, 217, 219, 224
Photoelastic
– coatings, 215
– dispersion, 203
– model, 197, 204–208, 214, 216, 227, 236
Photoelasticity, 37, 44, 197, 199–201, 203–205, 207, 209, 211, 213, 215–217, 219, 221, 223, 225–227, 229–231, 295, 307, 317, 319
Photographic subtraction, 132
Photosensitivity, 248
Piezoelectric, 110, 164, 166, 262
– transducer, 110
Planar
– Doppler velocimetry, 391
– surfaces, 323, 367
Plane
– polariscope, 200, 204, 205, 207, 217, 218, 223, 225
– polarized light, 207
– strain, 40, 44–46, 175
– stress, 35, 40, 44, 45, 297, 298, 315
Plasma diagnostics, 121
Plateau Honing, 418, 444
Ply-by-ply deformation, 166, 167
Pneumatic methods, 433
Point-wise reading, 131
Poisson's ratio, 298, 315
Polariscope, 200, 204, 205, 207, 209, 210, 213, 216–218, 222, 223, 225, 226, 300, 304, 307
Polarization fading, 238, 242, 250, 267, 268
Polarization-stepping, 222, 227
Polarizer, 200, 201, 204, 206, 207, 210, 223, 224, 305
Polishing, 415, 418, 422, 423
Polyimide coating, 282
Power
– budget, 241, 246
– spectral density, 439, 447
Principal stress, 37, 48, 197, 203, 204, 206, 207, 209, 210, 216, 217, 222–227, 229
– difference, 203, 206, 209
– direction, 207, 216, 217, 222–224, 227, 229
Probability density, 446, 447
Process control, 152, 415, 439, 458
Profile trace, 439
Profilometry, 57, 89
Pulsed laser velocimetry, 373–375, 377, 379, 381, 383, 385, 387, 389, 391, 393, 395, 397, 399, 401, 403, 405, 407–409, 411

Quality control, 414, 415
Quarter-wave plate, 202, 207–209, 216, 304
Quasi-heterodyne, 183, 184

Radius of curvature, 3, 4, 50, 433
Raman scattering, 286
Random process, 173, 446, 447, 455
Ratiometric demodulation, 266
Real-time holographic interferometry, 105, 113
Reconstructed image, 104–106, 120, 124, 403
Reference grating, 151, 154–156, 160, 162, 163, 165, 166, 177, 180–184, 189, 192, 193
Reference speckles, 134
Reference wave, 103, 106, 110, 135, 384, 404
Reflection, 2, 3, 198, 241, 246, 288, 297–301, 307, 309, 311
Refractive index, 2, 11, 28, 121, 122, 165, 166, 246, 280, 286, 287, 297, 298, 302, 313, 377, 389
Relative phase shift, 202, 204
Relative retardation, 200, 201, 203, 207, 216
Relief variations, 121
Replication, 157, 170, 171, 176
Resolved part of displacement, 109
Resonance states, 119
Response time, 426, 427

Retardation, 200, 201, 203, 206, 207, 216, 217, 305, 306
Reverse engineering, 323, 354, 357
Rigid body rotation, 111
Ronchi
- ruling,26, 27, 58
- test, 25, 26, 58
Rotating structures, 120
Rotation compensation, 120
Roughness, 413–419, 421–425, 427–429, 431, 433–437, 439–441, 443, 445, 447, 449–451, 453, 455, 457, 459, 461

Sagittal rays, 11
Sagittal surface, 10
Sagnac, 276
Sampling, 28, 277, 324, 436, 439, 441, 454, 455
- frequency, 59, 277, 324
- functions, 71, 72
- range, 66
- spacing, 84
Sandwich holography, 106
Scanning, 227, 251, 262, 263, 279, 288, 325, 367, 389, 390, 401, 413, 428, 430, 432, 433, 455
Scatter method, 429
Schlieren Techniques, 27, 28
Secondary speckles, 128
Seeding particles, 373, 375, 376, 382, 383, 389, 395
Sensitivity, 28, 116, 117, 120, 124, 129, 131, 138, 139, 144, 151, 152, 154, 156, 162, 164, 182, 184, 185, 189, 191, 193, 216, 237, 238, 243, 247, 266–268, 276, 277, 280, 289, 296, 303, 319, 333, 383, 393
Sensor packaging, 233, 280, 289, 290
Sensors, 49, 183, 233–244, 246, 247, 249–254, 256, 257, 259–262, 265–268, 273–277, 279–281, 283–287, 289–291, 293, 324, 327, 328, 393
Serrodyne, 259, 260
Shape, 4, 8, 23, 27, 31, 103–105, 118, 121–123, 140, 141, 298, 309, 311, 313, 326, 330, 334, 354, 415, 418, 420, 422, 425, 433, 434, 437, 438, 440, 443, 446–450
- contouring, 140

Shear
- strain, 37, 41, 160, 173, 178, 180
- stress, 34, 36, 37, 40, 48, 50–52, 226
Shrinkage, 112, 114, 283, 284
Signal-to-noise ratio, 84, 250, 375, 388
Single mode, 14
Size arguments, 416
Skew, 440, 447, 454, 457
Skid, 420–422, 431
Slope, 23, 28, 29, 52, 116, 117, 129, 142–144, 160, 162, 163, 216, 298, 350, 422, 423, 429, 445, 457
Smooth reference, 135
SOFO sensor, 282
Spacing
- information, 445, 447
- parameters, 415, 442, 444
Spatial
- division multiplexing (SDM), 267
- phase-stepping, 57–59, 81, 93
- resolution, 81, 94, 151, 228, 286–289, 373, 393, 400, 401, 408, 409
Specimen grating, 152–158, 162, 166–168, 170–172, 177
Speckle
- interferometry, 125, 131, 132, 141, 295
- metrology, 103, 105, 107, 109, 111, 113, 115, 117, 119, 121, 123–125, 127, 129, 131, 133, 135–137, 139, 141, 143, 145, 147, 149
- noise, 110, 124
- pattern, 125, 128, 132–135, 333
- photography, 125–129, 131, 395
- shearing interferometry, 125, 140
- shearing photography, 129
Speckle-shearogram, 131
Spectral
- contents, 215
- interrogation, 263–265
Speed of light, 1, 2, 438
Spherical
- aberration, 7, 8, 19, 22, 25
- stress, 41
- surface, 3, 4
Spontaneous Brillouin scattering, 287, 288
Stable crack growth, 346

Stiffness effects, 416
Stokes, 286, 376
– drag, 376
Strain, 33, 37–49, 52, 53, 131, 159–161, 163, 164, 167–169, 171, 173, 175, 177, 178, 180, 187, 197, 233, 235–239, 241–247, 249–261, 263–269, 271, 273–277, 280, 286–291, 324–326, 330–333, 344, 348, 349, 352–354, 361, 366, 390
– gradients, 247, 257, 353
– tuning, 249
Stress, 33–37, 39–53, 103, 105, 177, 197, 200, 203, 204, 206, 207, 209, 210, 215–217, 220, 222–227, 229, 236, 244, 255, 295–300, 303, 309, 311, 312, 315–317, 319, 344
– optic coefficient, 203
– optic law, 202–204, 206, 220, 226
– strain Equations, 33, 40, 41, 43, 45, 52
Stroboscopic holography, 120
Stylus methods, 414, 419, 424, 425, 428, 432, 433, 450, 458
Surface characterization, 413, 415, 417, 419, 421, 423, 425, 427, 429, 431, 433, 435, 437, 439, 441, 443, 445, 447, 449, 451, 453, 455, 457, 459, 461
– material ratio, 457
– of localization, 111
– slope, 163
– strains, 129
Symmetric optical systems, 8
System calibration, 286, 366

Talbot interferometer, 26
Tandem speckle photography, 129
Tangential
– rays, 10, 11
– surface, 10
Tardy method, 210
Temperature, 33, 109, 121, 151, 156–158, 171, 191, 194, 233, 236–238, 242, 243, 246, 250–252, 254, 257, 259, 260, 267, 273, 276–278, 284, 286–291, 348, 367, 391
Temporal phase-stepping, 57, 59, 91–93
Temporally coherent, 14

Texture, 132, 415, 416, 418, 419, 439, 449, 457, 459
Thermal
– apparent strain sensitivity, 237, 238, 267
– deformation, 151, 157
– expansion, 14, 158, 171, 237, 238, 243, 244, 260, 267, 277
– loading, 157, 172
Thermoplastic camera, 107
Thick hologram, 30
Thinning, 213
Three-dimensional
– digital image correlation, 334
– displacement, 326, 334, 338, 361
– vector displacement, 109
Time-average holography, 119, 120
Tolerance, 19, 409, 434
Transducer, 110, 234, 427, 428, 438
Transformation, 35, 36, 39, 48, 163, 212, 213, 327, 332, 336, 337, 356, 357, 385
Transients, 434
Tribology, 441, 455, 459
Tunable filter, 263
Tuning requirement, 58
Twist, 47, 117, 343
Two-aperture speckle interferometer, 138, 143, 144
Two-dimensional digital image correlation, 358
Two-wave interferometers, 15
Two-wavelength, 28, 219, 227
– interferometry, 28
Twyman-Green interferometer, 19–21

Uniaxial tension, 115, 137
Unwrapped phase, 277

Vertex, 4
Vibration
– analysis, 119, 120, 135
– modes, 110
– slideway error, 415
Vibrations, 14, 58, 59, 119, 120, 156, 204, 207, 380
Video rates, 110
Video-based PIV, 402

Virtual
- grating, 154
- image, 119
- reference grating, 154, 155, 162, 163, 165, 166

Volume strain, 41

Walls of interference, 154, 162, 163
Warpage, 191–194
Wave-plate, 305
Wavelength-division multiplexing, 251, 253, 268
Wavelet, 58
Waviness, 434–438
WDFT algorithm, 66–68, 86, 95
Wear, 413, 414, 418, 422, 445, 447–449, 458

Wedge splitting test, 116
White-light cross-correlation, 263
Whole field, 211
Workpiece, 414, 417, 423, 434, 435, 439, 447, 450

Young's
- double slit, 12
- fringe analysis, 396
- fringes, 128, 130, 397
- modulus, 40, 45, 49, 298

Zero-order
- Bessel function, 119
- fringe, 119, 158, 219

Topics in Applied Physics

60 **Ultrashort Laser Pulses**
    Generation and Applications
    By W. Kaiser (Ed.) 2nd edn. 1993. 197 figs. XVII, 492 pages

61 **Photorefractive Materials and Their Applications I**
    Fundamental Phenomena
    By P. Günter and J.-P. Huignard (Eds.) 1988. 134 figs. XVI, 295 pages

62 **Photorefractive Materials and Their Applications II**
    Survey of Applications
    By P. Günter and J.-P. Huignard (Eds.) 1989. 209 figs. XVIII, 367 pages

63 **Hydrogen in Intermetallic Compounds I**
    Electronic, Thermodynamic, and Crystallographic Properties, Preparation
    By L. Schlapbach (Ed.) 1988. 118 figs. 12 tabs. XIV, 350 pages

64 **Sputtering by Particle Bombardment III**
    Characteristics of Sputtered Particles, Technical Applications
    By R. Behrisch and K. Wittmaack (Eds.) 1991. 190 figs. 6 tabs. XV, 410 pages

65 **Laser Spectroscopy of Solids II**
    By W. M. Yen 1989. 144 figs. XI, 307 pages

66 **Light Scattering in Solids V**
    Superlattices and Other Microstructures
    By M. Cardona and G. Güntherodt (Eds.) 1989. 184 figs. XIII, 351 pages

67 **Hydrogen in Intermetallic Compounds II**
    Surface and Dynamic Properties, Applications
    By L. Schlapbach (Ed.) 1992. 126 figs. XIV, 328 pages

68 **Light Scattering in Solids VI**
    Recent Results, Including High-T_c Superconductivity
    By M. Cardona and G. Güntherodt (Eds.) 1991. 267 figs., 31 tabs., XIV, 526 pages

69 **Unoccupied Electronic States**
    Fundamentals for XANES, EELS, IPS and BIS
    By J. C. Fuggle and J. E. Inglesfield (Eds.) 1992. 175 figs. XIV, 359 pages

70 **Dye Lasers: 25 Years**
    By M. Stuke (Ed.) 1992. 151 figs. XVI, 247 pages

71 **The Monte Carlo Method in Condensed Matter Physics**
    By K. Binder (Ed.) 2nd edn. 1995. 83 figs. XX, 418 pages

72 **Glassy Metals III**
    Amorphization Techniques, Catalysis, Electronic and Ionic Structure
    By H. Beck and H.-J. Güntherodt (Eds.) 1994. 145 figs. XI, 259 pages

73 **Hydrogen in Metals III**
    Properties and Applications
    By H. Wipf (Ed.) 1997. 117 figs. XV, 348 pages

74 **Millimeter and Submillimeter Wave Spectroscopy of Solids**
    By G. Grüner (Ed.) 1998. 173 figs. XI, 286 pages

75 **Light Scattering in Solids VII**
    Christal-Field and Magnetic Excitations
    By M. Cardona and G. Güntherodt (Eds.) 1999. 96 figs. X, 310 pages

76 **Light Scattering in Solids VIII**
    C60, Semiconductor Surfaces, Coherent Phonons
    By M. Cardona and G. Güntherodt (Eds.) 1999. 86 figs. XII, 228 pages

77 **Photomechanics**
    By P. K. Rastogy (Ed.) 2000. 314 Figs. XVI, 472 pages

Printing: Saladruck, Berlin
Binding: H. Stürtz AG, Würzburg